Thermal Physics: Concepts and Practice

Thermodynamics has benefited from nearly 100 years of parallel development with quantum mechanics. As a result, thermal physics has been considerably enriched in concepts, technique and purpose, and now has a dominant role in developments of physics, chemistry and biology. This unique book explores the meaning and application of these developments using quantum theory as the starting point.

The book links thermal physics and quantum mechanics in a natural way. Concepts are combined with interesting examples, and entire chapters are dedicated to applying the principles to familiar, practical and unusual situations. Together with end-of-chapter exercises, this book gives advanced undergraduate and graduate students a modern perception and appreciation for this remarkable subject.

Allen L. Wasserman is Professor (Emeritus) in the Department of Physics, Oregon State University. His research area is condensed matter physics.

Thermal Physics: Concepts and Practice

Thermal Physics

Concepts and Practice

ALLEN L. WASSERMAN

Oregon State University

CAMBRIDGE
UNIVERSITY PRESS

University Printing House, Cambridge CB2 8BS, United Kingdom

One Liberty Plaza, 20th Floor, New York, NY 10006, USA

477 Williamstown Road, Port Melbourne, VIC 3207, Australia

314-321, 3rd Floor, Plot 3, Splendor Forum, Jasola District Centre, New Delhi 110025, India

79 Anson Road, #06-04/06, Singapore 079906

Cambridge University Press is part of the University of Cambridge.

It furthers the University's mission by disseminating knowledge in the pursuit of education, learning and research at the highest international levels of excellence.

www.cambridge.org
Information on this title: www.cambridge.org/9781107006492

First published 2012

A catalogue record for this publication is available from the British Library

Library of Congress Cataloging in Publication data
Wasserman, Allen L.
Thermal physics : concepts and practice / Allen L. Wasserman.
p. cm.
Includes bibliographical references and index.
ISBN 978-1-107-00649-2 (Hardback)
1. Thermodynamics. 2. Entropy. 3. Statistical mechanics. I. Title.
QC311.W37 2011
536′.7–dc23

2011036379

ISBN 978-1-107-00649-2 Hardback

Contents

Preface

In the preface to his book Statistical Mechanics Made Simple Professor Daniel Mattis writes:

> My own experience in thermodynamics and statistical mechanics, a half century ago at M.I.T., consisted of a single semester of Sears, skillfully taught by the man himself. But it was a subject that seemed as distant from "real" physics as did poetry or French literature.[1]

This frank but discouraging admission suggests that thermodynamics may not be a course eagerly anticipated by many students – not even physics, chemistry or engineering majors – and at completion I would suppose that few are likely to claim it was an especially inspiring experience. With such open aversion, the often disappointing performance on GRE[2] questions covering the subject should not be a surprise. As a teacher of the subject I have often conjectured on reasons for this lack of enthusiasm.

Apart from its subtlety and perceived difficulty, which are probably immutable, I venture to guess that one problem might be that most curricula resemble the thermodynamics of nearly a century ago.

Another might be that, unlike other areas of physics with their epigrammatic equations – Newton's, Maxwell's or Schrödinger's, which provide accessibility and direction – thermal physics seems to lack a comparable unifying principle.[3] Students may therefore fail to see conceptual or methodological coherence and experience confusion instead.

With those assumptions I propose in this book alternatives which try to address the disappointing experience of Professor Mattis and undoubtedly others.

Thermodynamics, the set of rules and constraints governing interconversion and dissipation of energy in macroscopic systems, can be regarded as having begun with Carnot's (1824) pioneering paper on heat-engine efficiency. It was the time of the industrial revolution, when the caloric fluid theory of heat was just being questioned and steam-engine efficiency was, understandably, an essential preoccupation. Later in that formative period Rudolf Clausius introduced a *First Law of Thermodynamics* (1850), formalizing the principles governing macroscopic energy conservation.

[1] Daniel Mattis, *Statistical Mechanics Made Simple*, World Scientific Publishing, Singapore (2003).
[2] Graduate Record Examination: standardized graduate school admission exam.
[3] R. Baierlein, "A central organizing principle for statistical and thermal physics?", *Am. J. Phys.* **63**, 108 (1995).

Microscopic models were, at the time, largely ignored and even regarded with suspicion, to the point where scientific contributions by some proponents of such interpretations were roundly rejected by the editors of esteemed journals. Even with the intercession in support of kinetic models by the respected Clausius (1857), they stubbornly remained regarded as over-imaginative, unnecessary appeals to invisible, unverifiable detail – even by physicists. A decade later when Maxwell (1866) introduced probability into physics bringing a measure of statistical rigor to kinetic (atomic) gas models there came, at last, a modicum of acceptance.

Within that defining decade the already esteemed Clausius (1864) invented a novel, abstract quantity as the centerpiece of a *Second Law of Thermodynamics*, a new principle – which he named *entropy* – to change forever our understanding of thermal processes and, indeed, all natural processes. Clausius offered no physical interpretation of entropy, leaving the matter open to intense speculation. Ludwig Boltzmann, soon to be a center of controversy, applied Maxwell's microscopic probability arguments to postulate a statistical model of entropy based on counting discrete "atomic" configurations referred to, both then and now, as "microstates".[4] However, Boltzmann's ideas on entropy, which assumed an atomic and molecular reality, were far from universally embraced – a personal disappointment which some speculate led to his suicide in 1906.

Closing the book on 19th-century thermal physics, J. W. Gibbs reconciled Newtonian mechanics with thermodynamics by inventing *statistical mechanics*[5] based on the still mistrusted presumption that atoms and molecules were physical realities. In this indisputably classic work, novel statistical "ensembles" were postulated to define thermodynamic averages, a statistical notion later adopted in interpreting quantum theories. Shortcomings and limited applicability of this essentially Newtonian approach notwithstanding, it provided prescient insights into the *quantum mechanics*, whose full realization was still a quarter century in the future.

Quantum mechanics revolutionized physics and defines the modern scientific era. Developing in parallel with it, and synergistically benefiting from this reshaped scientific landscape, thermal physics has come to occupy a rightful place among the pillars of modern physics.

Quantum mechanics' natural, internally consistent unification of statistics with microscopic mechanics immediately suggests the possibility of a thermodynamics derived, in some way, from microscopic quantum averages and quantum probabilities. But thermodynamic systems are not simply the isolated quantum systems familiar from most quantum mechanics courses. Thermodynamics is about *macroscopic* systems, i.e. many-particle quantum systems that are never perfectly isolated from the remainder of the universe. This interaction with the "outside" has enormous

[4] Boltzmann's microstates suggested to Planck (1900) what eventually became the quantization he incorporated into his theory of electromagnetic radiation.

[5] J. W. Gibbs, *The Elementary Principles of Statistical Mechanics*, C. Scribner, New York (1902).

consequences which, when taken into account quantitatively, clarifies the essence of thermodynamics.

Many thermal variables may then be approached as macroscopic quantum averages and their associated thermal probabilities as macroscopic quantum probabilities, the micro-to-macro translation achieved in part by an entropy postulate. This approach gives rise to a practical organizing principle with a clear pedagogical path by which thermodynamics' structure attains the epigrammatic status of "real physics".

Thermal physics is nevertheless frequently taught in the spirit of its utile 19th-century origins, minimizing both 20th- and 21st-century developments and, for the most part, disregarding the beauty, subtlety, profundity and laboratory realities of its modern rebirth – returning us to Professor Mattis' reflections. In proposing a remedy for his justifiable concerns, the opening chapter introduces a moderate dose of quantum-based content, both for review and, hopefully, to inspire interest in and, eventually, better understanding of thermodynamics. The second chapter develops ideas that take us to the threshold of a thermodynamics that we should begin to recognize. In Chapter 6 thermodynamics flies from a quantum nest nurtured, ready to take on challenges of modern physics.

Students and practitioners of thermodynamics come from a variety of disciplines. Engineers, chemists, biologists and physicists all use thermodynamics, each with practical or scientific concerns that motivate different emphases, stress different legacies and demand different pedagogical objectives. Since contemporary university curricula in most of these disciplines integrate some modern physics, i.e. quantum mechanics – if not its mathematical details at least its primary concepts and aims – the basic thermodynamic ideas as discussed in the first two chapters should lie within the range of students of science, engineering and chemistry. After a few chapters of re-acquaintance with classic thermodynamic ideas, the book's remaining chapters are dedicated to applications of thermodynamic ideas developed in Chapter 6 in practical and model examples for students and other readers.

Parts of this book first appeared in 1997 as notes for a course in thermal physics designed as a component of the revised undergraduate physics curriculum at Oregon State University. An objective of this revision was to create paradigmatic material stressing ideas common to modern understandings and contemporary problems. Consequently, concepts and dynamic structures basic to quantum mechanics – such as hamiltonians, eigen-energies and quantum degeneracy – appear and play important roles in this view of thermal physics. They are used to maintain the intended "paradigm" spirit by avoiding the isolation of thermal physics from developments of the past 100 years while, hopefully, cultivating in students and teachers alike a new perception of and appreciation for this absolutely remarkable subject.

This work has been funded in part by NSF Grants: DUE 9653250, 0231194, 0837829.

1 Introducing thermodynamics

The atomistic nature of matter as conceptualized by the Greeks had, by the 19th century, been raised by scientists to a high probability. But it was Planck's law of radiation that yielded the first exact determination of the absolute size of atoms. More than that, he convincingly showed that in addition to the atomistic structure of matter there is a kind of atomistic structure to energy, governed by the universal constant h.

This discovery has almost completely dominated the development of physics in the 20th century. Without this discovery a workable theory of molecules and atoms and the energy processes that govern their transformations would not have been possible. It has, moreover, shaken the whole framework of classical mechanics and electrodynamics and set science the fresh task of finding a new conceptual basis for all of physics. Despite partial success, the problem is still far from solved.

Albert Einstein, "Max Planck memorial service" (1948).
Original image, Einstein Archives Online, Jerusalem
(trans. A. Wasserman)

1.1 The beginning

Thermodynamics has exceeded the scope and applicability of its utile origins in the industrial revolution to a far greater extent than other subjects of physics' classical era, such as mechanics and electromagnetism. Unquestionably this results from over a century of synergistic development with quantum mechanics, to which it has given and from which it has gained clarification, enhancement and relevance, earning for it a vital role in the modern development of physics as well as chemistry, biology, engineering, and even aspects of philosophy.

The subject's fascinating history is intertwined with seminal characters who contributed much to its present form – some colorful and famous with others lesser known, maligned or merely ignored.

"Atomism" (i.e. molecular models), a source of early conflict, seems to have had Daniel Bernoulli as its earliest documented proponent when in 1738 he hypothesized a kinetic theory of gases.[1] Although science – thermodynamics in particular – is now unthinkable without such models, Bernoulli's ideas were largely ignored and for nearly a century little interest was shown in matter as based on microscopic constituents. Despite a continuing atmosphere of suspicion, particle models occasionally reappeared[2,3] and evolved, shedding some of their controversy (Joule 1851) while moving towards firm, accepted hypotheses (Kronig 1856). Adoption of kinetic models by the respected Rudolf Clausius (1857) began to erode the skeptics' position and encourage the new statistical theories of Maxwell (1859) and Boltzmann (1872). At about the same time van der Waals (1873), theorizing forces between *real* atoms and molecules, developed a successful equation of state for *non-ideal* gases.[4] Nevertheless, as the 20th century dawned, controversy continued only slightly abated. J. J. Thomson's discovery of the electron (1897) should have finally convinced remaining doubters of matter's microscopic essence – but it didn't. The argument continued into the 20th century stilled, finally, by the paradigm shift towards "quantized" models.

It all started with Max Planck who in 1900[5,6] introduced quantized energy into theories of electromagnetic radiation – and Einstein[7] who used quantized lattice vibrations in his ground-breaking heat capacity calculation. But it would be another 25 years before the inherently probabilistic, microscopic theory of matter – quantum mechanics – with its quantum probabilities and expectation values – would completely reshape the scientific landscape and permanently dominate most areas of physics, providing a basis for deeper understanding of particles, atoms and nuclei while grooming thermodynamics for its essential role in modern physics.

Thermodynamics is primarily concerned with mechanical, thermal and electromagnetic interactions in *macroscopic* matter, i.e. systems with huge numbers of microscopic constituents ($\sim 10^{23}$ particles). Although thermodynamic descriptions are generally in terms of largely intuitive macroscopic variables, most macroscopic behaviors are, at their root, quantum mechanical. Precisely how the classical

[1] D. Bernoulli, *Hydrodynamica* (1738).

[2] J. Herapath,"On the causes, laws and phenomena of heat, gases, gravitation", *Annals of Philosophy* **9** (1821). Herapath's was one of the early papers on kinetic theory, but rejected by the Royal Society, whose reviewer objected to the implication that there was an absolute zero of temperature at which molecular motion ceased.

[3] J.J.Waterston,"Thoughts on the mental functions" (1843). This peculiar title was a most likely cause of its rejection by the Royal Society as "nothing but nonsense". In recognition of Waterston's unfairly maligned achievement, Lord Rayleigh recovered the original manuscript and had it published as "On the physics of media that are composed of free and perfectly elastic molecules in a state of motion", *Philosophical Transactions of the Royal Society A* **183**, 1 (1892), nearly 10 years after Waterston's death.

[4] Van der Waals' work also implied a molecular basis for critical points and the liquid–vapor phase transition.

[5] Max Planck,"Entropy and temperature of radiant heat", *Ann. der Physik* **1**, 719 (1900).

[6] Max Planck, "On the law of distribution of energy in the normal spectrum", *Ann. der Physik* **4**, 553 (1901).

[7] A. Einstein, "Planck's theory of radiation and the theory of specific heat", *Ann. der Physik* **22**, 180 (1907).

measurement arises from quantum behavior has been a subject of some controversy ever since quantum theory's introduction. But it now seems clear that macroscopic systems are quantum systems that are particularly distinguished by always being entangled (however weakly) with an environment (sometimes referred to as a reservoir) that is also a quantum system. Although environmental coupling may be conceptually simple and even uninteresting in detail, it has enormous consequences for quantum-based descriptions of macroscopic matter, i.e. thermodynamics.[8]

1.2 Thermodynamic vocabulary

A few general, large-scale terms are used in describing objects and conditions of interest in thermodynamics.

- **System:** A macroscopic unit of particular interest, especially one whose thermal properties are under investigation. It may, for example, be a gas confined within physical boundaries or a rod or string of elastic material (metal, rubber or polymer). It can also be matter that is magnetizable or electrically polarizable.
- **Surroundings:** Everything physical that is not the system or which lies outside the system's boundaries is regarded as "surroundings". This may be external weights, an external atmosphere, or static electric and magnetic fields. The system plus surroundings comprise, somewhat metaphorically, "the universe".
- **Thermal variables:** A set of macroscopic variables that describe the state of the system. Some variables are intuitive and familiar, such as pressure, volume, elongation, tension, etc. Others may be less intuitive and even abstract – such as temperature – but, nevertheless, also play important roles in thermodynamics. These will be discussed in detail in this and later chapters.
- **Thermal equilibrium:** The final state attained in which thermal state variables that describe the macroscopic system (pressure, temperature, volume, etc.) no longer change in time.[9] It is only at thermal equilibrium that thermodynamic variables are well defined. The time elapsed in attaining equilibrium is largely irrelevant.

[8] Macroscopic behavior can be quite different from the behavior of individual constituents (atoms, molecules, nuclei, etc.) As an example, the appearance of spontaneous bulk magnetism in iron (at temperatures below some critical temperature T_C) is not a property of individual iron atoms but arises from large numbers of interacting iron atoms behaving collectively.

[9] There are, nevertheless, small departures from equilibrium averages, referred to as *fluctuations*, whose values are also part of any complete thermodynamic description.

1.3 Energy and the First Law

James Joule's classic contribution on the mechanical equivalent of heat and his theory of energy reallocation between a system and its surroundings[10] (referred to as energy conservation) led Rudolph Clausius[11] to the historic *First Law of Thermodynamics*:

$$\Delta\mathcal{U} = \mathcal{Q} - \mathcal{W}. \tag{1.1}$$

Here $\mathcal{W}$ is *mechanical work* done *by* the system and $\mathcal{Q}$ is *heat* (thermal energy) *added to* the system, both of which are classical quantities associated with surroundings. In Clausius' time controversy about reality in atomic models left $\mathcal{U}$ with no definitive interpretation. But being the maximum *work* which could be theoretically extracted from a substance it was initially called "intrinsic energy". As kinetic (atomic) models gained acceptance (Clausius having played an influential role) $\mathcal{U}$ became the *mean kinetic energy* of the system's microscopic constituents or, more generally, as *internal energy*. Although the *change* in internal energy, $\Delta\mathcal{U}$, is brought about by mechanical and thermal, i.e. classical, interactions, quantum mechanics provides a clear and specific meaning to $\Delta\mathcal{U}$ as an *average* change in energy of the *macroscopic system as determined from kinetic, potential and interaction energies of its microscopic constituents*, clearly distinguishing it from other energy contributions. The precise meaning and interrelation of this with similar *macroscopic averages* provides the basis for thermodynamics.[12]

1.3.1 Thermodynamic variables defined

Some thermodynamic concepts and macroscopic variables are familiar from classical physics while others arise simply from operational experience. Moving beyond this, quantum mechanics provides definitions and context for not only *internal* energy $\mathcal{U}$, but for other *macroscopic* (thermodynamic) variables, placing them within a microscopic context that adds considerably to their meaning and their role within thermal physics. The First Law, in arraying $\mathcal{Q}$ and $\mathcal{W}$ (both classical) against $\Delta\mathcal{U}$ (quantum mechanical), highlights this intrinsic partitioning of macroscopic variables into "classical"(C) vis-à-vis quantum (Q).

Examples of *Q-variables* – macroscopic variables having microscopic origins[13] – are:

[10] James P. Joule, "On the existence of an equivalent relation between heat and the ordinary forms of mechanical power", *Phil. Mag.* **27**, 205 (1850).

[11] R. Clausius, "On the moving force of heat, and the laws regarding the nature of heat", *Phil. Mag.* **2**, 1–21, 102–119 (1851).

[12] Macroscopic "averages" will be discussed in Chapter 2.

[13] These are defined by quantum operators.

- internal energy: $\mathbf{h}_{op}$ (energy of the system's microscopic constituents – kinetic plus potential);
- pressure: $\boldsymbol{p}_{op} = -\left(\frac{\partial\, \mathbf{h}_{op}}{\partial V}\right)$ (pressure arising from a system's internal constituents);
- electric polarization: $\mathcal{P}_{op}$;
- magnetization: $\boldsymbol{M}_{op}$;
- elongation (length): $\boldsymbol{\chi}_{op}$;
- particle number: $\mathcal{N}_{op}$.[14]

Examples of *C-variables* – classical (macroscopic) variables that exist apart from microscopic mechanics[15] – are:

- temperature: T;
- volume: V;
- *static* magnetic fields: $\mathcal{B}$ or H;
- *static* electric fields: $\mathcal{E}$ or $\mathcal{D}$;[16]
- elastic tension: τ;
- chemical potential: μ.[17]

Interaction energy

Most Q-variables listed above appear from interaction terms added to $\mathbf{h}_{op}$. The following are examples of such variables and their interactions.

a. Tension $\boldsymbol{\tau}$ applied to an elastic material produces a "conjugate" elongation $\boldsymbol{\chi}$, as described by an interaction operator[18]

$$\mathcal{H}_{op}^{\chi} = -\boldsymbol{\tau} \cdot \boldsymbol{\chi}_{op}. \tag{1.2}$$

b. A static magnetic induction field $\boldsymbol{\mathcal{B}}_0$ contributes an interaction operator (energy)

$$\mathcal{H}_{op}^{\mathcal{M}} = -\boldsymbol{m}_{op} \cdot \boldsymbol{\mathcal{B}}_0. \tag{1.3}$$

[14] Variable particle number is essential for thermodynamic descriptions of phase transitions, chemical reactions and inhomogeneous systems. However, the particle number operator is not a part of Schrödinger's fixed particle number theory, though it appears quite naturally in quantum field theories. Implementing variable particle number requires operators to create and destroy them and operators to count them.

[15] These are not defined by quantum operators.

[16] Electromagnetic *radiation* fields are, on the other hand, representable by quantum field operators $\mathcal{B}_{op}$ and $\mathcal{E}_{op}$ obtained from a quantum electromagnetic vector potential operator $\mathcal{A}_{op}$. This will be discussed in Chapter 14, on radiation theory.

[17] μ is an energy per particle and is associated with processes having varying particle number.

[18] The tension $\boldsymbol{\tau}$ is said to be *conjugate* to the elongation $\boldsymbol{\chi}$. The variable and its conjugate comprise a thermodynamic energy.

Here $\boldsymbol{m}_{op}$ represents a magnetic moment *operator* for elementary or composite particles. The sum[19,20]

$$\boldsymbol{M}_{op} = \sum_i \boldsymbol{m}_{op}(i), \tag{1.4}$$

represents the total magnetization operator.

c. A static electric field $\boldsymbol{\mathcal{E}}_0$ can contribute an interaction operator (energy)

$$\mathcal{H}_{op}^{\mathcal{P}} = -\mathcal{P}_{op} \cdot \boldsymbol{\mathcal{E}}_0, \tag{1.5}$$

where p_{op} represents an electric dipole moment *operator* and

$$\mathcal{P}_{op} = \sum_i p_{op}(i) \tag{1.6}$$

represents the total polarization.[21]

d. An energy associated with creating "space" for the system is

$$\mathcal{H}_{op}^{\mathcal{W}} = \boldsymbol{p}_{op} V, \tag{1.7}$$

with $\boldsymbol{p}_{op}$ representing the system pressure and V the displaced volume.

e. An "open system" energy associated with particle creation and/or destruction is

$$\mathcal{H}_{op}^{\mathcal{N}} = -\mu \mathcal{N}_{op}, \tag{1.8}$$

where the chemical potential μ (an energy per particle) is conjugate to a particle number *operator* $\mathcal{N}_{op}$. Chemical reactions, phase transitions and other cases with variable numbers of particles of different species are examples of open systems.[22]

1.4 Quantum mechanics, the "mother of theories"

As out of place as rigorous quantum ideas might seem in this introduction to thermodynamics, it is the author's view that they are an essential topic for an approach that strives to bring unity of structure and calculable meaning to the subject.

[19] The field $\boldsymbol{\mathcal{B}}_0$ conjugate to $\boldsymbol{m}_{op}$ is the field present prior to the insertion of matter. Matter itself may be magnetized and contribute to an effective $\mathcal{B}$.

[20] There are ambiguities in the thermodynamic roles of static fields, e.g. Maxwell's local magnetic average $\boldsymbol{\mathcal{B}}$ vs. an external $\boldsymbol{\mathcal{B}}_0$ and Maxwell's local electric average $\boldsymbol{\mathcal{E}}$ vs. external $\boldsymbol{\mathcal{E}}_0$.

[21] The field $\boldsymbol{\mathcal{E}}_0$ is conjugate to the electric dipole operator p_{op}.

[22] Variable particle number is intrinsic to thermal physics even though it may be suppressed for simplicity when particle number is assumed constant.

1.4.1 Introduction

Macroscopic variables such as pressure p, magnetization $\boldsymbol{M}$, elastic elongation χ, particle number N, etc. – any (or all) of which can appear in thermodynamic descriptions of matter – have microscopic origins and should be obtainable from quantum models. The question is: "How?"

Investigating the means by which microscopic models eventually lead to macroscopic quantities is the aim of Chapters 2 and 6. Preliminary to that goal, this chapter reviews postulates, definitions and rules of quantum mechanics for typical *isolated* systems.[23] Particular attention is given to probabilities and expectation values of dynamical variables (*observables*).[24] This provides the basic rules, language and notation to carry us into Chapter 2 where the question "What is thermodynamics?" is raised, and then to Chapter 6 where the ultimate question "How does it arise?" is addressed.

In achieving this goal Chapter 2 will diverge somewhat from the familiar "wave mechanics" of introductory courses and focus on a less-familiar but closely related quantum mechanical object called the *density operator*, ρ_{op}. This quantity, although derived from a quantum state function, goes beyond state function limitations by adding the breadth and flexibility critical for answering most of our questions about thermodynamics. It is a "Yellow Brick Road" that will guide us from the land of the Munchkins to the Emerald City[25] – from the microscopic to the macroscopic – from quantum mechanics to thermodynamics.

1.4.2 A brief review

The review starts with Schrödinger's famous linear equation of non-relativistic quantum mechanics:

$$\mathcal{H}_{op}\psi\left(\mathbf{x},t\right) = i\hbar\frac{\partial\psi\left(\mathbf{x},t\right)}{\partial t}. \tag{1.9}$$

The solution to Eq. 1.9 is the time-evolving, complex, scalar *wavefunction* $\psi\left(\mathbf{x},t\right)$, which describes the *quantum* dynamics of an *isolated, fixed particle number, microscopic* system.

In the widely used (and preferred) Dirac notation, Eq. 1.9 is identical with

$$\mathcal{H}_{op}\left|\Psi\right\rangle = i\hbar\frac{\partial}{\partial t}\left|\Psi\right\rangle, \tag{1.10}$$

[23] These are the systems described by Schrödinger theory.

[24] Most of this review material is likely to be familiar from an earlier course in modern physics. If not, it is recommended that you work alongside one of the many books on introductory quantum mechanics.

[25] L. Frank Baum, *Wizard of Oz*, Dover Publications, New York (1996).

where $|\Psi\rangle$ is the time-dependent state function, or "ket", corresponding to the wavefunction $\psi(\mathbf{x}, t)$, i.e.

$$\langle \mathbf{x} \mid \Psi \rangle \equiv \psi(\mathbf{x}, t) . \tag{1.11}$$

Here $|\mathbf{x}\rangle$ is an "eigen-ket" of the position operator with $\mathbf{x} \equiv \mathbf{x_1}, \mathbf{x_2}, \ldots, \mathbf{x_N}$ the coordinates of an N-particle system. As indicated in Eq. 1.11, the wavefunction $\psi(\mathbf{x}, t)$ is merely the "ket" $|\Psi\rangle$ in a coordinate representation.[26] $\mathcal{H}_{op}$ is the hamiltonian operator (inspired by classical dynamics). A hamiltonian operator for a case with only internal particle dynamics may, for example, be written

$$\mathbf{h}_{op} = \mathcal{T}_{op} + \mathcal{V}_{op}, \tag{1.12}$$

where $\mathcal{T}_{op}$ and $\mathcal{V}_{op}$ are kinetic energy and potential energy operators, respectively.[27] Additional interaction terms $\mathcal{H}_{op}^{int}$ such as Eqs. 1.2 → 1.8 may appear, depending on the physical situation, in which case

$$\mathcal{H}_{op} = \mathbf{h}_{op} + \mathcal{H}_{op}^{int} . \tag{1.13}$$

The quantum state function $|\Psi\rangle$ (see Eq. 1.11) has no classical counterpart and is, moreover, not even a measurable! But it is nevertheless interpreted as the generating function for a statistical description of the quantum system – including probabilities, statistical averages (expectation values) and fluctuations about these averages. It contains, in principle, all that is knowable about the isolated, microscopic system.[28]

Devising an arbitrary but reasonable quantum knowledge scale, say

$$1 \Rightarrow \textit{all knowledge}, \tag{1.14}$$

$$0 \Rightarrow \textit{no knowledge}, \tag{1.15}$$

and using this scale to calibrate "information", Schrödinger's quantum state function has information value 1.[29] Unless there is some quantum interaction with the surroundings (i.e. the system is no longer isolated) $|\Psi\rangle$ will retain information value 1 indefinitely.[30,31]

[26] The "ket" $|\Psi\rangle$ is an abstract, representation-independent object. Its only dependence is t, time.

[27] $\mathcal{V}_{op}$ is assumed to also include particle–particle interactions.

[28] Whereas probabilities are derivable from the system state function, the reverse is not the case – state functions $|\Psi\rangle$ cannot be inferred from measurement.

[29] In Chapter 2 we will introduce a more solidly based mathematical measure of "information", extending this arbitrary scale.

[30] The isolated system state function is a quantity of maximal information. The system's isolation assures that no information will "leak" in or out.

[31] Macroscopic complexity has relaxed absolute reliance on a detailed many-body quantum description of a macroscopic system. Even at this early point we can foresee that exact (or even approximate) knowledge of the macroscopic (many-body) quantum state function (even if that were possible, which it is not) is unnecessary for macroscopic descriptions.

An energy expectation value – an observable – is defined (in terms of the wavefunction) by

$$\langle \mathcal{H}(t)\rangle = \int d\mathbf{x}\psi^*(\mathbf{x},t)\,\mathcal{H}_{op}\psi(\mathbf{x},t)\,, \tag{1.16}$$

where $\psi^*(\mathbf{x},t)$ is the complex conjugate of the wavefunction $\psi(\mathbf{x},t)$. In the more convenient Dirac notation

$$\langle \mathcal{H}\rangle = \langle \Psi|\mathcal{H}_{op}|\Psi\rangle, \tag{1.17}$$

where $\langle\Psi|$, the conjugate to $|\Psi\rangle$, is called a "bra". The energy expectation value (sometimes called average value) is often a measurable[32] of physical interest.

But neither Eq. 1.16 nor Eq. 1.17 represent thermodynamic state variables. In particular, $\langle \mathbf{h}\rangle = \langle\Psi|\mathbf{h}_{op}|\Psi\rangle$ is *not* macroscopic internal energy $\mathcal{U}$, the centerpiece of the First Law. This is a *crucial* point that will be further addressed in Chapter 2 when we inquire specifically about thermodynamic variables.

In quantum mechanics, each of nature's dynamical observables (e.g. momentum, position, energy, current density, magnetic moment, etc.) is represented by an hermitian operator. (An *hermitian operator* satisfies the condition $\Omega_{op}^{\dagger} = \Omega_{op}$, where $\dagger$ is the symbol for hermitian conjugation. In matrix language *hermiticity* means $\Omega^{\dagger} = (\Omega^*)^T = \Omega$, where $(\Omega^*)^T$ is the complex conjugate-transpose of the matrix Ω.)[33]

The quantum theory postulates an auxiliary eigenvalue problem for Ω_{op} (which represents a typical hermitian operator)

$$\Omega_{op}\,|\omega_n\rangle = \omega_n\,|\omega_n\rangle\,, \tag{1.18}$$

with $n = 1, 2, 3, \ldots$, from which the spectrum of allowed quantum observables ω_n (called eigenvalues) is derived. The ω_n are real numbers[34] and $|\omega_n\rangle$ are their corresponding set of complex eigenfunctions. (For simplicity we assume the eigenvalues are discrete.) The specific forms required for dynamical hermitian operators will be introduced as the models require.

1.5 Probabilities in quantum mechanics

According to the Great Probability Postulate of quantum mechanics, the Schrödinger state function $|\Psi\rangle$ is a generator of *all probability information* about an isolated

[32] The value of this measurable is the average of a very large number of identical measurements on identical systems with identical apparatus.

[33] In the following discussion Ω_{op} is used as a generic symbol for operators representing observables.

[34] The hermitian property of quantum mechanical operators assures that the eigenvalues will be real numbers.

quantum system with fixed number of particles.[35] Using the complete orthonormal set of eigenfunctions determined from Eq. 1.18, a state function $|\Psi\rangle$ can be expressed as a linear coherent superposition[36]

$$|\Psi\rangle = \sum_n |\omega_n\rangle\langle\omega_n|\Psi\rangle, \tag{1.19}$$

where the coefficients

$$p(\omega_n, t) = \langle\omega_n|\Psi\rangle \tag{1.20}$$

are complex probability amplitudes. *Coherence* as used here implies phase relations among the terms in the sum which leads to characteristic quantum interference effects.

The probability that a measurement outcome will be ω_n is given by

$$\mathcal{P}(\omega_n, t) = |\langle\omega_n|\Psi\rangle|^2 \tag{1.21}$$

$$= \langle\omega_n|\Psi\rangle\langle\Psi|\omega_n\rangle\,. \tag{1.22}$$

The eigenfunction $|\omega_n\rangle$ acts, therefore, as a statistical projector in the sense that with some appropriate measuring device it projects out from the state function $|\Psi\rangle$ the statistical probability of measuring ω_n. The probabilities $\mathcal{P}(\omega_n, t)$ are normalized in the usual probabilistic sense, i.e.

$$\sum_n \mathcal{P}(\omega_n, t) = 1, \tag{1.23}$$

which corresponds to state function normalization

$$\langle\Psi|\Psi\rangle = 1\,. \tag{1.24}$$

1.5.1 Expectation values

If the probability of observing the measurable value ω_n is $\mathcal{P}(\omega_n, t)$, then its expectation (average) value $\langle\omega\rangle$ is, as in ordinary statistics,

$$\langle\omega\rangle = \sum_n \omega_n \mathcal{P}(\omega_n, t) \tag{1.25}$$

$$= \sum_n \omega_n \langle\omega_n|\Psi\rangle\langle\Psi|\omega_n\rangle\,. \tag{1.26}$$

For the important special case of energy of an isolated (microscopic) system, where

$$\mathcal{H}_{op}|E_n\rangle = E_n|E_n\rangle, \tag{1.27}$$

[35] The state function $|\Psi\rangle$ depends on time t so that all quantities calculated from it will – in principle – also depend on time.

[36] This is the famous *linear superposition theorem.*

with E_n the allowed eigen-energies (eigenvalues) and $|E_n\rangle$ the corresponding eigen-functions, the energy expectation value for the state $|\Psi\rangle$ is

$$\langle E \rangle = \sum_n E_n \langle E_n | \Psi \rangle \langle \Psi | E_n \rangle \tag{1.28}$$

$$= \sum_n E_n \mathcal{P}(E_n, t) , \tag{1.29}$$

which is equivalent to Eq. 1.17.

The expectation value of any hermitian observable Ω_{op} is, for a microscopic system,

$$\langle \Omega \rangle = \langle \Psi | \, \Omega_{op} \, | \Psi \rangle . \tag{1.30}$$

1.6 Closing comments

The main points in this chapter are as follows.

1. Schrödinger's wavefunctions yield quantum probabilities and quantum expectation values (averages). But, contrary to expectations, these measurables are *not* macroscopic state variables of thermodynamics. Temporarily putting aside this letdown, the issue will be discussed and resolved in the next chapters.
2. Schrödinger state functions are capable of generating *all* dynamical information that can be known about the physical system. They can be said to have "information content" 1 and are called *pure* states. From these few clues it should be apparent that "information" is a term with relevance in quantum mechanics and, ultimately, in thermodynamics.

The reader is encouraged to consult the ever-growing number of fine monographs on quantum mechanics (far too many to list here).

2 A road to thermodynamics

If someone says that he can think or talk about quantum physics without becoming dizzy, that shows only that he has not understood anything whatever about it.
Murray Gell-Mann, Thinking Aloud: The Simple and the Complex",
Thinking Allowed Productions, Oakland, California (1998)

2.1 The density operator: pure states

Results from Chapter 1 can, alternatively, be expressed in terms of a *density operator* ρ_{op}, which is defined (in Dirac notation) as

$$\rho_{op} = |\Psi\rangle\langle\Psi| \tag{2.1}$$

where the Dirac "ket" $|\Psi\rangle$ is a pure Schrödinger state function, and the Dirac "bra" $\langle\Psi|$ is its conjugate. Density operator ρ_{op} is clearly hermitian with matrix elements – the *density matrix* – in some complete basis $|\omega_i\rangle$

$$\rho_{ij} = \langle\omega_i|\Psi\rangle\langle\Psi|\omega_j\rangle, \tag{2.2}$$

whose diagonal elements

$$\rho_{ii} = |\langle\Psi|\omega_i\rangle|^2 \tag{2.3}$$

are identical with the probabilities of Eq.1.21

$$\rho_{ii} = \mathcal{P}(\omega_i). \tag{2.4}$$

An introduction to properties of the density operator[1] (matrix) will demonstrate that this object's value is not merely that of an alternative to the wave equation description in the previous chapter, but that it is capable of defining and accommodating a broader class of quantum states – in particular, states which are not pure, i.e. states

[1] Also called the statistical operator.

for which information[2] content is *less than* 1. This generalization is a huge stride towards realizing thermodynamics.[3]

2.1.1 Traces, expectations and information

If $\mathcal{A}_{op}$ is an operator and its matrix representation is

$$A_{ij} = \langle \phi_i | \mathcal{A}_{op} | \phi_j \rangle, \tag{2.5}$$

then its *trace* $Tr\mathcal{A}_{op}$ is defined as the sum of its diagonal matrix elements in the orthonormal basis $|\phi_i\rangle$,[4]

$$Tr\mathcal{A}_{op} = \sum_i \langle \phi_i | \mathcal{A}_{op} | \phi_i \rangle. \tag{2.6}$$

The expectation value $\langle \Omega \rangle$ in terms of the pure state density operator ρ_{op} is

$$\langle \Omega \rangle = \frac{\sum\limits_\alpha \langle \phi_\alpha | \rho_{op} \Omega_{op} | \phi_\alpha \rangle}{\sum\limits_\alpha \langle \phi_\alpha | \rho_{op} | \phi_\alpha \rangle}, \tag{2.7}$$

where $|\phi_\alpha\rangle$ is any complete set of states. Using Eq. 2.6, this can be condensed to

$$\langle \Omega \rangle = \frac{Tr\,\rho_{op}\Omega_{op}}{Tr\,\rho_{op}}. \tag{2.8}$$

Since $|\Psi\rangle$ is normalized,

$$Tr\,\rho_{op} = \sum_\alpha \langle \phi_\alpha | \rho_{op} | \phi_\alpha \rangle \tag{2.9}$$

$$= \sum_\alpha \langle \phi_\alpha | \Psi \rangle \langle \Psi | \phi_\alpha \rangle \tag{2.10}$$

$$= \sum_\alpha \langle \Psi | \phi_\alpha \rangle \langle \phi_\alpha | \Psi \rangle \tag{2.11}$$

$$= \langle \Psi | \Psi \rangle \tag{2.12}$$

$$= 1, \tag{2.13}$$

reducing Eq.2.8 to

$$\langle \Omega \rangle = Tr\,\rho_{op}\Omega_{op}, \tag{2.14}$$

[2] A brief discussion about information appeared in Chapter 1.

[3] Moreover, the density operator (matrix) can be a measurable. See: R. Newton and B. Young, "Measurability of the spin density matrix", *Annals of Physics* **49**, 393 (1968); Jean-Pierre Amiet and Stefan Weigert, "Reconstructing the density matrix of a spin *s* through Stern-Gerlach measurements: II", *J. Phys. A* **32**, L269 (1999).

[4] The trace of a matrix is independent of the orthonormal basis in which the matrix is expressed.

where completeness of $|\phi_\alpha\rangle$ has been applied. Using as a basis $|\omega_n\rangle$ (the eigenfunctions of Ω_{op}) Eq. 2.14 becomes, as expected,

$$\langle\Omega\rangle = \sum_n \omega_n \left|\langle\Psi|\omega_n\rangle\right|^2, \tag{2.15}$$

where, as discussed in Chapter 1,

$$\mathcal{P}(\omega_n) = \left|\langle\Psi|\omega_n\rangle\right|^2 \tag{2.16}$$

is the probability of finding the system in an eigenstate $|\omega_n\rangle$.

The probability $\mathcal{P}(\omega_n)$ is easily shown to be expressible as

$$\mathcal{P}(\omega_n) = \mathcal{T}r\left\{\rho_{op}|\omega_n\rangle\langle\omega_n|\right\}. \tag{2.17}$$

The brief introduction to *information* in Chapter 1 is now carried a step further by defining a new quantity, *purity* $\mathcal{I}$,

$$\mathcal{I} = \mathcal{T}r\,{\rho_{op}}^2, \tag{2.18}$$

as a way to measure information.[5] The pure state, Eq. 2.1, now has purity (information)

$$\mathcal{I} = \mathcal{T}r\,\rho_{op}\rho_{op} = \sum_\alpha \langle\phi_\alpha|\Psi\rangle\langle\Psi|\Psi\rangle\langle\Psi|\phi_\alpha\rangle \tag{2.19}$$

$$= \sum_\alpha \langle\Psi|\phi_\alpha\rangle\langle\phi_\alpha|\Psi\rangle\langle\Psi|\Psi\rangle \tag{2.20}$$

$$= \langle\Psi|\Psi\rangle \tag{2.21}$$

$$= 1, \tag{2.22}$$

in agreement with the idea of *maximal* information on the arbitrary scale devised in Chapter 1. The pure state density operator ρ_{op} carries complete quantum information ($\mathcal{I} = 1$) about an isolated microscopic system, in the same way as does its corresponding state function $|\Psi\rangle$, but within a more useful framework. (The best is still to come!)

The density operator has an equation of motion found by using the time derivative of the state operator,

[5] This is not the only way "information" can be defined. A more thermodynamically relevant definition will be expanded upon in Chapter 6.

$$\frac{\partial \rho_{op}}{\partial t} = \left(\frac{\partial}{\partial t}|\Psi\rangle\right)\langle\Psi| + |\Psi\rangle\left(\langle\Psi|\frac{\partial}{\partial t}\right) \tag{2.23}$$

$$= \left(-\frac{i}{\hbar}\mathcal{H}_{op}|\Psi\rangle\right)\langle\Psi| + |\Psi\rangle\left(\langle\Psi|\frac{i}{\hbar}\mathcal{H}_{op}\right)$$

$$= \frac{i}{\hbar}(\rho_{op}\mathcal{H}_{op} - \mathcal{H}_{op}\rho_{op})$$

$$= \frac{i}{\hbar}\left[\rho_{op}, \mathcal{H}_{op}\right], \tag{2.24}$$

which, in practical terms, is equivalent to Schrödinger's equation.[6]

2.2 Mixed states

At this point it might seem that not much has been gained by introducing the density operator. So why bother?

Keeping in mind that our interest is in *macroscopic* quantum systems whose states are inextricably entangled with those of a quantum environment, the density operator provides a logical foundation for describing environmental "mixing" and its consequences.[7] Although it seems that mixing can only irreparably complicate matters, in fact it ultimately reveals the essence of thermodynamics, bringing its meaning into focus.

Pure quantum states, which evolve according to Schrödinger's equation, can be individually prepared and then classically combined by some (mechanical) process or apparatus.[8] This combined configuration *cannot* be a solution to any Schrödinger equation and is no longer a pure state described by a state function, say $|\Psi^{(\xi)}\rangle$. It can, however, be described by a density operator – in particular a *mixed state density operator* ρ_{op}^{M} – defined by

$$\rho_{op}^{M} = \sum_{j} \left|\psi^{(j)}\right\rangle w_j \left\langle\psi^{(j)}\right| \tag{2.25}$$

with

$$\sum_{j} w_j = 1, \tag{2.26}$$

where the w_j are real, classical, probabilities that the jth pure state density operator $|\psi^{(j)}\rangle\langle\psi^{(j)}|$ contributes to Eq. 2.25. The sum in Eq. 2.25 is *not* an example of the superposition principle. It is not equivalent to or related to any linear coherent

[6] The hamiltonian $\mathcal{H}_{op}$ is the full system hamiltonian.

[7] Even if at some initial time a state starts out *pure* with maximal information, interactions with the environment quickly destroy that purity – the larger the system the more rapid the loss of purity.

[8] A Stern-Gerlach "machine" is such an apparatus. See Example 2.2.1.

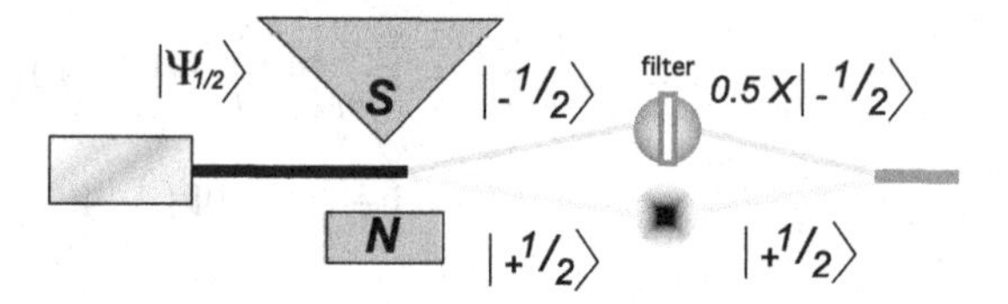

Fig. 2.1 Stern-Gerlach apparatus with filter.

superpositions (with phase relationships) that exhibit quantum interference. Eq. 2.25 represents an ensemble of systems, $|\psi^{(j)}\rangle$, combined according to classical probabilities w_j with no phase relationships.

2.2.1 Example

Consider ions with spin $+1/2$ in a pure Schrödinger state $|\Psi_{1/2}\rangle$ which can be expressed as a linear coherent superposition (the superposition principle)

$$|\Psi_{1/2}\rangle = |\varphi_{+1/2}\rangle\langle\varphi_{+1/2}|\Psi_{1/2}\rangle + |\varphi_{-1/2}\rangle\langle\varphi_{-1/2}|\Psi_{1/2}\rangle \tag{2.27}$$

where $|\varphi_{+1/2}\rangle$ and $|\varphi_{-1/2}\rangle$ are the complete set of spin-$1/2$ eigenstates quantized along the z-axis.

The quantum probability an ion will be found in a $+1/2$-spin eigenstate is

$$\mathcal{P}(+1/2) = \left|\langle\varphi_{+1/2} \mid \Psi_{1/2}\rangle\right|^2 \tag{2.28}$$

while the quantum probability an ion will be found in a $-1/2$-spin eigenstate is

$$\mathcal{P}(-1/2) = \left|\langle\varphi_{-1/2} \mid \Psi_{1/2}\rangle\right|^2. \tag{2.29}$$

After passing the beam through a magnetic field gradient, (Stern-Gerlach apparatus) ions in the pure $|\varphi_{\pm(1/2)}\rangle$ states will emerge deflected along a pair of divergent trajectories with probabilities $w_{+1/2} = 1/2$ and $w_{-1/2} = 1/2$ with $w_{+1/2}+w_{-1/2} = 1$.

Following this, each of the separated ion beams is itself deflected by, for example, a uniform static magnetic field, with one of the beams, say the $|\varphi_{-1/2}\rangle$, first passing through a purpose-designed filter that diminishes its intensity to one-half its initial value. Then, by incoherently (classically) mixing the redirected pure $|\varphi_{+1/2}\rangle$ and $|\varphi_{-1/2}\rangle$ states, a new single beam is reconstituted. Clearly there is no single Schrödinger state function for the reconstituted beam. But a mixed state density operator, as in Eq. 2.25, is possible with

$$\rho_{op}^{\pm 1/2} = \frac{2}{3}|\varphi_{+1/2}\rangle\langle\varphi_{+1/2}| + \frac{1}{3}|\varphi_{-1/2}\rangle\langle\varphi_{-1/2}|. \tag{2.30}$$

This is an *incoherent* (non-interfering) ensemble sum. There are no phase relationships between different pure state contributions.

2.2.2 Mixed state properties

The mixed state density operator has several properties that coincide with those of the pure state density operator. For example:

1. Using Eqs. 2.25 and 2.26 it is easy to show that the mixed state density operator has the normalization property

$$\mathcal{T}r\,\rho_{op}^{M} = 1. \tag{2.31}$$

2. The average value of the observable Ω_{op} for the mixed state is

$$\langle \Omega \rangle = \mathcal{T}r\,\rho_{op}^{M}\Omega_{op}. \tag{2.32}$$

3. In the mixed state the probability of measuring the observable ω_s is

$$\mathcal{P}(\omega_s) = \mathcal{T}r\left\{\rho_{op}^{M}|\omega_s\rangle\langle\omega_s|\right\}. \tag{2.33}$$

4. Following the steps for deriving Eq. 2.24, the mixed state density operator has an equation of motion

$$\frac{\partial \rho_{op}^{M}}{\partial t} = \frac{i}{\hbar}\left[\rho_{op}^{M}, \mathcal{H}_{op}\right]. \tag{2.34}$$

2.2.3 Macroscopic consequences

Macroscopic systems are *always* quantum-coupled to their surroundings, i.e. many-particle quantum systems entangled with their quantum environment. It is here that distinctive characteristics of the mixed state density operator cause it to emerge as the correct object for describing macroscopic (thermodynamic) systems.

Expanding the pure state $|\Psi\rangle$ in a linear coherent superposition of system eigenstates $|\omega_i\rangle$,

$$|\Psi\rangle = \sum_i |\omega_i\rangle\langle\omega_i|\Psi\rangle, \tag{2.35}$$

its corresponding density operator is

$$\rho_{op} = \sum_i |\langle\omega_i|\Psi\rangle|^2\,|\omega_i\rangle\langle\omega_i| + \sum_{i\neq j}|\omega_i\rangle\rho_{ij}\langle\omega_j| \tag{2.36}$$

whose off-diagonal terms are typical of quantum interference.

The density operator formulation accommodates an approach in which a system, such as defined by Eq. 2.35, interacts with a quantum environment resulting in an entangled system–environment density operator. Later, the uninteresting and unmeasurable environmental degrees of freedom are identified and summed (traced) out, leaving a reduced density operator consisting of two components.

- The first component is a mixed state density operator ρ_{op}^{s} of the system alone, independent of the environment,

$$\rho_{op}^{s}=\sum_{i}|\langle\omega_i|\Psi\rangle|^2|\omega_i\rangle\langle\omega_i|. \tag{2.37}$$

  Here $\mathcal{P}(\omega_i)=|\langle\omega_i|\Psi\rangle|^2$ are the probabilities of Eq. 2.16 for which

$$\sum_{i}\mathcal{P}(\omega_i)=1. \tag{2.38}$$

  Clearly ρ_{op}^{s} has less than complete system information.
- The second component is a sum of off-diagonal interference terms containing a system–environment interaction dependent factor that rapidly decays with time. This term represents information "lost" to the environment.

The density operator sum of Eqs. 2.37 represents an ensemble of pure system states having all the mixed state properties exhibited in Eqs. 2.31–2.34.

With the loss of non-diagonal interference terms, the reduced density operator has crossed a boundary from pure quantum results (displaying interference) into a classical regime with no quantum interference. This is the meaning and consequence of *entangled environmental decoherence*.[9–12] Using ρ_{op}^{s} macroscopic system averages can now – in principle – be calculated.

The equation of motion for the density operator ρ_{op}^{s} is found by the method leading to Eq. 2.34, and gives

$$\frac{\partial\rho_{op}^{s}}{\partial t}=\frac{i}{\hbar}\left[\rho_{op}^{s},\mathcal{H}_{op}\right]. \tag{2.39}$$

Since we are interested in thermal equilibrium, i.e. thermodynamic variables are not on average changing in time, we also require

$$\frac{\partial\rho_{op}^{s}}{\partial t}=0. \tag{2.40}$$

It therefore follows that at thermal equilibrium

$$\left[\rho_{op}^{s},\mathcal{H}_{op}\right]=0, \tag{2.41}$$

i.e. the mixed state density operator ρ_{op}^{s} and the quantum mechanical hamiltonian $\mathcal{H}_{op}$ commute, guaranteeing that the density operator and the hamiltonian have the

[9] E. Joos and H. D. Zeh, "The emergence of classical properties through interaction with the environment", *Z. Phys. B* **59**, 223 (1985).

[10] E. Joos, H. D. Zeh, et al. *Decoherence and the Appearance of a Classical World in Quantum Theory*, Springer, Berlin (2003).

[11] W. H. Zurek, "Decoherence and the transition to the classical", *Phys. Today* **44**, 36–44 (1991).

[12] Maximilian Schlosshauer, *Decoherence and the Quantum to Classical Transition*, Springer, New York (2007).

same set of macroscopic eigenstates. Specifically basing Eq. 2.35 on the linear superposition generated from the eigenstates of

$$\mathcal{H}_{op}|E_s\rangle = E_s|E_s\rangle, \tag{2.42}$$

where E_s are macroscopic system eigen-energies and $|E_s\rangle$ are corresponding eigenstates, entangled environmental coupling reduces Eq.2.37 to a specific thermal density operator

$$\rho^{\tau}_{op} = \sum_s \mathcal{P}(E_s)|E_s\rangle\langle E_s|, \tag{2.43}$$

which commutes with $\mathcal{H}_{op}$ as required and where $\mathcal{P}(E_s)$ are probabilities the macroscopic system has eigen-energies E_s.[13]

2.2.4 Thermodynamic state functions(?)

Applying the hamiltonian of Eq.1.12 the spectrum of macroscopic internal eigen-energies[14] can be found from

$$\mathbf{h}_{op}|\epsilon_s\rangle = \epsilon_s|\epsilon_s\rangle. \tag{2.44}$$

Furthermore with these eigenstates a thermal density operator can be constructed

$$\rho^{\tau}_{op} = \sum_s \mathcal{P}(\epsilon_s)|\epsilon_s\rangle\langle\epsilon_s| \tag{2.45}$$

from which, in principle, we can find the following.

1. *Internal energy* $\mathcal{U}$

$$\mathcal{U} = Tr\,\rho^{\tau}_{op}\mathbf{h}_{op} \tag{2.46}$$

$$= \sum_s \epsilon_s \mathcal{P}(\epsilon_s). \tag{2.47}$$

2. *Internal pressure* p

$$p = Tr\,\rho^{\tau}_{op}\left(-\frac{\partial \mathbf{h}_{op}}{\partial V}\right). \tag{2.48}$$

3. *Internal energy fluctuations* $\langle(\Delta\mathcal{U})^2\rangle$

$$\langle(\Delta\mathcal{U})^2\rangle = Tr\,\rho^{\tau}_{op}\big(\mathbf{h}_{op} - \langle\mathbf{h}_{op}\rangle\big)^2 \tag{2.49}$$

$$= \langle\mathbf{h}^2_{op}\rangle - \langle\mathbf{h}_{op}\rangle^2. \tag{2.50}$$

[13] In non-equilibrium thermodynamics $\partial\rho^{\tau}_{op}/\partial t \neq 0$, which requires evaluation of Eq. 2.39. This equation can be studied by a systematic thermodynamic perturbation theory.

[14] Internal energy is generally taken to be the kinetic and potential energy of constituent particles. Total energy, on the other hand, reflects energy contributions from all interactions.

Thermodynamics is beginning to emerge from quantum theory. However, there remains an obstacle – no measurement or set of independent measurements can reveal ρ_{op}^{τ}, suggesting that ρ_{op}^{τ} are unknowable! Is this fatal? The question is addressed in Chapter 6.

2.3 Thermal density operator ρ_{op}^{τ} and entropy

Macroscopic quantum systems are entangled with an environment. As a result:

1. They are not isolated.
2. They are not representable as pure Schrödinger states.
3. They "leak" information to their environment.

As defined in Eq. 2.18[15,16] purity covers the range of values

$$0 < \mathcal{I} = \mathcal{T}r \left(\rho_{op}^{\tau} \right)^2 \leq 1 \tag{2.51}$$

(see Appendix A).

Now, beginning an adventurous progression, purity is replaced by an alternative logarithmic scale

$$s = -\ln \mathcal{T}r \left(\rho_{op}^{\tau} \right)^2, \tag{2.52}$$

which, according to Eq. 2.51, now covers the range $0 \leq s < \infty$, where

$$s = 0 \Rightarrow \text{complete information,} \tag{2.53}$$

$$s > 0 \Rightarrow \text{missing information.} \tag{2.54}$$

Finally, a more physically based alternative to purity (see Appendix B) is defined as

$$\mathcal{F} = -\kappa \, \mathcal{T}r \, \rho_{op}^{\tau} \ln \rho_{op}^{\tau}, \tag{2.55}$$

where κ is a real, positive scale constant.[17] It covers the same range as s but with different scaling:

$$\mathcal{F} = 0 \equiv \text{complete information (biased probabilities)} \Rightarrow \text{pure state,}$$

$$\mathcal{F} > 0 \equiv \text{incomplete information (unbiased probabilities)} \Rightarrow \text{mixed state.}$$

Although as introduced here $\mathcal{F}$ appears nearly identical to the quantity Gibbs called entropy, it is not yet the entropy of thermodynamics. As in Example 2.2 below (and

[15] The following result is proved in Appendix A.

[16] Zero "purity" means that all outcomes are equally likely.

[17] In thermodynamics $\kappa \to k_B T$ (Boltzmann's constant $\times$ temperature.)

everywhere else in this book), only the maximum value assumed by $\mathcal{F}$, i.e. $\mathcal{F}_{max}$, can claim distinction as *entropy*.

Even quantum mechanics' exceptional power offers no recipe for finding ρ_{op}^{τ} (see Eq. 2.43). Overcoming this obstacle requires that we look outside quantum mechanics for a well-reasoned postulate whose results accord with the measurable, macroscopic, physical universe. This is the subject of Chapter 6.

2.3.1 Examples

Example 2.1

Consider a density matrix

$$\rho = \begin{pmatrix} 1 & 0 \\ 0 & 0 \end{pmatrix}. \tag{2.56}$$

After matrix multiplication

$$(\rho)^2 = \begin{pmatrix} 1 & 0 \\ 0 & 0 \end{pmatrix}\begin{pmatrix} 1 & 0 \\ 0 & 0 \end{pmatrix} = \begin{pmatrix} 1 & 0 \\ 0 & 0 \end{pmatrix}, \tag{2.57}$$

so its corresponding purity $\mathcal{I} = \mathcal{T}r\,(\rho)^2$ is

$$\mathcal{I} = \mathcal{T}r\,(\rho)^2 \tag{2.58}$$

$$= 1, \tag{2.59}$$

corresponding to total information – a pure state (totally biased probabilities.) As introduced in Eq. 2.55,

$$\mathcal{F}/\kappa = -\mathcal{T}r\,\rho \ln \rho, \tag{2.60}$$

whose evaluation requires a little analysis.

If $f(\mathbf{M})$ is a function of the operator or matrix $\mathbf{M}$, then its trace $\mathcal{T}r\,f(\mathbf{M})$ can be evaluated by first assuming that the function $f(\mathbf{M})$ implies a power series in that matrix. For example,

$$e^{\mathbf{M}} = 1 + \mathbf{M} + \frac{1}{2}\mathbf{M}^2 + \cdots. \tag{2.61}$$

Then, since the trace of an operator (matrix) is independent of the basis in which it is taken, the trace can be written

$$\mathcal{T}r\,f(\mathbf{M}) = \sum_m \langle m| f(\mathbf{M}) |m\rangle, \tag{2.62}$$

where the matrix $\langle m| f(\mathbf{M}) |m\rangle$ is in the basis that diagonalizes $\mathbf{M}$,

$$\mathbf{M}|m\rangle = \mu_m |m\rangle, \tag{2.63}$$

with μ_m the eigenvalues of $\mathbf{M}$ and $|m\rangle$ its eigenfunctions. Referring to Eq. 2.56 we see ρ is already diagonal with eigenvalues $\{0, 1\}$. Therefore[18]

$$\frac{\mathcal{F}}{\kappa} = -\{1 \times \ln(1) + 0 \times \ln(0)\} \tag{2.64}$$

$$= 0, \tag{2.65}$$

corresponding to a pure state – maximal information (totally biased probabilities).

Example 2.2

Now consider the density matrix

$$\rho = \begin{pmatrix} \frac{1}{2} & 0 \\ 0 & \frac{1}{2} \end{pmatrix} \tag{2.66}$$

for which matrix multiplication gives

$$(\rho)^2 = \begin{pmatrix} \frac{1}{2} & 0 \\ 0 & \frac{1}{2} \end{pmatrix} \begin{pmatrix} \frac{1}{2} & 0 \\ 0 & \frac{1}{2} \end{pmatrix}$$

$$= \begin{pmatrix} \frac{1}{4} & 0 \\ 0 & \frac{1}{4} \end{pmatrix}. \tag{2.67}$$

Therefore its purity $\mathcal{I}$ is

$$\mathcal{T}r\,(\rho)^2 = \frac{1}{2}, \tag{2.68}$$

which corresponds to a mixed state with less than maximal information. The matrix of Eq. 2.66, which is already diagonal with eigenvalues $\left\{\frac{1}{2}, \frac{1}{2}\right\}$, has

$$\frac{\mathcal{F}}{k} = -\left\{\frac{1}{2} \times \ln\left(\frac{1}{2}\right) + \frac{1}{2} \times \ln\left(\frac{1}{2}\right)\right\} \tag{2.69}$$

$$= \ln 2, \tag{2.70}$$

also implying missing information – not a pure state. Moreover, for a two-dimensional density matrix $\mathcal{F}/k = \ln 2$ is the highest value $\mathcal{F}/k$ can attain. This maximal value, $\mathcal{F}_{max}/k = \ln 2$, describes *least biased* probabilities. Furthermore,

[18] $\lim_{x\to 0} x \ln x = 0$.

$\mathcal{F}_{max}$ is now revealed as the entropy of thermodynamics – the quantity invented by Clausius,[19] interpreted by Boltzmann[20] and formalized by Gibbs and especially von Neumann[21,22] (see Appendix B).

Problems and exercises

2.1 In a two-dimensional space a density matrix (hermitian with $\mathcal{T}r\rho = 1$) is

$$\rho = \begin{pmatrix} \alpha & X \\ X^* & 1-\alpha \end{pmatrix}.$$

a. Calculate $\mathcal{F}/\kappa = -\mathcal{T}r\,\rho \ln \rho$ as a function of α and X.
b. Find α and X that minimize $\mathcal{F}/\kappa$. What is $\mathcal{F}_{min}/\kappa$?
c. Find α and X that maximize $\mathcal{F}/\kappa$. What is the entropy $\mathcal{F}_{max}/\kappa$? (See Section 2.3.1.)

2.2 In a three-dimensional space a density matrix (hermitian with $\mathcal{T}r\rho = 1$) is

$$\rho = \begin{pmatrix} \alpha_1 & X & Y \\ X^* & \alpha_2 & Z \\ Y^* & Z^* & 1-\alpha_1-\alpha_2 \end{pmatrix}.$$

Find X, Y, Z, α_1 and α_2 that maximize $\mathcal{F}/k_B$. What is the entropy $\mathcal{F}_{max}/k_B$?

2.3 Consider two normalized pure quantum states:

$$|\psi_1\rangle = \frac{1}{\sqrt{3}}|+\tfrac{1}{2}\rangle + \imath\sqrt{\frac{2}{3}}|-\tfrac{1}{2}\rangle, \tag{2.71}$$

$$|\psi_2\rangle = \frac{1}{\sqrt{5}}|+\tfrac{1}{2}\rangle - \frac{2}{\sqrt{5}}|-\tfrac{1}{2}\rangle. \tag{2.72}$$

where $|+\frac{1}{2}\rangle$ and $|-\frac{1}{2}\rangle$ are orthonormal eigenstates of the operator S_z. Let them be incoherently mixed by some machine in equal proportions, i.e. $w_1 = w_2 = \frac{1}{2}$.

a. Find the two-dimensional density matrix (in the basis $|+\frac{1}{2}\rangle, |-\frac{1}{2}\rangle$) corresponding to this mixed state.

[19] R. Clausius, *The Mechanical Theory of Heat, with its Applications to the Steam-Engine and to the Physical Properties of Bodies,* John Van Voorst, London (1867).

[20] Ludwig Boltzmann, "Über die Mechanische Bedeutung des Zweiten Hauptsatzes der Wärmetheorie", *Wiener Berichte* **53**, 195–220 (1866).

[21] John von Neumann, *Mathematical Foundations of Quantum Mechanics,* Princeton University Press (1996).

[22] Entropy was also introduced in a context of digital messaging *information* by Claude Shannon, "A mathematical theory of communication", *Bell System Technical Journal* **27**, 379–423, 623–656 (1948).

b. Find the average value $\langle S_z \rangle$ in a measurement of the mixed state's S_z spin components.
c. Find $\mathcal{F}/\kappa$ for this mixed state.

2.4 Consider two normalized pure quantum states:

$$|\psi_1\rangle = \frac{1}{\sqrt{3}}|+\tfrac{1}{2}\rangle + \imath\sqrt{\frac{2}{3}}|-\tfrac{1}{2}\rangle \tag{2.73}$$

$$|\psi_2\rangle = \frac{1}{\sqrt{5}}|+\tfrac{1}{2}\rangle - \frac{2}{\sqrt{5}}|-\tfrac{1}{2}\rangle. \tag{2.74}$$

where $|+\frac{1}{2}\rangle$ and $|-\frac{1}{2}\rangle$ are orthonormal eigenstates of the operator S_z. Let them be incoherently mixed by some machine such that $w_1 = \frac{1}{3}$, $w_2 = \frac{2}{3}$.

a. Find the two-dimensional density matrix (in the basis $|+\frac{1}{2}\rangle$, $|-\frac{1}{2}\rangle$) corresponding to this mixed state.
b. Find the average value $\langle S_z \rangle$ in a measurement of the mixed state's S_z spin components.
c. Find $\mathcal{F}/\kappa$ for this mixed state.

3 Work, heat and the First Law

> ...there is in the physical world one agent only, and this is called *Kraft* [energy]. It may appear, according to circumstances, as motion, chemical affinity, cohesion, electricity, light and magnetism; and from any one of these forms it can be transformed into any of the others.
>
> Karl Friedrich Mohr, *Zeitschrift für Physik* (1837)

3.1 Introduction

Taking respectful note of Professor Murray Gell-Mann's epigraphic remark in Chapter 2, a respite from quantum mechanics seems well deserved. Therefore, the discussion in Chapter 1 is continued with a restatement of the First Law of Thermodynamics:

$$\Delta\mathcal{U} = \mathcal{Q} - \mathcal{W}, \tag{3.1}$$

where $\mathcal{U}$ is internal energy, a thermodynamic state function with quantum foundations. In this chapter the "classical" terms $\mathcal{W}$ (work) and $\mathcal{Q}$ (heat) are discussed, followed by a few elementary applications.

3.1.1 Work, $\mathcal{W}$

Work $\mathcal{W}$ is energy transferred between a system and its surroundings by mechanical or electrical changes at its boundaries. It is a notion imported from Newton's mechanics with the definition

$$\mathcal{W} = \int \mathbf{F} \cdot d\mathbf{r}, \tag{3.2}$$

where $\mathbf{F}$ is an applied force and $d\mathbf{r}$ is an infinitesimal displacement. The integral is carried out over the path of the displacement.

Work is neither stored within a system nor does it describe any state of the system. Therefore $\Delta\mathcal{W} \stackrel{?}{=} \mathcal{W}_{final} - \mathcal{W}_{initial}$ has absolutely *no* thermodynamic meaning!

Incremental work, on the other hand, can be defined as $đ\mathcal{W}$, with finite work being the integral

$$\mathcal{W} = \int đ\mathcal{W}. \tag{3.3}$$

The bar on the d is a reminder that $đ\mathcal{W}$ is *not* a true mathematical differential and that the integral in Eq. 3.3 depends on the path.[1]

Macroscopic parameters are further classified as:

- extensive – proportional to the size or quantity of matter (e.g. energy, length, area, volume, number of particles, polarization, magnetization);
- intensive – independent of the quantity of matter (e.g. force, temperature, pressure, density, tension, electric field, magnetic field).

Incremental work has the generalized meaning of energy expended by/on the system due to an equivalent of force (*intensive*) acting with an infinitesimal equivalent of displacement (*extensive*) which, in accord with Eq. 3.2, is written as the "conjugate" pair,

$$đ\mathcal{W} = \sum_j \underbrace{\mathrm{F}_j}_{\text{intensive}} \overbrace{\mathrm{d}x_j}^{\text{extensive}} , \tag{3.4}$$

where F_j is generalized force and $\mathrm{d}x_j$ is its conjugate differential displacement. If work takes place in infinitesimal steps, continually passing through equilibrium states, the work is said to be *quasi-static* and is distinguished by the representation $đ\mathcal{W}_{QS}$.

Examples of conjugate "work" pairs, intensive $\leftrightarrow$ d(extensive) are:

- pressure $(p) \leftrightarrow$ volume $(\mathrm{d}V)$;
- tension $(\boldsymbol{\tau}) \leftrightarrow$ elongation $(\mathrm{d}\boldsymbol{\chi})$;
- stress $(\boldsymbol{\sigma}) \leftrightarrow$ strain $(\mathrm{d}\boldsymbol{\epsilon})$;
- surface tension $(\boldsymbol{\Sigma}) \leftrightarrow$ surface area $(\mathrm{d}\boldsymbol{A}_S)$;
- electric field $(\mathcal{E}) \leftrightarrow$ polarization $(\mathrm{d}\mathcal{P})$;
- magnetic field $(\mathcal{B}) \leftrightarrow$ magnetization $(\mathrm{d}\boldsymbol{M})$;[2]
- chemical potential $(\mu) \leftrightarrow$ particle number $(\mathrm{d}N)$.

(It is often helpful in thermodynamics to think of mechanical work as the equivalent of raising and lowering weights external to the system.)

Finite work, as expressed by Eq. 3.3, depends on the details of the process and not just on the process end-points.

The sign convention adopted here[3] is

$$đ\mathcal{W} > 0 \Rightarrow \textit{system performs work} \quad \left(đ\mathcal{W}^{gas}_{by} = p\,\mathrm{d}V\right),$$

$$đ\mathcal{W} < 0 \Rightarrow \textit{work done on the system} \quad \left(đ\mathcal{W}^{gas}_{on} = -p\,\mathrm{d}V\right).$$

[1] Work is not a function of independent thermal variables so that $đ\mathcal{W}$ cannot be an exact differential.
[2] The fields used in accounting for electric and magnetic work will be further examined in Chapter 11.
[3] In the case of pressure and volume as the conjugate pair.

3.1.2 Heat, $\mathcal{Q}$

Heat $\mathcal{Q}$ is energy transferred between a system and any surroundings with which it is in diathermic contact. $\mathcal{Q}$ is not stored within a system and is not a state of a macroscopic system. Therefore $\Delta\mathcal{Q} \stackrel{?}{=} \mathcal{Q}_{final} - \mathcal{Q}_{initial}$ has *no* thermodynamic meaning! On the other hand, *incremental heat* is defined as $đ\mathcal{Q}$, with finite heat $\mathcal{Q}$ as the integral

$$\mathcal{Q} = \int đ\mathcal{Q}. \tag{3.5}$$

Finite heat depends on the details of the process and not just the integral's end-points.

The bar on the *d* is an essential reminder that $đ\mathcal{Q}$ is not a function of temperature or any other independent variables. It is *not* a true mathematical differential so "derivatives" like $\frac{đ\mathcal{Q}}{\mathrm{d}T}$ have dubious meaning.

Processes for which $\mathcal{Q} = 0$, either by thermal isolation or by extreme rapidity (too quickly to exchange heat), are said to be *adiabatic*.

3.1.3 Temperature, T

Temperature is associated with a subjective sense of "hotness" or "coldness". However, in thermodynamics temperature has empirical (objective) meaning through thermometers – e.g. the height of a column of mercury, the pressure of an ideal gas in a container with fixed volume, the electrical resistance of a length of wire – all of which vary in a way that allows a temperature scale to be defined. Since there is no quantum hermitian "temperature operator", a formal, consistent definition of *temperature* is elusive. But it is achievable in practice through thermal measurements[4] showing, with remarkable consistency, how and where it thermodynamically appears and that it measures the same property as the common thermometer.

3.1.4 Internal energy, $\mathcal{U}$

As discussed in Chapter 1, *internal energy* $\mathcal{U}$ *is* a state of the system. Therefore $\Delta\mathcal{U}$ *does* have meaning and $\mathrm{d}\mathcal{U}$ *is* an ordinary mathematical differential[5] with

$$\Delta\mathcal{U} = \mathcal{U}(B) - \mathcal{U}(A) = \int_A^B \mathrm{d}\mathcal{U}, \tag{3.6}$$

where A and B represent different macroscopic equilibrium states. Clearly, $\Delta\mathcal{U}$ depends *only* on the end-points of the integration and *not* on any particular thermodynamic process (path) connecting them.

[4] This subject will be discussed in Chapter 6.

[5] It is called an exact differential.

3.2 Exact differentials

All thermodynamic state variables are true (exact) differentials with change in value as defined in Eq. 3.6. Moreover, state variables are not independent and can be functionally expressed in terms of other state variables. Usually only a few are needed to completely specify any state of a system. The precise number required is the content of an important result which is stated here without proof:

> The number of independent parameters needed to specify a thermodynamic state is equal to the number of distinct conjugate pairs (see Section 3.1.1) of quasi-static work needed to produce a differential change in the internal energy – *plus ONE*.[6]

For example, if a gas undergoes mechanical expansion or compression, the only quasi-static work is mechanical work, $đ\mathcal{W}_{QS} = p\,\mathrm{d}V$. Therefore only two variables are required to specify a state. We could then write the internal energy of a system as a function of any two other state parameters. For example, we can write $\mathcal{U} = \mathcal{U}(T, V)$ or $\mathcal{U} = \mathcal{U}(p, V)$. To completely specify the volume of a gas we could write $V = V(p, T)$ or even $V = V(p, \mathcal{U})$. The choice of functional dependences is determined by the physical situation and computational objectives.

If physical circumstances dictate particle number changes, say by evaporation of liquid to a gas phase or condensation of gas to the liquid phase, or chemical reactions where atomic or molecular species are lost or gained, a thermodynamic description requires "chemical work" $đ\mathcal{W}_{QS} = -\mu\,\mathrm{d}N$, where $\mathrm{d}N$ is an infinitesimal change in particle number and μ is the (conjugate) chemical potential. Chemical potential μ is an intensive variable denoting energy per particle while particle number N is obviously an extensive quantity. For cases where there is both mechanical and chemical work,

$$đ\mathcal{W}_{QS} = p\,\mathrm{d}V - \mu\,\mathrm{d}N, \tag{3.7}$$

three variables are required and, for example, $\mathcal{U} = \mathcal{U}(T, V, N)$ or $\mathcal{U} = \mathcal{U}(p, V, \mu)$.

Incorporating the definitions above, an incremental form of the First Law of Thermodynamics is

$$\mathrm{d}\mathcal{U} = đ\mathcal{Q} - đ\mathcal{W}. \tag{3.8}$$

3.3 Equations of state

Equations of state are functional relationships among a system's equilibrium state parameters – usually those that are most easily measured and controlled in the

[6] The *ONE* of the *plus ONE* originates from the conjugate pair T and $d\mathcal{S}$, where $\mathcal{S}$ is entropy, about which much will be said throughout the remainder of this book. $T\ d\mathcal{S}$ can be thought of as quasi-static thermal work.

laboratory. Determining equations of state, by experiment or theory, is often a primary objective in practical research, for then thermodynamics can fulfill its role in understanding and predicting state parameter changes for particular macroscopic systems.

The following are samples of equations of state for several different types of macroscopic systems. Their application is central in the Examples subsections below.

1. Equations of state for gases (often called gas laws) take the functional form $f(p, V, T) = 0$, where p is the equilibrium internal gas pressure, V is the gas volume, T is the gas temperature and f defines the functional relationship. For example, a gas at low pressure and low particle density, $n = N/V$, where N is the number of gas particles (atoms, molecules, etc.), has the equation of state

$$p - (N/V)\, k_B T = 0 \tag{3.9}$$

or, more familiarly (the ideal gas law),

$$pV = N k_B T, \tag{3.10}$$

where k_B is Boltzmann's constant.[7,8]

2. An approximate equation of state for two- and three-dimensional metallic materials is usually expressed as $g(\sigma, \epsilon, T) = 0$, where $\sigma = \tau/A$ is the stress, $\epsilon = (\ell - \ell_0)/\ell_0$ is the strain,[9] T is the temperature, and A is the cross-sectional area. An approximate equation of state for a metal rod is the Hookian behavior

$$\epsilon = \kappa\, \sigma \tag{3.11}$$

where κ is a material-specific constant.

3. An equation of state for a uniform one-dimensional elastomeric (rubber-like) material has the form $\Lambda(\tau, \chi, T) = 0$, where τ is the equilibrium elastic tension, χ is the sample elongation, T is the temperature and Λ is the functional relationship. For example, a rubber-like polymer that has not been stretched beyond its elastic limit can have the simple equation of state

$$\tau - \mathscr{K} T \chi = 0, \tag{3.12}$$

or

$$\sigma - \mathscr{K}' T \epsilon = 0, \tag{3.13}$$

where $\mathscr{K}$ and $\mathscr{K}'$ are material-specific elastic constants.[10,11]

[7] For a derivation see Chapter 7.

[8] For gases under realistic conditions Eq. 3.10 may be inadequate requiring more complicated equations of state. Nevertheless, because of its simplicity and "zeroth order" validity, the ideal gas "law" is pervasive in models and exercises.

[9] ℓ is the stretched length and ℓ_0 is the unstretched length.

[10] H. M. James and E. Guth, "Theory of elastic properties of rubber", *J. Chem. Phys.* **11**, 455–481 (1943).

[11] Comparing these equations of state with Eq. 3.11 suggests a physically different origin for rubber elasticity. This will be explored in Chapter 10.

4. Dielectrics with no permanent polarization in *weak* polarizing electric fields have an equation of state

$$\mathcal{P}_i = \sum_j \alpha_{ij}(T)\,\mathcal{E}_j, \tag{3.14}$$

where $\mathcal{P}_i$ is an electric polarization vector component, $\alpha_{ij}(T)$ is the temperature-dependent electric susceptibility tensor and $\mathcal{E}_j$ is an electric field component.[12] In stronger fields equations of state may acquire non-linear terms,[13] such as

$$\mathcal{P}_i = \sum_j \alpha_{ij}^{(1)}(T)\mathcal{E}_j + \sum_{j,k} \alpha_{ijk}^{(2)}(T)\mathcal{E}_j\mathcal{E}_k. \tag{3.15}$$

5. Non-permanently magnetized materials which are uniformly magnetized in weak magnetic fields are described by the equation of state

$$\boldsymbol{M} = \chi_M(T)\,\boldsymbol{\mathcal{B}}_0, \tag{3.16}$$

where $\boldsymbol{M}$ is the magnetization vector, $\chi_M(T)$ is a temperature-dependent magnetic susceptibility tensor and $\mathcal{B}_0$ is the existing magnetic induction field prior to insertion of the sample.[14,15]

3.3.1 Examples I

Quasi-static reversible work

A gas is confined in a cylinder by a frictionless piston of negligible mass and cross-sectional area A. The gas pressure p is counteracted by a pile of fine sand of total mass m resting on the piston (see Figure 3.1) so that it is initially at rest. Neglecting any atmospheric (external) force acting on the piston, the downward force exerted by the sand is initially $\mathcal{F}_0^{sand} = mg$. Therefore the initial equilibrium gas pressure p_0^{gas} is

$$p_0^{gas} = \frac{mg}{A}. \tag{3.17}$$

Sand is now removed "one grain at a time", allowing the gas to expand in infinitesimal steps, continually passing through equilibrium states. Moreover, it is conceivable that at any point a single grain of sand could be returned to the pile and exactly restore the previous equilibrium state. Such a real or idealized infinitesimal process is called *quasi-static*. A quasi-static process, by definition, takes place so slowly that the system (in this example a gas) always passes through equilibrium states with an equation of state always relating system pressure, volume and temperature.

[12] Is this an external electric field or some average local internal electric field? How do Maxwell's fields fit in here? This will be addressed in Chapter 11. For the present, we assume that the sample is ellipsoidal in which case the internal and external electric fields are the same.

[13] Robert W. Boyd, *Nonlinear Optics*, Academic Press, San Diego CA (2003).

[14] Magnetism and magnetic work will be discussed in Chapter 11.

[15] Solid-state physics textbooks often express Eq. 3.16 in terms of an external magnetic field $\mathcal{H}$.

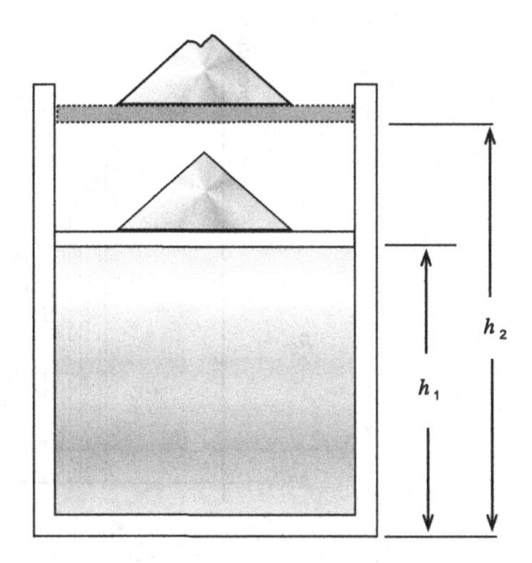

Fig. 3.1 Quasi-static reversible work. A piston loaded down with a pile of fine sand.

Note: If, as in this case, the previous state can be restored by an infinitesimal displacement – with no dissipation of energy – the process is also said to be *reversible*.[16]

Question 3.1 Assuming the quasi-static expansion also takes place isothermally at temperature T_R, what work is done by the gas in raising the piston from h_1 to h_2?

Solution

At all stages of the quasi-static expansion equilibrium is maintained with pressure p^{gas} and temperature T_R related by, say, the ideal gas law. Therefore incremental work done by the gas is

$$\text{đ}\mathcal{W}_{QS} = p^{gas}\, \mathrm{d}V \tag{3.18}$$

$$= \frac{N\, k_B\, T_R}{V}\, \mathrm{d}V. \tag{3.19}$$

Integrating Eq. 3.19 along an isothermal path (as pictured in Figure 3.2) gives

$$\mathcal{W}_{QS} = N\, k_B\, T_R \int_{V_1=A\,h_1}^{V_2=A\,h_2} \frac{\mathrm{d}V}{V} \tag{3.20}$$

$$= N\, k_B\, T_R\, \ln\left(V_2/V_1\right) \tag{3.21}$$

$$= N\, k_B\, T_R\, \ln\left(p^{gas}_{-f}/p^{gas}_{-o}\right). \tag{3.22}$$

[16] Not all quasi-static processes are reversible but all reversible processes are quasi-static. Reversibility implies no friction or other energy dissipation in the process.

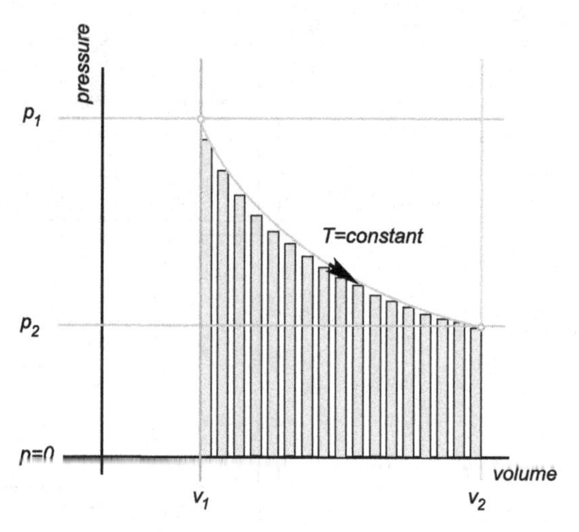

Fig. 3.2 $p - V$ diagram for an isothermal quasi-static process.

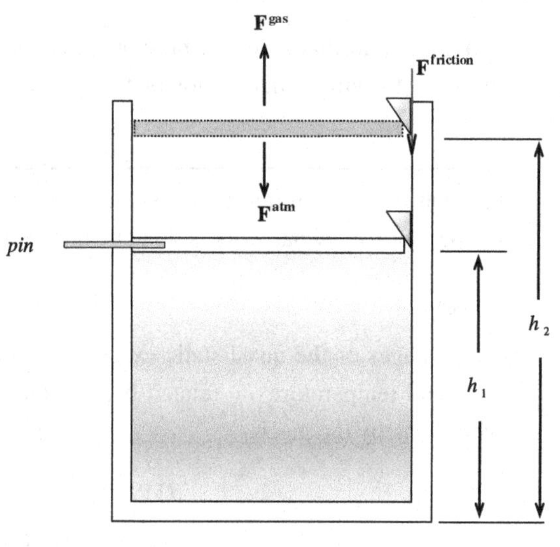

Fig. 3.3 Irreversible work: a gas-filled piston initially pinned at height h_1.

Irreversible work

Instead of a restraining sand pile, the piston is now pinned a distance h_1 above the bottom of the cylinder (see Figure 3.3). At the same time a constant atmospheric pressure P^{atm} exerts an external force $\mathcal{F}^{atm}$ on the piston,

$$\mathcal{F}^{atm} = P^{atm} A.$$

Quickly extracting the pin frees the piston, allowing the gas to expand rapidly and *non-uniformly*. After things quieten down the gas reaches an equilibrium state, the piston having moved upward a distance $\ell = h_2 - h_1$ with a final gas pressure

$p_f^{gas} = p^{atm}$. This isothermal process is characterized by a finite expansion rate and system (gas) non-uniformity. It is neither quasi-static nor reversible.[17]

Question 3.2 What isothermal work is done by the rapidly expanding gas in raising the piston a distance ℓ?

Solution

The gas expands irreversibly so that during expansion a gas law does not apply. But initially the gas has equilibrium pressure p_0^{gas} with volume V_0,

$$V_0 = \frac{N k_B T}{p_0^{gas}}. \tag{3.23}$$

After the piston ceases its erratic motion it returns to rest with the gas in equilibrium at atmospheric pressure,

$$p_f^{gas} = p^{atm}, \tag{3.24}$$

with volume

$$V_f = \frac{N k_B T}{p^{atm}}. \tag{3.25}$$

During expansion the gas lifts the piston against a constant force $\mathcal{F}^{atm}$ (equivalent to raising a weight $mg = \mathcal{F}^{atm}$ in the Earth's gravitational field).[18] The incremental, irreversible work done by the confined gas (the system) is

$$đ\mathcal{W} = p^{atm}\, dV \tag{3.26}$$

which, with constant p^{atm}, is integrated to give

$$\mathcal{W} = p^{atm}\left(V_f - V_0\right). \tag{3.27}$$

With Eqs. 3.23 and 3.25

$$\mathcal{W} = N k_B T\left[1 - \frac{p^{atm}}{p_0^{gas}}\right]. \tag{3.28}$$

---

[17] All real processes in nature are irreversible. Reversibility is a convenient idealization for thermodynamic modeling.

[18] The piston starts at rest and returns to rest. There is no change in the piston's kinetic energy.

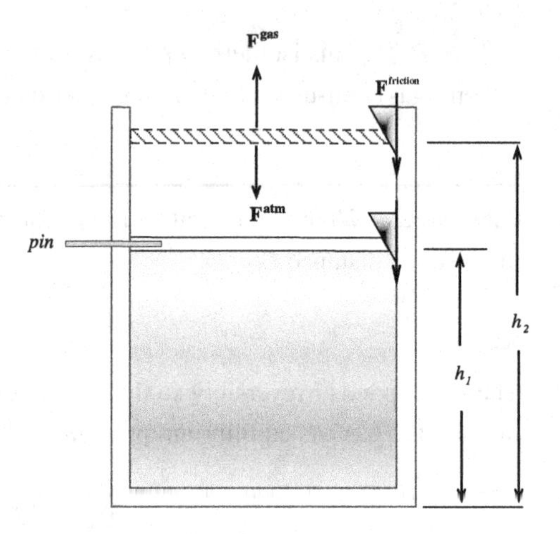

Fig. 3.4 Quasi-static irreversible work: a piston slowed by a frictional wedge.

Quasi-static irreversible work

A gas is confined by a piston whose position is again secured by a pin. The piston is exposed to atmospheric pressure P^{atm} and the initial gas pressure is $p_0 > P^{atm}$. Removing the pin allows the gas to expand with the piston's upward movement opposed by both the atmosphere and a wedge of variable friction (see Figure 3.4). The wedge "conspires" with the external atmosphere to create an approximately quasi-static expansion. However, as a result of friction the expansion is irreversible.

Question 3.3 Regarding the gas alone as the system and assuming the process is isothermal, what work is done by the quasi-statically expanding gas?

Solution

Gas pressure p^{gas} quasi-statically lifts the piston against the combined force $F^{atm} + F^{friction}$. Incremental work done by the gas, assumed ideal, is

$$\begin{aligned} đ\mathcal{W}_{QS} &= p^{gas}\,\mathrm{d}V \\ &= \frac{Nk_BT}{V}\,\mathrm{d}V, \end{aligned} \tag{3.29}$$

which after integration becomes

$$\mathcal{W}_{QS} = Nk_BT\ln\left(\frac{V_f}{V_0}\right) \tag{3.30}$$

$$= Nk_BT\ln\left(\frac{p_0}{p_{atm}}\right). \tag{3.31}$$

Examining the same process from another point of view, take as the system the combined piston, gas and frictional wedge. The sole external force acting on this

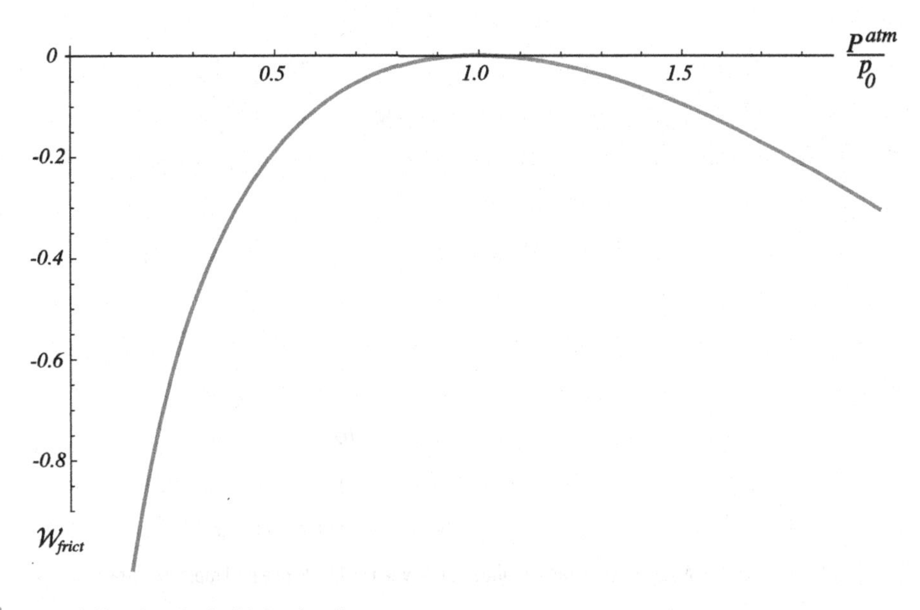

Fig. 3.5 Work $\mathcal{W}_{frict}$ done by the frictional wedge.

system is the constant atmospheric pressure P^{atm}. Therefore the work done by the system against the atmosphere is

$$\mathcal{W}_{sys} = P^{atm}\left(V_f - V_0\right) \tag{3.32}$$

or

$$\mathcal{W}_{sys} = Nk_B T\left(1 - \frac{p_{atm}}{p_0}\right). \tag{3.33}$$

The work done by the frictional wedge is the difference

$$\mathcal{W}_{friction} = Nk_B T\left[\left(1 - \frac{p_{atm}}{p_0}\right) - \ln\left(\frac{p_0}{p_{atm}}\right)\right], \tag{3.34}$$

which is negative for either upward ($p_0 > P^{atm}$) or downward ($p_0 < P^{atm}$) movement of the piston (as shown in Figure 3.5), underscoring the irreversibility of the process.

3.3.2 Examples II

A loaded rubber band

A mass $2\,m$, affixed to the end of a vertically stretched massless rubber band, rests a distance h_1 (rubber band length ℓ_1) above a table. When half the mass is suddenly

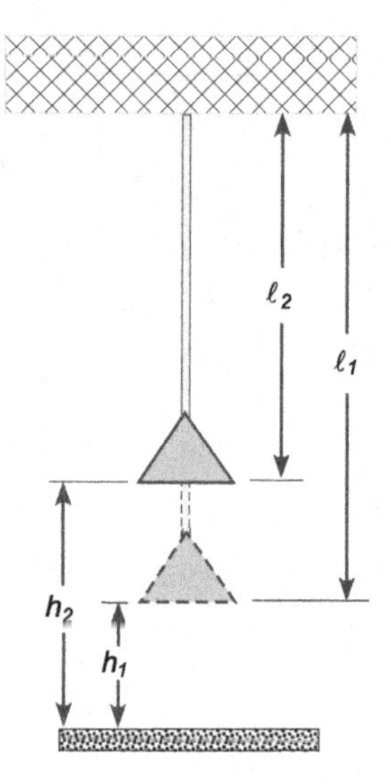

Fig. 3.6 A loaded, constrained rubber band when suddenly released rapidly contracts.

removed the rubber band rapidly contracts causing the remaining mass m to rise. After oscillating for a while it finally comes to rest a distance h_2 ($h_2 > h_1$) (rubber band length ℓ_2) above the table. (See Figure 3.6.)

Question 3.4 What is the work done by the rubber band?

Solution

After returning to rest the work done (not quasi-static) *by* the rubber band (work done *by* the system) in raising the weight m (against gravity) is

$$\mathcal{W} = -mg\Delta\ell, \tag{3.35}$$

where $\Delta\ell = \ell_2 - \ell_1 < 0$.

Question 3.5 What is the change in the rubber band's internal energy?

Solution

Since the process takes place rapidly it is irreversible. Moreover, there is not enough time for the rubber band to lose heat to the surroundings and the process can be regarded as adiabatic. Using the First Law

$$\Delta\mathcal{U} = +mg\,\Delta\ell. \tag{3.36}$$

The rubber band loses internal energy.

Question 3.6 This time, to counteract rubber elasticity, the total mass $2\,m$ is restrained to rise slowly (quasi-statically) but isothermally. What is the work done by the contracting rubber band?

Solution

We can safely use an equilibrium tension (an elastic state parameter) which acts to raise the weight. The incremental quasi-static elastic work done by the system is

$$đ\mathcal{W}_{QS} = -\tau \; d\chi, \tag{3.37}$$

where $d\chi$ is the differential of elastic extension. Using the "rubber" equation of state, Eq. 3.12, and integrating over the range of extension χ,

$$\mathcal{W}_{QS} = -\mathcal{K}'\,T \int_{-\Delta\ell}^{0} \chi \; d\chi \tag{3.38}$$

$$= \mathcal{K}'\,T\,(\Delta\ell)^2 . \tag{3.39}$$

A compressed metal rod

A metal rod, cross-sectional area A, is suddenly struck by a sledgehammer, subjecting it to a transient but constant compressive (external) force $\mathcal{F}$ which momentarily changes the rod's length by $\Delta\ell = \ell_f - \ell_0 < 0$.

Question 3.7 What is the work done on the rod during the interval that its length decreases by $|\Delta\ell|$?

Solution

Sudden compression of the rod implies that it does not pass through equilibrium states so an equilibrium internal stress in the bar $\sigma = f/A$ is not defined. But the mechanical work done *on the rod* by the constant compressive force is, however,

$$\mathcal{W} = -\mathcal{F}\,\Delta\ell. \tag{3.40}$$

Question 3.8 On the other hand, if the rod could be compressed infinitely slowly (perhaps by a slowly cranked vise) the rod's internal stress is a well-defined equilibrium property and the quasi-static incremental work done by the system (i.e. the rod) is

$$đ\mathcal{W}_{QS} = -V\sigma \; d\epsilon, \tag{3.41}$$

where V is the rod's volume, σ is the internal stress and $d\epsilon = (d\ell/\ell)$ defines the differential strain. What is the work done by the rod?

Solution

We could now apply an approximate equation of state, such as Eq. 3.11, and get

$$\mathcal{W}_{QS} = -V \int_{\epsilon_0}^{\epsilon_f} (\epsilon/\kappa)\ d\epsilon \tag{3.42}$$

$$= -(V/2\kappa)\left(\epsilon_f^2 - \epsilon_0^2\right). \tag{3.43}$$

3.4 Heat capacity

An infinitesimal amount of thermal energy $đ\mathcal{Q}_{QS}$ quasi-statically entering or leaving a system generally produces an infinitesimal change in temperature dT,

$$đ\mathcal{Q}_{QS} = \mathcal{C}_\alpha\ dT \quad \text{(constant } \alpha\text{)}. \tag{3.44}$$

(Exceptions are certain first-order phase transitions, such as boiling of water or melting of ice.) The proportionality constant between the two infinitesimals, $\mathcal{C}_\alpha$, together with the list of state variables, α, that are fixed during the quasi-static process, is called *heat capacity at constant* α.[19] The units of heat capacity are energy K^{-1}. *Specific heats* are alternative intensive quantities, having units energy K^{-1} per unit mass or energy K^{-1} per mole.

3.4.1 Heat transfer at constant volume

If heat is transferred at constant volume,[20] by the "definition" in Eq. 3.44,

$$đ\mathcal{Q}_{QS} = \mathcal{C}_V\ dT \quad \text{(constant volume)}. \tag{3.45}$$

Heat capacities can be equivalently expressed in terms of partial derivatives of state parameters. For example, if a gas performs only quasi-static mechanical work, the infinitesimal First Law for the process is

$$d\mathcal{U} = đ\mathcal{Q}_{QS} - p\ dV, \tag{3.46}$$

[19] The term *heat capacity* is a legacy from early thermodynamics when heat was regarded as a stored quantity. Although we now know better, the language seems never to have similarly evolved.
[20] And constant particle number.

where $p\ \mathrm{d}V$ is quasi-static incremental work done by the gas (system). Since for this case $\mathcal{U}$ can be expressed in terms of two other thermal parameters, choose $\mathcal{U} = \mathcal{U}(T, V)$ so that its total differential is

$$d\mathcal{U} = \left(\frac{\partial \mathcal{U}}{\partial V}\right)_T \mathrm{d}V + \left(\frac{\partial \mathcal{U}}{\partial T}\right)_V \mathrm{d}T \tag{3.47}$$

and Eq. 3.46 can then be rewritten

$$đ\mathcal{Q}_{QS} = \left[\left(\frac{\partial \mathcal{U}}{\partial V}\right)_T + p\right] \mathrm{d}V + \left(\frac{\partial \mathcal{U}}{\partial T}\right)_V \mathrm{d}T. \tag{3.48}$$

Thus, for an isochoric (constant volume) path, i.e. $\mathrm{d}V = 0$,

$$đ\mathcal{Q}_{QS} = \left(\frac{\partial \mathcal{U}}{\partial T}\right)_V \mathrm{d}T \tag{3.49}$$

and from Eq. 3.45

$$\mathcal{C}_V = \left(\frac{\partial \mathcal{U}}{\partial T}\right)_V. \tag{3.50}$$

3.4.2 Heat transfer at constant pressure

Thermodynamic experiments, especially chemical reactions, are often carried out in an open atmosphere where the pressure is constant. In such cases we define

$$đ\mathcal{Q}_{QS} = \mathcal{C}_p\ \mathrm{d}T \quad \text{(constant pressure)}. \tag{3.51}$$

Expressing $\mathcal{C}_p$ as the partial derivative of a state parameter requires introducing a new, abstract, state variable. Here again the infinitesimal First Law description of the quasi-static process is

$$d\mathcal{U} = đ\mathcal{Q}_{QS} - p\ \mathrm{d}V. \tag{3.52}$$

But this time, using

$$\mathrm{d}(pV) = p\ \mathrm{d}V + V\ \mathrm{d}p, \tag{3.53}$$

Eq. 3.52 can be rewritten as

$$đ\mathcal{Q}_{QS} = \mathrm{d}(\mathcal{U} + pV) - V\ \mathrm{d}p. \tag{3.54}$$

Defining

$$H = \mathcal{U} + pV, \tag{3.55}$$

where H – also a state function – is called *enthalpy*,[21] permits recasting a "new" differential "law":

$$đ\mathcal{Q}_{QS} = dH - V\ \mathrm{d}p. \tag{3.56}$$

[21] Enthalpy is one of several defined state variables that are introduced to simplify specific thermodynamic constraints. In this case the constraint is constant pressure. Others will be introduced as needed.

Then, expressing H in terms of T and p, i.e. $H = H(T, p)$ so that a total differential is

$$\mathrm{d}H = \left(\frac{\partial H}{\partial p}\right)_T \mathrm{d}p + \left(\frac{\partial H}{\partial T}\right)_p \mathrm{d}T, \tag{3.57}$$

for an isobaric (constant pressure) path, i.e. $\mathrm{d}p = 0$,

$$đQ_{QS} = \left(\frac{\partial H}{\partial T}\right)_p \mathrm{d}T \tag{3.58}$$

and we have the definition of C_p in terms of state variables

$$C_p = \left(\frac{\partial H}{\partial T}\right)_p . \tag{3.59}$$

Heat capacity measurements play a surprisingly important role in exploring microscopic properties of matter. The unexpectedly low temperature heat capacity of diamond led to a quantum theory of the solid state. Equations of state, together with heat capacities, can offer a complete thermal picture of most matter.

3.4.3 Examples III

Ideal monatomic gas

An ideal monatomic gas, volume V_0, temperature T_0, is confined by a frictionless, movable piston to a cylinder with diathermic walls. The external face of the piston is exposed to the atmosphere at pressure P_0. A transient pulse of heat is injected through the cylinder walls causing the gas to expand against the piston until it occupies a final volume V_f.

Question 3.9 What is the change in temperature of the gas?

Solution

At the beginning and end of the expansion the system is in equilibrium so that an equation of state applies, i.e. initially $P_0 V_0 = N k_B T_0$ and finally $P_0 V_f = N k_B T_f$. Therefore

$$\Delta T = \frac{P_0 \Delta V}{N k_B}, \tag{3.60}$$

where $\Delta T = T_f - T_0$ and $\Delta V = V_f - V_0$.

Question 3.10 How much heat $\mathcal{Q}$ was in the pulse?

Solution

The gas will expand turbulently, i.e. irreversibly, against constant atmospheric pressure P_0 (an external "weight") so that incremental work done by the expanding gas is

$$đ\mathcal{W} = P_0 \, dV. \tag{3.61}$$

Applying the First Law

$$\mathcal{Q} = \Delta\mathcal{U} + P_0 \, \Delta V \tag{3.62}$$

and using Eqs. 3.47, 3.50 and 3.60, together with the ideal gas property $(\partial u/\partial V) = 0$, gives the internal energy change

$$\Delta\mathcal{U} = \mathcal{C}_V \, \Delta T \tag{3.63}$$

$$= \frac{\mathcal{C}_V P_0 \Delta V}{N k_B}. \tag{3.64}$$

Thus the injected heat is

$$\mathcal{Q} = \left(\frac{\mathcal{C}_V}{N k_B} + 1 \right) P_0 \Delta V, \tag{3.65}$$

where Eqs. 3.62 and 3.64 have been used. Using another ideal gas property[22]

$$\mathcal{C}_p - \mathcal{C}_V = N k_B \tag{3.66}$$

and the definition

$$\gamma = \mathcal{C}_p / \mathcal{C}_V, \tag{3.67}$$

Eq. 3.65 can be restated as

$$\mathcal{Q} = \frac{\gamma}{\gamma - 1} P_0 \, \Delta V. \tag{3.68}$$

A steel wire

A steel wire, length L and radius ρ, is held with zero tension τ between two firmly planted rigid vertical posts, as shown in Figure 3.7. The wire, initially at a temperature T_H, is then cooled to a temperature T_L.

Question 3.11 Find an expression for the tension τ in the cooled wire.

[22] This well-known result for ideal gases is stated, for now, without proof.

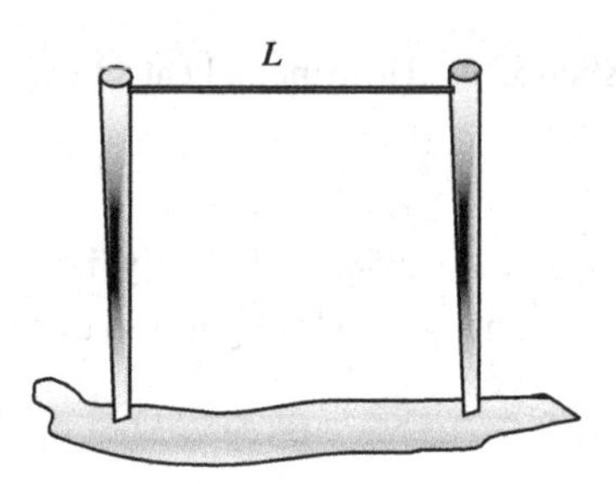

Fig. 3.7 A wire held between firmly planted rigid vertical posts.

Solution

The thermodynamic state variables used for describing elastic behavior in the steel wire (or rod) are stress $\sigma = \tau/A$, where A is the wire's cross-sectional area, and *strain* $\epsilon = (\ell - \ell_0)/\ell_0$, where ℓ and ℓ_0 are the wire's stretched and unstretched lengths. These are the same variables referred to in elastic work (see Eq. 3.41) and in the elastic equation of state (see Eq. 3.16).

In the present situation the rigid posts constrain the length of the wire to remain constant so that a decrease in temperature is expected to increase the strain (tension) in the wire. Referring to Section 3.2, consider a functional dependence of strain $\epsilon = \epsilon(T, \sigma)$, which are the physically relevant state variables for the wire, and take the total differential

$$\mathrm{d}\epsilon = \left(\frac{\partial \epsilon}{\partial T}\right)_\sigma \mathrm{d}T + \left(\frac{\partial \epsilon}{\partial \sigma}\right)_T \mathrm{d}\sigma. \tag{3.69}$$

The partial derivatives represent physical properties of the system under investigation for which extensive numerical tables can be found in the literature or on the web.[23] In Eq. 3.69

$$\left(\frac{\partial \epsilon}{\partial T}\right)_\sigma = \alpha_L, \tag{3.70}$$

where α_L is the coefficient of linear thermal expansivity, and

$$\left(\frac{\partial \epsilon}{\partial \sigma}\right)_T = \frac{1}{E_T}, \tag{3.71}$$

where E_T is the isothermal Young's modulus.

Thus we can write

$$\mathrm{d}\epsilon = \alpha_L\, \mathrm{d}T + \frac{1}{E_T}\, \mathrm{d}\sigma, \tag{3.72}$$

and since the length of the wire is constant, i.e. $\mathrm{d}\epsilon = 0$, we have

$$0 = \alpha_L\, \mathrm{d}T + \frac{1}{E_T}\, \mathrm{d}\sigma. \tag{3.73}$$

[23] See, for example, http://www.periodictable.com/Properties/A/YoungModulus.html.

Finally, assuming α_L and E_T are average constant values, integration gives

$$\Delta\sigma = \sigma(T_L) - \sigma(T_H) = -\alpha_L\, E_T\,(T_L - T_H)\,, \tag{3.74}$$

and from the definition of stress given above, $\tau(T_H) = 0$,

$$\tau(T_L) = -\, A\alpha_L E_T\,(T_L - T_H) \tag{3.75}$$

$$= \;\pi r^2 \alpha_L E_T\,(T_H - T_L)\,. \tag{3.76}$$

A rapidly expanding gas

A gas is allowed to expand rapidly from an initial state (p_0, V_0, T_0) to a final state f.

Question 3.12 What is the final state (p_f, V_f, T_f)?

Solution

This rapid process can be regarded as taking place adiabatically, $\mathcal{Q} = 0$. Applying the infinitesimal First Law we have

$$đ\mathcal{Q} = 0 = \,\mathrm{d}\mathcal{U} + đ\mathcal{W}. \tag{3.77}$$

Since $\mathcal{U}$ is a state variable the integral

$$\Delta\mathcal{U} = \int d\mathcal{U} \tag{3.78}$$

is path (i.e. process) independent. Therefore Eq. 3.77 is integrated to become

$$0 = \Delta\mathcal{U} + \int đ\mathcal{W} \tag{3.79}$$

which depends only on equilibrium end-points. So the equilibrium end-points can be connected by a unique, quasi-static process

$$0 = \int_{p_0,\,V_0,\,T_0}^{p_f,V_f,T_f} \left[\mathrm{d}\mathcal{U} + p\,\mathrm{d}V\right]. \tag{3.80}$$

Then with $\mathrm{d}\mathcal{U}$ as in Eq. 3.47

$$0 = \int_{p_0,\,V_0,\,T_0}^{p_f,V_f,T_f} \left\{\left[\left(\frac{\partial\mathcal{U}}{\partial V}\right)_T + p\right]\mathrm{d}V + \left(\frac{\partial\mathcal{U}}{\partial T}\right)_V \mathrm{d}T\right\}. \tag{3.81}$$

So far the result is general (any equation of state). But for simplicity assume the gas is ideal. Using Eq. 3.49 and applying the ideal gas property[24]

$$\left(\frac{\partial \mathcal{U}}{\partial V}\right)_T = 0 \tag{3.82}$$

together with the ideal gas equation of state (see Eq. 3.10) gives

$$0 = \int_{p_0, V_0, T_0}^{p_f, V_f, T_f} \left\{ \left[\frac{N k_B T}{V}\right] \mathrm{d}V + \mathcal{C}_V \, \mathrm{d}T \right\}, \tag{3.83}$$

which implies[25]

$$0 = \left[\frac{N k_B T}{V}\right] \mathrm{d}V + \mathcal{C}_V \, \mathrm{d}T. \tag{3.84}$$

This differential equation separates quite nicely to give upon integration

$$N k_B \ln\left(\frac{V_f}{V_0}\right) = \mathcal{C}_V \ln\left(\frac{T_0}{T_f}\right) \tag{3.85}$$

or

$$\left(\frac{V_f}{V_0}\right)^{N k_B} = \left(\frac{T_0}{T_f}\right)^{\mathcal{C}_V} \tag{3.86}$$

which with Eqs. 3.66 and 3.67 becomes[26]

$$\left(\frac{V_f}{V_0}\right)^{\gamma - 1} = \left(\frac{T_0}{T_f}\right). \tag{3.87}$$

The filling problem: atmospheric gas filling an evacuated vessel

An insulated, rigid, very narrow-necked vessel is evacuated and sealed with a valve. The vessel rests in an atmosphere with pressure P_0 and temperature T_0 (Figure 3.8). The valve is *suddenly* opened and air quickly fills the vessel until the pressure inside the vessel also reaches P_0.

Question 3.13 If the air is treated as an ideal gas, what is the temperature T_f of the air inside the vessel immediately after equilibrium is attained?

Solution

In this problem we introduce the strategy of replacing an essentially open system (the number of molecules inside the vessel is not constant) with a closed system (the

[24] This useful ideal gas result is the basis of a later problem.
[25] Note that the integral still represents Q, which is *not* a state variable.
[26] This ideal gas expression is not an equation of state.

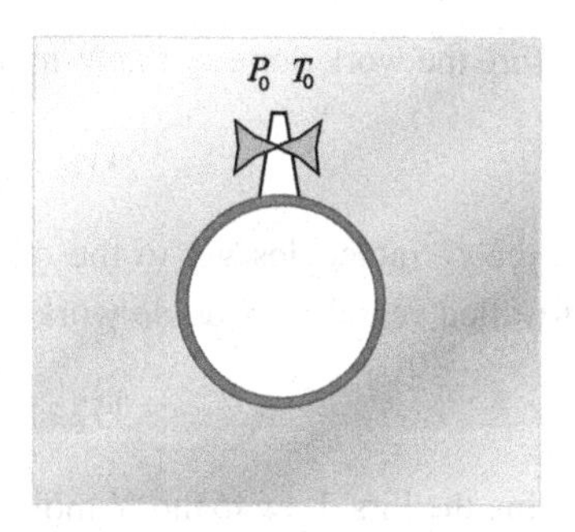

Fig. 3.8 Evacuated, insulated vessel in an atmosphere with pressure P_0 and temperature T_0.

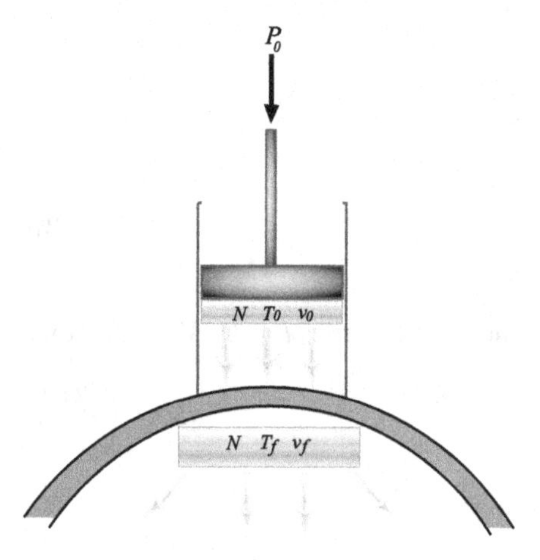

Fig. 3.9 Evacuated vessel filling under pressure P_0.

gas is conceptually bounded by a movable piston that seals the neck and, ultimately, the closed vessel below, as in Figure 3.8). This notional replacement is a common trick in thermodynamics and we take the opportunity to illustrate it here.

Physically, the narrow neck separates the air inside the vessel from the air outside the vessel, the latter maintained at constant pressure P_0 by the infinite atmosphere. The strategy is to replace the external atmosphere with a piston that forces outside air across the narrow neck into the vessel by applying the constant external pressure P_0, as shown in Figure 3.9.

Focus attention on a small, fixed amount of air being forced into the vessel by the conjectured piston, say N molecules at pressure P_0 and temperature T_0 occupying a small volume v_0. In forcing these molecules through the valve the piston displaces the volume v_0. The work done *on* the N molecules *by* the constant pressure piston is

$$\mathcal{W}_{by\ the\ piston} = \int đW \tag{3.88}$$

$$= P_0 \Delta v \tag{3.89}$$

$$= P_0 v_0. \tag{3.90}$$

Therefore the work done *by* the N molecules of gas is

$$\mathcal{W}_{by\ the\ gas} = -P_0 v_0. \tag{3.91}$$

Once the N molecules get to the other side of the valve and enter the evacuated rigid-walled vessel, they do *no* work.[27] Therefore the total work done *by* the gas is

$$\mathcal{W}_{total} = -P_0\, v_0 + 0. \tag{3.92}$$

Applying the First Law to the N molecules while approximating this sudden process as adiabatic[28]

$$0 = \Delta\mathcal{U} + \mathcal{W}_{total} \tag{3.93}$$

or

$$0 = -P_0 v_0 + \int_0^f d\mathcal{U}. \tag{3.94}$$

Choosing to write $\mathcal{U} = \mathcal{U}(V, T)$ the total differential is

$$d\mathcal{U} = \left(\frac{\partial \mathcal{U}}{\partial T}\right)_V dT + \left(\frac{\partial \mathcal{U}}{\partial V}\right)_T dV. \tag{3.95}$$

Since

$$\left(\frac{\partial \mathcal{U}}{\partial T}\right)_V = C_V \tag{3.96}$$

and for an ideal gas

$$\left(\frac{\partial \mathcal{U}}{\partial V}\right)_T = 0 \tag{3.97}$$

we have

$$P_0 v_0 = \mathcal{C}_V \left(T_f - T_0\right). \tag{3.98}$$

Then with the ideal gas equation of state

$$N k_B T_0 = \mathcal{C}_V \left(T_f - T_0\right) \tag{3.99}$$

followed by Eqs. 3.66 and 3.67, this becomes

$$T_f = \gamma T_0. \tag{3.100}$$

[27] This is evident since no weights are raised or lowered.

[28] Too fast to allow heat exchange with the external atmosphere or heat to leak outside the vessel.

At room temperature ($T_0 = 300\,\mathrm{K}$) and for an ideal monatomic gas ($\gamma = 5/3$), Eq. 3.100 gives $T_f = 500\,\mathrm{K}$, indicating a substantial rise in temperature for the gas streaming into the vessel.

3.5 Concluding remarks

In this hiatus from quantum mechanics we have resumed the discussion of classical thermodynamics, emphasizing the distinction between $\mathcal{U}$ and purely classical work $\mathcal{W}$ and "heat" $\mathcal{Q}$.

Chapter 4 pauses to develop some mathematical skills essential for thermodynamic applications.

Problems and exercises

3.1 Isothermal compressibility κ_T is defined as

$$\kappa_T = -\frac{1}{V}\left(\frac{\partial V}{\partial p}\right)_T. \tag{3.101}$$

a. Describe in words the meaning of κ_T.

b. Find κ_T for an ideal gas.

c. A more descriptive gas law is the van der Waals equation

$$\left(p + \frac{aN^2}{V^2}\right)(V - Nb) = Nk_BT, \tag{3.102}$$

where the constant b describes volume excluding short-range repulsion between gas molecules and the constant a takes account of long-range intermolecular attraction. Find κ_T for a van der Waals gas.

3.2 A thick rubber band diameter d is fastened between two rigid vertical posts a distance L apart. When the temperature is T_0 the rubber band has positive tension τ_0.

Assuming rubber has an equation of state

$$\sigma = \mathscr{K}T\epsilon \tag{3.103}$$

at what temperature T_1 will the tension be $\tau = 0.9\tau_0$?

3.3 An ideal gas expands adiabatically from the state (V_0, p_0) to the state (V_f, p_f). Show that the work done by the expanding gas is

$$\mathcal{W} = \frac{(p_0 V_0 - p_f V_f)}{\gamma - 1}, \tag{3.104}$$

where

$$\gamma = \frac{\mathcal{C}_p}{\mathcal{C}_v}. \tag{3.105}$$

3.4 The constant volume heat capacity for a hypothetical nearly ideal gas is, at low temperature,

$$\mathcal{C}_V = a + bT, \tag{3.106}$$

where a and b are constants. Show that for an adiabatic, reversible expansion carried out at low temperature

$$pV^{\gamma} \approx \exp\{(b/a)\,T\} \tag{3.107}$$

with

$$\gamma = \frac{\mathcal{C}_p}{\mathcal{C}_v}. \tag{3.108}$$

3.5 Although the heat capacity at constant volume is

$$\mathcal{C}_V = \left(\frac{\partial \mathcal{U}}{\partial T}\right)_V, \tag{3.109}$$

show that for the heat capacity at constant pressure

$$\mathcal{C}_p \neq \left(\frac{\partial \mathcal{U}}{\partial T}\right)_p \tag{3.110}$$

by proving that

$$\mathcal{C}_p = \left(\frac{\partial \mathcal{U}}{\partial T}\right)_p + pV\alpha, \tag{3.111}$$

where α is the volume thermal expansivity

$$\alpha = \frac{1}{V}\left(\frac{\partial V}{\partial T}\right)_p. \tag{3.112}$$

Note: The heat capacity at constant pressure is

$$\mathcal{C}_p = \left(\frac{\partial H}{\partial T}\right)_p \tag{3.113}$$

where enthalpy H is defined as

$$H = \mathcal{U} + pV. \tag{3.114}$$

4 A mathematical digression

To those who do not know Mathematics it is difficult to get across a real feeling as to the beauty, the deepest beauty of nature. ... If you want to learn about nature, to appreciate nature, it is necessary to understand the language that she speaks in.

Richard Feynman, *The Character of Physical Law*, MIT Press (1967)

4.1 Thermodynamic differentials

In thermodynamics the equilibrium state of a system is described by macroscopic variables, say $\mathcal{U}, p, T, V, \ldots$, which can roughly be classified as:

- variables that are readily measurable in a laboratory such as $p, V, T, \mathcal{C}_v$, etc.;
- variables that are less accessible to direct measurement, e.g. $\mathcal{U}$, H, and others, soon to be introduced.

The choice of variables is generally from among those that appear in thermodynamic laws, those that can be measured or controlled in laboratories (such as appear in equations of state) and those defining process constraints, i.e. isothermal, isobaric or adiabatic. Since physics of interest involves changes in equilibrium macroscopic states, we must study differential relationships among them. Assuming a state variable $\phi\,(x, y, z)$, where x, y, z are also (independent) state variables, these relationships have the mathematical form

$$\mathrm{d}\phi\,(x, y, z, \ldots) = \left(\frac{\partial \phi}{\partial x}\right)_{y,z,\ldots} \mathrm{d}x + \left(\frac{\partial \phi}{\partial y}\right)_{x,z,\ldots} \mathrm{d}y + \left(\frac{\partial \phi}{\partial z}\right)_{x,y,\ldots} \mathrm{d}z \ldots, \tag{4.1}$$

where

$$\left(\frac{\partial \phi}{\partial x}\right)_{y,z,\ldots}, \quad \left(\frac{\partial \phi}{\partial y}\right)_{x,z,\ldots}, \quad \left(\frac{\partial \phi}{\partial z}\right)_{x,y,\ldots} \tag{4.2}$$

are partial derivatives, i.e. derivatives with respect to $x, y, z, \ldots$, with the remaining independent variables held constant. In thermodynamics the variables that are held constant dictate specific experimental conditions and must be retained in all expressions. They are not merely inconvenient baggage that one is free to carelessly misplace.

Several mathematical rules associated with differentials and partial differentiation especially useful in thermodynamics are introduced in this chapter and then applied in sample scenarios.

4.2 Exact differentials

Finite changes in thermal state parameters, e.g. $\Delta\mathcal{U}$, depend only on final and initial state values and not on the path or process driving the change. This independence defines "exactness" of a differential. More generally, exactness of any thermal parameter, say $\mathcal{Y}$, means that the closed circuit integral $\oint \mathrm{d}\mathcal{Y}$ satisfies the closed path condition

$$\oint \mathrm{d}\mathcal{Y} = 0. \tag{4.3}$$

Even though integrals corresponding to finite work $\int đ\mathcal{W} = \mathcal{W}$ and finite heat $\int đ\mathcal{Q} = \mathcal{Q}$ have macroscopic meaning, as stressed earlier $\mathcal{W}$ and $\mathcal{Q}$ are not state parameters and do not have exact differentials, i.e. $\oint đ\mathcal{W} \neq 0$ and $\oint đ\mathcal{Q} \neq 0$. That is the meaning of the bar on the *d*.

4.2.1 Exactness

A class of thermodynamic identities called *Maxwell relations* play an important role in problem solving strategies by replacing "unfriendly" partial derivatives – i.e. those that do not obviously represent accessible measurables – with equivalent "friendly" partial derivatives that contain heat capacities or state variables that are part of an equation of state. For example, measured values of the following partial derivatives, called *response functions*, are found in tables of physical properties of materials[1] and can be placed in the "friendly" category:

$$\kappa_T = -V\left(\frac{\partial p}{\partial V}\right)_T, \quad \text{isothermal bulk modulus,} \tag{4.4}$$

$$\beta_T = -\frac{1}{V}\left(\frac{\partial V}{\partial p}\right)_T, \quad \text{isothermal compressibility,} \tag{4.5}$$

$$\alpha_p = \frac{1}{V}\left(\frac{\partial V}{\partial T}\right)_p, \quad \text{coefficient of thermal expansion.} \tag{4.6}$$

Several elastic coefficients previously encountered in Chapter 3 may also be found in tables. Partial derivatives in the "unfriendly" category will appear in our analyses

[1] See, for example, D.R. Lide (ed.), *Chemical Rubber Company Handbook of Chemistry and Physics*, CRC Press, Boca Raton, Florida, USA (79th edition, 1998).

and will certainly be recognized. Methods for dealing with them are introduced later in the chapter.

4.2.2 Euler's criterion for exactness

If $M = M(x, y)$ is a differentiable function of the independent variables x and y whose total differential is

$$\mathrm{d}M = A\,(x, y)\ \mathrm{d}x + B\,(x, y)\ \mathrm{d}y, \tag{4.7}$$

where by definition

$$A\,(x, y) = \left(\frac{\partial M}{\partial x}\right)_y \tag{4.8}$$

and

$$B\,(x, y) = \left(\frac{\partial M}{\partial y}\right)_x, \tag{4.9}$$

then according to Green's theorem

$$\oint \mathrm{d}M = \oint \left[A\,(x, y)\ \mathrm{d}x + B\,(x, y)\ \mathrm{d}y\right] \tag{4.10}$$

$$= \iint_R \left[\left(\frac{\partial B}{\partial x}\right)_y - \left(\frac{\partial A}{\partial y}\right)_x\right] \mathrm{d}x\ \mathrm{d}y, \tag{4.11}$$

where R is a two-dimensional region completely enclosed by the integration path. Then if

$$\left(\frac{\partial B}{\partial x}\right)_y = \left(\frac{\partial A}{\partial y}\right)_x \tag{4.12}$$

it follows that

$$\oint \mathrm{d}M = 0 \tag{4.13}$$

i.e., M is exact. Conversely, if M is exact, i.e. is a thermodynamic state function, then

$$\left(\frac{\partial B}{\partial x}\right)_y = \left(\frac{\partial A}{\partial y}\right)_x. \tag{4.14}$$

4.2.3 Entropy: a thermal introduction

Consider the case of a gas expanding quasi-statically, for which the First Law says

$$\int đ\,\mathcal{Q}_{QS} = \int \mathrm{d}\mathcal{U} + \int p\ \mathrm{d}V \tag{4.15}$$

where p is the gas pressure. In Chapter 1 it was argued, as a matter of definition, that $\int đ\mathcal{Q}_{QS}$ must depend on the process path, i.e. $\oint đ\mathcal{Q}_{QS} \neq 0$. Delving deeper, this claim is tested with Euler's criterion for exactness.

Beginning with Eq. 3.48, as applied to an ideal gas,

$$đ\mathcal{Q}_{QS} = \left(\frac{\partial \mathcal{U}}{\partial T}\right)_V \mathrm{d}T + p\,\mathrm{d}V \tag{4.16}$$

$$= \mathcal{C}_V\,\mathrm{d}T + \frac{Nk_BT}{V}\,\mathrm{d}V. \tag{4.17}$$

Then applying Euler's criterion we find

$$\left(\frac{\partial \mathcal{C}_V}{\partial V}\right)_T = 0 \tag{4.18}$$

and

$$\left(\frac{\partial}{\partial T}\frac{Nk_BT}{V}\right)_V \neq 0. \tag{4.19}$$

So for this basic test case $đ\mathcal{Q}_{QS}$ fails the exactness test and, revisiting old news, $đ\mathcal{Q}_{QS}$ is not a true differential.

What may be a surprise however, is that exactness can be "induced" by the expedient of introducing an integrating factor $1/T$, in which case Eq. 4.17 becomes

$$\frac{đ\mathcal{Q}_{QS}}{T} = \left(\frac{\mathcal{C}_V}{T}\right)\mathrm{d}T + \left(\frac{Nk_B}{V}\right)\mathrm{d}V. \tag{4.20}$$

Now applying Euler's criterion

$$\left(\frac{\partial}{\partial T}\left[\frac{Nk_B}{V}\right]\right)_V = 0 \tag{4.21}$$

and

$$\left(\frac{\partial \mathcal{C}_V/T}{\partial V}\right)_T = 0. \tag{4.22}$$

The "revised" integrand $đ\mathcal{Q}_{QS}/T$ *does* pass Euler's test,

$$\oint \frac{đ\mathcal{Q}_{QS}}{T} = 0, \tag{4.23}$$

and can therefore be integrated to yield the same result over *any* path in the T, V plane. The integrating factor $1/T$ generates a "new" thermodynamic state function $\mathcal{S}$, with

$$\mathrm{d}\mathcal{S} = \frac{đ\mathcal{Q}_{QS}}{T}. \tag{4.24}$$

$\mathcal{S}$ is *entropy*, the same quantity invoked in Chapter 2 but within a different narrative.[2] In addition, the quasi-static adiabatic process ($đ\, \mathcal{Q}_{QS}/T = 0$) is equivalent to $d\mathcal{S} = 0$, an *isenthalpic* process.

Having identified the new state function $\mathcal{S}$, the incremental First Law for a fixed number of particles becomes

$$\frac{đ\, \mathcal{Q}_{QS}}{T} = d\mathcal{S} = \frac{d\mathcal{U}}{T} + \frac{p}{T}\, dV \tag{4.25}$$

or equivalently

$$d\mathcal{U} = T\, d\mathcal{S} - p\, dV \tag{4.26}$$

which is called the *thermodynamic identity*. In the case of variable particle number the *identity* becomes

$$d\mathcal{U} = T\, d\mathcal{S} - p\, dV + \mu\, dN \tag{4.27}$$

from which

$$\left(\frac{\partial \mathcal{U}}{\partial \mathcal{S}}\right)_{V,N} = T, \tag{4.28}$$

$$\left(\frac{\partial \mathcal{U}}{\partial V}\right)_{\mathcal{S},N} = -p, \tag{4.29}$$

$$\left(\frac{\partial \mathcal{U}}{\partial N}\right)_{V,\mathcal{S}} = \mu. \tag{4.30}$$

Furthermore, writing $\mathcal{S} = \mathcal{S}(\mathcal{U}, V, N)$ and taking its total differential

$$d\mathcal{S} = \left(\frac{\partial \mathcal{S}}{\partial \mathcal{U}}\right)_{V,N} dU + \left(\frac{\partial \mathcal{S}}{\partial V}\right)_{\mathcal{U},N} dV + \left(\frac{\partial \mathcal{S}}{\partial N}\right)_{\mathcal{U},V} dN \tag{4.31}$$

gives, in association with Eq. 4.27,

$$\left(\frac{\partial \mathcal{S}}{\partial \mathcal{U}}\right)_{V,N} = \frac{1}{T}, \tag{4.32}$$

$$\left(\frac{\partial \mathcal{S}}{\partial V}\right)_{\mathcal{U},N} = \frac{p}{T}, \tag{4.33}$$

$$\left(\frac{\partial \mathcal{S}}{\partial N}\right)_{\mathcal{U},V} = -\frac{\mu}{T}. \tag{4.34}$$

[2] Admittedly, this formal emergence of entropy doesn't reveal "meaning" – a subject with a long and contentious history – and its thermal introduction is only for an ideal gas. But its generality, independent of the working substance, although initially assumed, will be established in Chapter 6 with an axiomatic rendering that assigns "meaning".

Eq. 4.32 may be used to define inverse temperature, even though it is not an especially intuitive or transparent definition.[3]

With the introduction of thermodynamic identities (Eqs. 4.26 and 4.27) "unfriendly" partial derivatives can now be "friended", i.e. recast into more practical forms involving only variables that appear in an equation of state or heat capacities, as a few examples will show.

The definition of *enthalpy*, Eq. 2.44, evolves into something resembling Eq. 4.26, with

$$d\mathcal{S} = \frac{đ\,\mathcal{Q}_{QS}}{T} = \frac{dH}{T} - \frac{V}{T}\,dp \tag{4.35}$$

or equivalently

$$dH = T\,d\mathcal{S} + V\,dp\,. \tag{4.36}$$

Finally, if the internal energy differential in Eq. 4.26 is replaced by its general First Law expression (which includes irreversibilty), i.e. $d\mathcal{U} = đ\,\mathcal{Q} - đ\,\mathcal{W}$, then the entropy differential is

$$d\mathcal{S} = \frac{1}{T}đ\,\mathcal{Q} + \left[\frac{đ\,\mathcal{W}_{QS} - đ\,\mathcal{W}}{T}\right]. \tag{4.37}$$

Since quasi-static (reversible) work is the maximum work available in any thermodynamic process, i.e. $đ\,\mathcal{W}_{QS} \geq đ\,\mathcal{W}$, for any *real* (irreversible) processes,

$$d\mathcal{S} \geq \frac{đ\,\mathcal{Q}}{T}, \tag{4.38}$$

which is, in essence, the Second Law of Thermodynamics.

4.3 Euler's homogeneous function theorem

A function $\varphi\,(x, y, z)$ is *homogeneous* of order n if

$$\varphi\,(\lambda x, \lambda y, \lambda z) = \lambda^n\,\varphi\,(x, y, z)\,. \tag{4.39}$$

Euler's homogeneous function theorem states: If $\varphi\,(x, y, z)$ is homogeneous of order n, then

$$x\left(\frac{\partial\varphi}{\partial x}\right)_{y,z} + y\left(\frac{\partial\varphi}{\partial y}\right)_{x,z} + z\left(\frac{\partial\varphi}{\partial z}\right)_{x,y} = n\,\varphi\,(x, y, z)\,. \tag{4.40}$$

[3] However, it is useful in defining "negative temperature", which will be discussed in Chapter 9.

This seemingly abstract theorem[4] has several important thermodynamic implications. For example, consider the internal energy $\mathcal{U}$ as a function of the extensive variables $\mathcal{S}$, V and N (see Eq. 3.7) and write $\mathcal{U} = \mathcal{U}(\mathcal{S}, V, N)$. Since $\mathcal{S}, \mathcal{U}, V$ and N are extensive they scale according to

$$\mathcal{U}(\lambda \mathcal{S}, \lambda V, \lambda N) = \lambda \mathcal{U}(\mathcal{S}, V, N), \tag{4.41}$$

i.e. $\mathcal{U}(\mathcal{S}, V, N)$ is a homogeneous function of order 1. Applying Euler's homogeneous function theorem (see Eq. 4.40),

$$\mathcal{S}\left(\frac{\partial \mathcal{U}}{\partial \mathcal{S}}\right)_{V,N} + V\left(\frac{\partial \mathcal{U}}{\partial V}\right)_{\mathcal{S},N} + \mathcal{N}\left(\frac{\partial \mathcal{U}}{\partial N}\right)_{\mathcal{S},V} = \mathcal{U}(\mathcal{S}, V, N). \tag{4.42}$$

Then, from Eqs. 4.28 and 4.29,

$$\mathcal{U} = \mathcal{S}T - Vp + \mu N, \tag{4.43}$$

where

$$\mu = \left(\frac{\partial \mathcal{U}}{\partial N}\right)_{\mathcal{S},V} \tag{4.44}$$

is the chemical potential. Note that μ, p and T are intensive state variables conjugate to the extensive state variables N, V and $\mathcal{S}$.

4.4 A cyclic chain rule

In addition to the familiar chain rule of partial differentiation, i.e.

$$\left(\frac{\partial z}{\partial x}\right)_s \left(\frac{\partial x}{\partial y}\right)_s = \left(\frac{\partial z}{\partial y}\right)_s \tag{4.45}$$

there is the following *cyclic chain rule*.

Consider a function $M(x, y)$ and its total differential

$$dM = \left(\frac{\partial M}{\partial x}\right)_y \mathrm{d}x + \left(\frac{\partial M}{\partial y}\right)_x \mathrm{d}y \tag{4.46}$$

from which, with the additional independent variables m and n, it follows that

$$\left(\frac{\partial M}{\partial m}\right)_n = \left(\frac{\partial M}{\partial x}\right)_y \left(\frac{\partial x}{\partial m}\right)_n + \left(\frac{\partial M}{\partial y}\right)_x \left(\frac{\partial y}{\partial m}\right)_n. \tag{4.47}$$

[4] A proof of the theorem is in Appendix C.

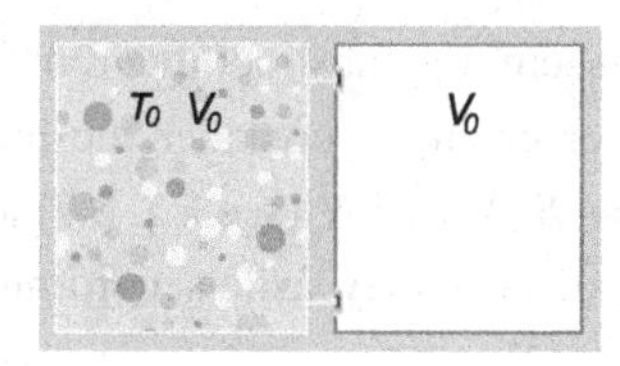

Fig. 4.1 Free expansion chamber.

With the replacements $m \to x$ and $n \to M$

$$0 = \left(\frac{\partial M}{\partial x}\right)_y + \left(\frac{\partial M}{\partial y}\right)_x \left(\frac{\partial y}{\partial x}\right)_M \tag{4.48}$$

which is reorganized as

$$\left(\frac{\partial x}{\partial y}\right)_M = -\frac{\left(\frac{\partial M}{\partial y}\right)_x}{\left(\frac{\partial M}{\partial x}\right)_y}. \tag{4.49}$$

This is often written as a "chain rule"[5]

$$\left(\frac{\partial x}{\partial y}\right)_M \left(\frac{\partial M}{\partial x}\right)_y \left(\frac{\partial y}{\partial M}\right)_x = -1. \tag{4.50}$$

4.4.1 Examples

Free expansion: the cyclic chain rule

A gas, assumed ideal with a fixed number of particles and initially at temperature T_0, is confined to volume V_0 in a partitioned insulated rigid cylinder. (See Figure 4.1.) The unoccupied part, also with volume V_0, is initially evacuated.

The partition spontaneously begins to leak and the gas surges into the formerly unoccupied part, eventually coming to equilibrium by uniformly occupying the entire cylinder.

Question 4.1 What is the change in temperature of the gas?

Solution

Since the cylinder is rigid the surging gas does no work. (The expanding gas does not raise or lower any "weights" in the surroundings.) Moreover, the cylinder is insulated

[5] This simple (but not obvious) bit of mathematics is useful in thermodynamic problem solving.

so that during the process, $\mathcal{Q} = 0$. (Regarding the expansion as rapid will lead to the same conclusion.) With

$$\mathcal{Q} = 0 \quad and \quad \mathcal{W} = 0 \tag{4.51}$$

the First Law, $\Delta\mathcal{U} = \mathcal{Q} - \mathcal{W}$, says

$$\Delta\mathcal{U} = 0; \tag{4.52}$$

i.e. internal energy is constant. This constant $\mathcal{U}$ process is called a *free expansion.* Since $\mathcal{U}$ is constant it follows that $d\mathcal{U} = 0$, so taking $\mathcal{U} = \mathcal{U}(T, V)$ its total differential is

$$d\mathcal{U} = \left(\frac{\partial\mathcal{U}}{\partial T}\right)_V dT + \left(\frac{\partial\mathcal{U}}{\partial V}\right)_T dV = 0 \tag{4.53}$$

and the temperature change we are looking for is

$$dT = -\frac{\left(\frac{\partial\mathcal{U}}{\partial V}\right)_T}{\left(\frac{\partial\mathcal{U}}{\partial T}\right)_V} dV. \tag{4.54}$$

But a little more work is needed to give ΔT in terms of practical thermal variables. Applying the cyclic chain rule (see Eq. 4.49) to this result we get

$$dT = \left(\frac{\partial T}{\partial V}\right)_{\mathcal{U}} dV. \tag{4.55}$$

The partial derivative is one of the "unfriendly" kind referred to earlier. The objective now is relating it to a "friendly" partial derivative that corresponds to laboratory measurables.

Using the thermodynamic identity, Eq. 4.26, apply the Euler exactness criterion to the differential $d\mathcal{S}$ (see Eq. 4.34), in the following way;

$$\left(\frac{\partial\left(\frac{1}{T}\right)}{\partial V}\right)_{\mathcal{U}} = \left(\frac{\partial\left(\frac{p}{T}\right)}{\partial\mathcal{U}}\right)_V. \tag{4.56}$$

After carrying out the differentiations it becomes

$$-\frac{1}{T^2}\left(\frac{\partial T}{\partial V}\right)_{\mathcal{U}} = -\frac{p}{T^2}\left(\frac{\partial T}{\partial\mathcal{U}}\right)_V + \frac{1}{T}\left(\frac{\partial p}{\partial\mathcal{U}}\right)_V. \tag{4.57}$$

Applying the "ordinary" chain rule to the last term on the right-hand side and then substituting Eq. 3.50 ($\mathcal{C}_V$), Eq. 4.57 becomes

$$\left(\frac{\partial T}{\partial V}\right)_{\mathcal{U}} = \frac{1}{\mathcal{C}_V}\left[p - T\left(\frac{\partial p}{\partial T}\right)_V\right], \tag{4.58}$$

which is now in terms of measurable quantities – our objective all along. Finally, applying the ideal gas equation of state,[6]

$$\frac{1}{C_V}\left[p - T\left(\frac{\partial p}{\partial T}\right)_V\right] = 0, \tag{4.59}$$

and we see from Eq. 4.55 that an ideal gas undergoing a free expansion remains at constant temperature.

Question 4.2 What is the change in entropy of the expanded gas?

Solution

Focusing directly on $\mathcal{S}$ we write $\mathcal{S} = \mathcal{S}(\mathcal{U}, V)$ and then take the total differential

$$\mathrm{d}\mathcal{S} = \left(\frac{\partial \mathcal{S}}{\partial \mathcal{U}}\right)_V \mathrm{d}U + \left(\frac{\partial \mathcal{S}}{\partial V}\right)_{\mathcal{U}} \mathrm{d}V. \tag{4.60}$$

Since this is a free expansion, $\mathcal{U}$ = constant ($\mathrm{d}\mathcal{U} = 0$) and

$$\mathrm{d}\mathcal{S} = \left(\frac{\partial \mathcal{S}}{\partial V}\right)_{\mathcal{U}} \mathrm{d}V. \tag{4.61}$$

Then from Eq. 4.33

$$\mathrm{d}\mathcal{S} = \frac{p}{T}\,\mathrm{d}V. \tag{4.62}$$

Then using the ideal gas equation of state

$$\mathrm{d}\mathcal{S} = \frac{Nk_B}{V}\,\mathrm{d}V. \tag{4.63}$$

Since the expansion is from V_0 to $2V_0$, this is integrated to give

$$\Delta\mathcal{S} = \mathcal{S}(2V_0) - \mathcal{S}(V_0) = Nk_B \int_{V_0}^{2V_0} \frac{\mathrm{d}V}{V} \tag{4.64}$$

$$= Nk_B \ln\frac{2V_0}{V_0} \tag{4.65}$$

$$= Nk_B \ln 2. \tag{4.66}$$

---

[6] The analysis can be applied to any gas equation of state.

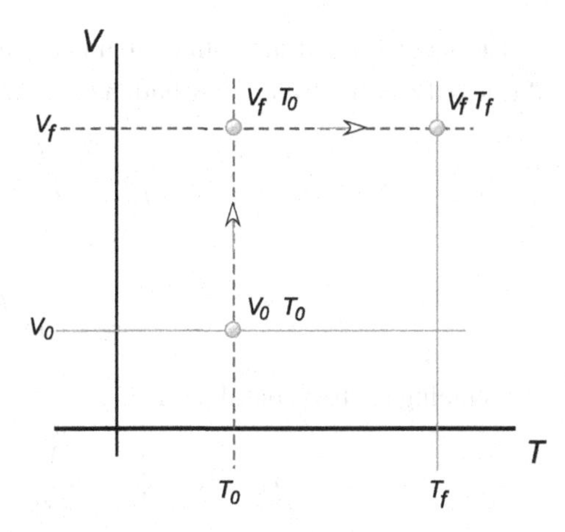

Fig. 4.2 Line paths of the "exact" integral.

If the gas was more realistically described by van der Waal's equation of state,

$$\left(p + \frac{N^2 a}{V^2}\right)(V - Nb) = Nk_B T, \tag{4.67}$$

where a and b are constants characteristic of the gas, would ΔT and $\Delta \mathcal{S}$ be different from the ideal gas?[7]

Exactness and line integration

The integral in Eq. 4.20, which arises from adiabatic expansion, is reproduced here for convenience:

$$0 = \int_{p_0,V_0,T_0}^{p_f,V_f,T_f} \left\{ \left[\frac{Nk_B}{V}\right] \mathrm{d}V + \frac{\mathcal{C}_V}{T}\,\mathrm{d}T \right\}. \tag{4.68}$$

This integral is exact. Since all paths connecting the end-points are equivalent (the virtue of exactness) this can now be line integrated along *any* path connecting the integration end-points ($V_0, T_0 \to V_f, T_f$). It is, however, good strategy to choose a simple pair of straight lines as shown in Figure 4.2 – integrating first along constant T and then integrating along constant V, as shown. Along constant T we have $\mathrm{d}T = 0$ and what survives is the integral (the first leg of the path)

$$I_1 = \int_{V_0,T_0}^{V_f,T_0} \frac{Nk_B}{V}\,\mathrm{d}V \tag{4.69}$$

$$= Nk_B \ln \frac{V_f}{V_0}. \tag{4.70}$$

[7] This is assigned as a problem.

The second leg of the path is along constant V as shown in the diagram. Along this leg $\mathrm{d}V = 0$, so this contribution is

$$I_2 = \int_{V_f,T_0}^{V_f,T_f} \frac{C_V}{T}\,\mathrm{d}T \tag{4.71}$$

$$= C_V \ln \frac{T_f}{T_0}. \tag{4.72}$$

Summing the two contributions

$$I_1 + I_2 = Nk_B \ln \frac{V_f}{V_0} + C_V \ln \frac{T_f}{T_0} = 0, \tag{4.73}$$

which is identical to Eq. 3.84, in Chapter 3.

4.5 Entropy and spontaneous processes

Example 1

As in Chapter 3 (Eqs. 3.18–3.20), consider an ideal gas, initially with volume V_0 and pressure p_0, confined in a cylinder by a massless, frictionless piston. The piston, although itself massless, is exposed to the atmosphere and loaded with a pile of fine sand that is slowly removed, allowing the piston to rise quasi-statically and attain final gas volume V_f and final pressure p^{atm}, identical to the surrounding atmosphere. The gas and surroundings are maintained at temperature T_0.

Question 4.3 What is the entropy change of the "universe"?

Solution

Apply the First Law to find $\mathcal{Q}_{QS}$, the heat absorbed quasi-statically from the surroundings by the gas. With pressure p_{gas}

$$\mathcal{Q}_{QS} = \int_{V_0}^{V_f} p_{gas} dV \tag{4.74}$$

$$= \int_{V_0}^{V_f} \frac{Nk_B T_0}{V} dV \tag{4.75}$$

$$= Nk_B T_0 \ln \frac{V_f}{V_0}. \tag{4.76}$$

Therefore the entropy change of the surroundings is

$$\Delta \mathcal{S}_{surr} = -\frac{\mathcal{Q}_{QS}}{T_0} \tag{4.77}$$

$$= -Nk_B \ln \frac{V_f}{V_0}. \tag{4.78}$$

To find the entropy change of the gas we write, at constant temperature,

$$\mathrm{d}\mathcal{S} = \left(\frac{\partial \mathcal{S}}{\partial V}\right)_T dV. \tag{4.79}$$

But since

$$\left(\frac{\partial S}{\partial V}\right)_T = \left(\frac{\partial p}{\partial T}\right)_V \tag{4.80}$$

the entropy change of the ideal gas is

$$\Delta \mathcal{S}_{gas} = \int_{V_0}^{V_f} \left(\frac{\partial p}{\partial T}\right)_V \mathrm{d}V \tag{4.81}$$

$$= Nk_B \ln \frac{V_f}{V_0}. \tag{4.82}$$

Therefore[8] the entropy change of the "universe" is, for this quasi-static reversible process,

$$\Delta \mathcal{S}_{universe} = \Delta \mathcal{S}_{gas} + \Delta \mathcal{S}_{surroundings} \tag{4.83}$$

$$= 0. \tag{4.84}$$

Example 2

An ideal gas is confined in a cylinder by a massless, frictionless piston. The piston is pinned some distance above the bottom of the cylinder while exposed to constant atmospheric pressure p^{atm} (see Figure 3.3). The initial state of the gas is $V_0, T_0, p_0 > p^{atm}$. The gas and surroundings are maintained at temperature T_0. (See Chapter 3, Eqs. 3.23–3.28.)

[8] A way to arrive at Eq. 4.80 starts from

$$\mathrm{d}\mathcal{U} = T\,\mathrm{d}\mathcal{S} - p\,\mathrm{d}V$$

followed by Euler's exact condition

$$\left(\frac{\partial p}{\partial \mathcal{S}}\right)_V = -\left(\frac{\partial T}{\partial V}\right)_{\mathcal{S}}$$

and an application of the cyclic chain rule.

Frictionlessly extracting the pin frees the piston and allows the gas to expand non-uniformly against the atmosphere, eventually reaching an equilibrium state with a final gas pressure $p_f^{gas} = p^{atm}$. The process – characterized by a finite expansion rate and system (gas) non-uniformity – is neither quasi-static nor reversible.[9]

Question 4.4 What is the entropy change of the "universe"?

Solution

Applying the First Law, the heat extracted from the surroundings by the isothermally expanding gas is

$$\mathcal{Q} = p^{atm} \Delta V, \tag{4.85}$$

where ΔV is the change in volume of the ideal gas,

$$\Delta V = N k_B T_0 \left(\frac{1}{p_{atm}} - \frac{1}{p_0} \right). \tag{4.86}$$

Therefore

$$\mathcal{Q} = N k_B T_0 p^{atm} \left(\frac{1}{p_{atm}} - \frac{1}{p_0} \right) \tag{4.87}$$

and the entropy change of the surroundings is

$$\Delta \mathcal{S}_{surroundings} = -\frac{\mathcal{Q}}{T_0} \tag{4.88}$$

$$= -N k_B \left(1 - \frac{p_{atm}}{p_0} \right). \tag{4.89}$$

Applying Eqs. 4.79–4.82 the entropy change of the expanding ideal gas is

$$\Delta \mathcal{S}_{gas} = -N k_B \ln \left(\frac{p_{atm}}{p_0} \right). \tag{4.90}$$

Therefore the entropy change of the "universe" is

$$\Delta \mathcal{S}_{universe} = \Delta \mathcal{S}_{gas} + \Delta \mathcal{S}_{surroundings} \tag{4.91}$$

$$= N k_B \left[\ln \left(\frac{p_0}{p_{atm}} \right) + \left(\frac{p_{atm}}{p_0} - 1 \right) \right]. \tag{4.92}$$

As shown in Figure 4.3 the entropy of the "universe" increases if $p^{atm} > p_0$, in which case the piston will drop, or if $p^{atm} < p_0$, in which case the piston will rise. Both are spontaneous events and for each $\Delta \mathcal{S}_{universe} > 0$.

---

[9] All real processes in nature are irreversible.

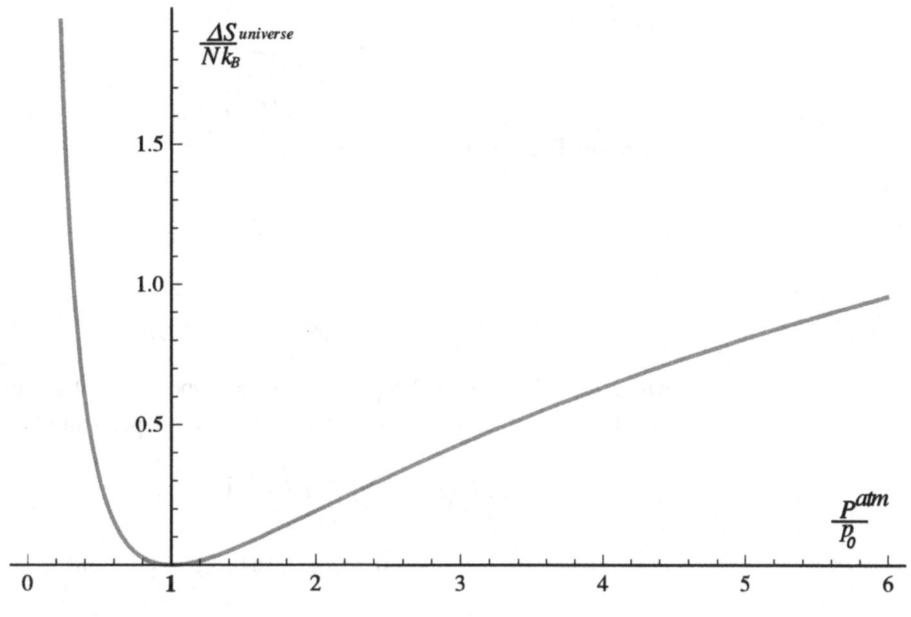

Fig. 4.3 $\Delta\mathcal{S}_{universe}$ after the "pin" is extracted.

Example 3

An uncovered mug holding 250 g of hot coffee at 95 ° C sits in a room with air temperature 20 ° C. The coffee slowly cools, eventually coming to room temperature.

Question 4.5 What is the change of entropy for the universe (coffee+surroundings)?

Solution

First, what is the change in entropy of the coffee?

Note that this (the coffee) entropy change takes place at constant pressure (the air in the room remains at constant pressure.) Knowing the initial coffee temperature and final equilibrium "universe" temperature, we wish to find $\Delta\mathcal{S}$ in this constant pressure process. With the typical strategy of expressing what we want to know, i.e. $\Delta\mathcal{S}$, in terms of what we can know, i.e. T and p, we choose to write $\mathcal{S} = \mathcal{S}(T, p)$ and take the total differential

$$d\mathcal{S} = \left(\frac{\partial \mathcal{S}}{\partial T}\right)_p \mathrm{d}T + \left(\frac{\partial \mathcal{S}}{\partial p}\right)_T \mathrm{d}p. \tag{4.93}$$

With constant pressure, i.e. $\mathrm{d}p = 0$, we only need to evaluate

$$\mathrm{d}\mathcal{S} = \left(\frac{\partial \mathcal{S}}{\partial T}\right)_p \mathrm{d}T, \tag{4.94}$$

where

$$T\left(\frac{\partial \mathcal{S}}{\partial T}\right)_p = \mathcal{C}_p. \tag{4.95}$$

Therefore Eq. 4.94 is

$$\mathrm{d}\mathcal{S} = \frac{1}{T}\mathcal{C}_p \, \mathrm{d}T \tag{4.96}$$

or

$$\mathrm{d}\mathcal{S} = \frac{1}{T}m\langle c_p\rangle \, \mathrm{d}T, \tag{4.97}$$

where m is the mass and $\langle c_p\rangle$ is an average specific heat per unit mass. Assuming that, for the coffee, $\langle c_p\rangle = 4184\,\mathrm{J\,kg^{-1}\,K^{-1}}$ is independent of temperature

$$\Delta\mathcal{S}_{coffee} = m\langle c_p\rangle \ln\left(\frac{T_f}{T_0}\right) \tag{4.98}$$

$$= 0.25\,\mathrm{kg} \cdot 4184\,\mathrm{J\,kg^{-1}\,K^{-1}} \cdot \ln\left(\frac{293\,\mathrm{K}}{368\,\mathrm{K}}\right) \tag{4.99}$$

$$= -238\,\mathrm{J\,kg^{-1}}. \tag{4.100}$$

The entropy of the coffee decreases.

Since coffee + surroundings = "the universe", having found the entropy change of the coffee we need to find the entropy change of the surroundings. Apart from receiving heat from the coffee, the surroundings are an infinite reservoir with volume, temperature and pressure remaining constant. So we can apply the First Law to the constant volume (zero work) surroundings:

$$\mathcal{Q}_{sur} = \Delta\mathcal{U}_{sur}. \tag{4.101}$$

Taking $\mathcal{Q}_{coffee}$ as the heat leaving the coffee, $-\mathcal{Q}_{coffee} = \mathcal{Q}_{sur}$. Then writing for the surroundings $\mathcal{U} = \mathcal{U}(V, \mathcal{S})$ and taking the total differential (with $\mathrm{d}V = 0$)

$$d\mathcal{U}_{sur} = \left(\frac{\partial \mathcal{U}_{sur}}{\partial \mathcal{S}_{sur}}\right)_V \mathrm{d}\mathcal{S}_{sur} \tag{4.102}$$

$$= T_{sur}\,\mathrm{d}\mathcal{S}_{sur} \tag{4.103}$$

where Eq. 4.32 has been used. After integration we have

$$\Delta\mathcal{S}_{sur} = \frac{-\mathcal{Q}_{coffee}}{T_{sur}}. \tag{4.104}$$

To find $\mathcal{Q}_{coffee}$ the First Law is applied to the coffee where we get, assuming negligible liquid volume change,

$$\mathcal{Q}_{coffee} = \Delta\mathcal{U}_{coffee} \tag{4.105}$$

$$= \mathcal{C}_p(T_0 - T_f) \tag{4.106}$$

$$= m\langle c_p\rangle(T_0 - T_f). \tag{4.107}$$

Finally, using this result in Eq. 4.104,

$$\Delta\mathcal{S}_{sur} = m\langle c_p\rangle \frac{T_f - T_0}{T_f} \tag{4.108}$$

$$= 0.25\,\text{kg} \cdot 4184\,\text{J}\,\text{kg}^{-1}\,\text{K}^{-1} \cdot 75\,\text{K}/293\,\text{K} \tag{4.109}$$

$$= 268\,\text{J}\,\text{K}^{-1} \tag{4.110}$$

so that

$$\Delta\mathcal{S}_{univ} = \Delta\mathcal{S}_{coffee} + \Delta\mathcal{S}_{sur} \tag{4.111}$$

$$= -238\,\text{J}\,\text{K}^{-1} + 268\,\text{J}\,\text{K}^{-1} \tag{4.112}$$

$$= 30\,\text{J}\,\text{K}^{-1}. \tag{4.113}$$

The entropy of the universe increases in this and any other spontaneous process. Cooling a mug of hot coffee by leaving it in a cool room is a spontaneous (irreversible) process, i.e. the entropy increase of the universe is a quantitative measure of its feasibility. The reverse process – the now cool coffee taking energy from the room to restore its original hot state – without some outside intervention – can "never" happen! That is the substance of the Second Law.

Examples 2 (Eq. 4.92) and 3 (Eq. 4.113) describe irreversible processes such as occur in nature whereas Example 1 (Eq. 4.84) describes the idealized quasi-static, reversible processes often invoked in thermodynamic models.
The Second Law,

$$\Delta\mathcal{S}_{universe} \geq 0, \tag{4.114}$$

is a measure of nature's propensity for processes to proceed, when there is no outside intervention, in a particular direction.

4.6 Thermal engines

A thermal engine is a device that operates in a cycle, during which:

1. a working substance absorbs heat $\mathcal{Q}_H$ from an energy source with which it is in thermal contact, resulting in thermodynamic changes to the working substance;
2. the working substance exhausts heat $\mathcal{Q}_L$ to an energy sink, also resulting in thermodynamic changes to the working substance;
3. external mechanical work $\mathcal{W}$ can be realized;

4. the working substance returns to its initial thermodynamic state, so the change in any state function per working substance cycle is zero, e.g. $\Delta\mathcal{U} = 0$.

Earliest 18th-century engines of various designs used steam exclusively as the working substance by condensing the steam to cause a partial vacuum, assisting external atmospheric pressure in driving a piston. Steam's only active role was to force the piston back to its starting position. In 1799 Richard Trevithick introduced a more efficient and compact high-pressure steam engine to drive the piston directly, which was the forerunner of the portable transportation engine. The steam-driven industrial revolution had begun.

However, engines of all designs were very inefficient, requiring large quantities of coal for little useful work output.

From the quest for improved efficiency evolved a new science, thermodynamics, created in 1824 by Sadi Carnot.[10,11]

Because of friction and turbulence, real engines are complicated irreversible devices. But their cycle can be modeled as ideal with accuracy sufficient to estimate their efficiency of operation η, defined as the ratio of mechanical work done $\mathcal{W}_{QS}$ to the heat absorbed from an energy source $\mathcal{Q}_H$, in each quasi-static engine cycle,

$$\eta = \frac{\mathcal{W}_{QS}}{\mathcal{Q}_H}. \tag{4.115}$$

Typical modern engines are of the combustion design in which burning fuel injects heat into a working substance which then expands against a piston to perform work. To return to its initial state, the expanded working substance must, at some point in the cycle, exhaust heat to a sink (usually the atmosphere or a nearby body of water). Exhausting heat to a sink is the step in the cycle which guarantees that the efficiency of any thermodynamic engine is less than 1.

4.6.1 Carnot's thermal engine

The ideal Carnot engine (see Figure 4.4) operates between an energy source, in which the working substance (in this case an ideal gas) absorbs heat $\mathcal{Q}_H$ isothermally at high temperature T_H (short dashed curve), and an energy sink, into which heat $\mathcal{Q}_L$ is isothermally exhausted at a low temperature T_L (solid curve). The long dash curves which close the Carnot cycle are adiabats in which no heat enters or leaves the working substance.

To calculate the engine efficiency (see Eq. 4.115) the work done in a cycle is determined by dividing the cycle into its four component paths:

[10] S. Carnot, *Reflections on the Motive Power of Fire*, Dover (1960).

[11] S. S. Wilson, "Sadi Carnot", *Scientific American* **245**, 102–114 (1981).

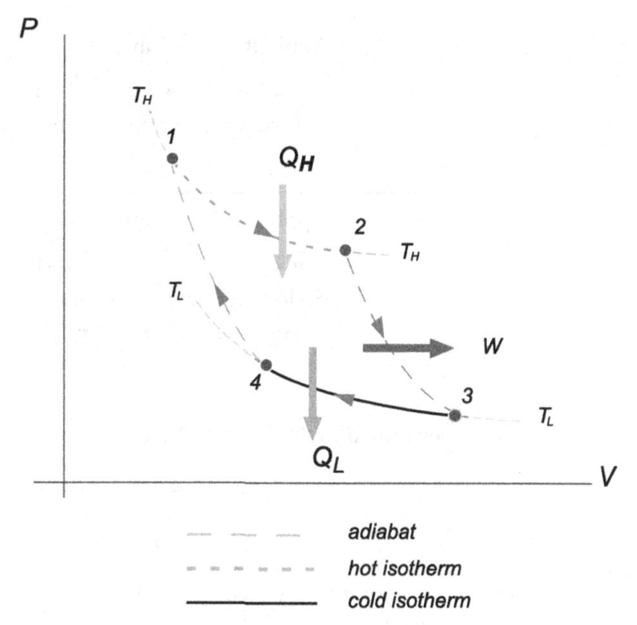

Fig. 4.4 An ideal Carnot engine cycle drawn in the P–V plane.

Work from an ideal gas Carnot cycle	
$1 \rightarrow 2$	Apply the First Law: $đ\mathcal{Q}_H = \mathrm{d}\mathcal{U} + đ\mathcal{W}_{1\rightarrow 2}$. Take $\mathrm{d}\mathcal{U} = \left(\frac{\partial \mathcal{U}}{\partial T}\right)_V \mathrm{d}T + \left(\frac{\partial \mathcal{U}}{\partial V}\right)_T \mathrm{d}V$. Along an isotherm $\mathrm{d}T = 0$ and for an ideal gas $\left(\frac{\partial \mathcal{U}}{\partial V}\right)_T = 0$. Therefore along the isotherm $\mathrm{d}\mathcal{U} = 0$ and $\mathcal{W}_{1\rightarrow 2} = \mathcal{Q}_H$.
$2 \rightarrow 3$	Apply the First Law: $đ\mathcal{Q}_{2\rightarrow 3} = \mathrm{d}\mathcal{U} + đ\mathcal{W}_{2\rightarrow 3}$. Along the adiabat $đ\mathcal{Q}_{2\rightarrow 3} = 0$. For the ideal gas $\mathrm{d}\mathcal{U} = \mathcal{C}_V\, \mathrm{d}T$. From this segment $\mathcal{W}_{2\rightarrow 3} = \mathcal{C}_V\,(T_H - T_L)$.

$3 \to 4$	Apply the First Law: $\bar{d}\mathcal{Q}_{3\to4} = \mathrm{d}\mathcal{U} + \bar{d}\mathcal{W}_{3\to4}$. Following the arguments used along $1 \to 2$ isotherm, $\mathcal{W}_{3\to4} = -\lvert\mathcal{Q}_L\rvert$.
$4 \to 1$	Apply the First Law: $\bar{d}\mathcal{Q}_{4\to1} = \mathrm{d}\mathcal{U} + \bar{d}\mathcal{W}_{4\to1}$. Following the $2 \to 3$ adiabat arguments, $\mathcal{W}_{4\to1} = \mathcal{C}_V\,(T_L - T_H)$.

Work performed in the cycle is therefore

$$\mathcal{W} = \mathcal{W}_{1\to2} + \mathcal{W}_{2\to3} + \mathcal{W}_{3\to4} + \mathcal{W}_{4\to1} \tag{4.116}$$

$$= \lvert\mathcal{Q}_H\rvert - \lvert\mathcal{Q}_L\rvert, \tag{4.117}$$

with the Carnot efficiency (see Eq. 4.115)

$$\eta = 1 - \frac{\lvert\mathcal{Q}_L\rvert}{\lvert\mathcal{Q}_H\rvert}. \tag{4.118}$$

In terms of working substance state variables (here assumed an ideal gas),

$$\lvert\mathcal{Q}_H\rvert = \int_{V_1}^{V_2} p\ \mathrm{d}V \tag{4.119}$$

$$= Nk_BT_H \int_{V_1}^{V_2} \frac{\mathrm{d}V}{V} \tag{4.120}$$

$$= Nk_BT_H \ln\frac{V_2}{V_1} \tag{4.121}$$

and

$$-\lvert\mathcal{Q}_L\rvert = \int_{V_3}^{V_4} p\ \mathrm{d}V \tag{4.122}$$

$$= Nk_BT_L \int_{V_3}^{V_4} \frac{\mathrm{d}V}{V} \tag{4.123}$$

$$= Nk_BT_L \ln\frac{V_4}{V_3}. \tag{4.124}$$

Therefore the efficiency is

$$\eta = 1 + \frac{T_L}{T_H}\frac{\ln\left(V_4/V_3\right)}{\ln\left(V_2/V_1\right)}. \tag{4.125}$$

But along the adiabats $2 \to 3$

$$\int_{T_H}^{T_L} C_V \frac{\mathrm{d}T}{T} = -Nk_B \int_{V_2}^{V_3} \frac{\mathrm{d}V}{V} \tag{4.126}$$

and $4 \to 1$

$$\int_{T_L}^{T_H} C_V \frac{\mathrm{d}T}{T} = -Nk_B \int_{V_4}^{V_1} \frac{\mathrm{d}V}{V}, \tag{4.127}$$

which are integrated to give

$$\frac{V_4}{V_1} = \frac{V_3}{V_2}. \tag{4.128}$$

Therefore the Carnot efficiency expressed in Eq. 4.125 becomes

$$\eta = 1 - \frac{T_L}{T_H}. \tag{4.129}$$

This result is consistent with the Kelvin–Planck statement of the Second Law of Thermodynamics:

- It is impossible to construct an engine such that heat is cyclically removed from a reservoir at high temperature with the sole effect of performing mechanical work. (In some part of the cycle heat must be exhausted and engine efficiency must be less than 1.)

A hypothetical Carnot engine with operating temperatures $T_H \sim 600\,\mathrm{K}$ and $T_L \sim 300\,\mathrm{K}$ would have an efficiency $\eta \sim 50\%$.[12]

4.6.2 The Carnot cycle – entropy

Any reversible engine cycle can be divided into infinitesimal Carnot segments (adiabats and isotherms) as shown in Figure 4.5.

Now, combining Eqs. 4.118 and 4.129, we have the result for a complete Carnot engine cycle

$$-\frac{|Q_H|}{T_H} + \frac{|Q_L|}{T_L} = 0, \tag{4.130}$$

[12] The Carnot engine is the most efficient engine operating between two fixed temperatures.

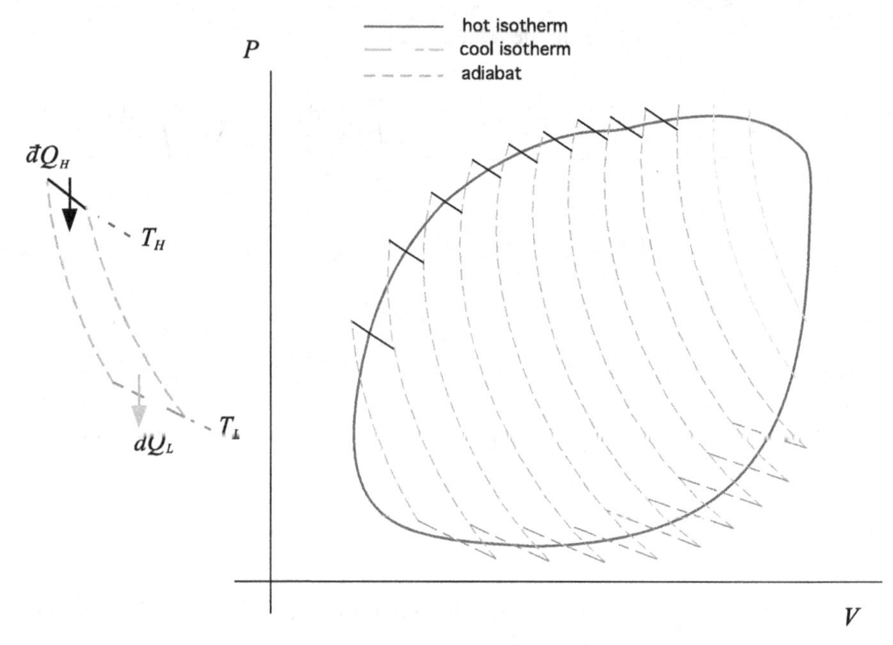

Fig. 4.5 An arbitrary quasi-static (reversible) engine cycle is divided into a series of infinitesimal Carnot cycles, a single one of which is shown on the left.

which applied to the infinitesimal Carnot engine cycle in the subdiagram of Figure 4.5 becomes

$$-\frac{|đQ_H|}{T_H} + \frac{|đQ_L|}{T_L} = 0. \tag{4.131}$$

Summing this expression for the sequence of infinitesimal Carnot cycles (which reconstruct the pictured arbitrary reversible cycle) gives

$$\oint \frac{đ\, Q_{QS}}{T} = 0. \tag{4.132}$$

Thus, $\frac{1}{T}$ is an integrating factor that makes $đQ_{QS}$ exact, confirming, even for an arbitrary substance, the redefinition

$$\frac{đ\, Q_{QS}}{T} = \mathrm{d}\mathcal{S}. \tag{4.133}$$

For the complete working substance cycle $\Delta\mathcal{S} = 0$.

Problems and exercises

4.1 A cube of lead 1 m on each side is to be slowly lowered from the sea surface 3×10^3 m down to the site of a deep drilling oil spill. The surface water

temperature is 23 °C while the sea water temperature at the oil well head is 4 °C. The mean density of sea water is 1035 kg · m^{-3}.
Find the volume of the lead cube when it reaches well-head depth.
Isothermal compressibility of Pb:

$$\kappa_T = -\frac{1}{V}\left(\frac{\partial V}{\partial p}\right)_T = 2.3 \times 10^{-6}\,\text{atm}^{-1}. \tag{4.134}$$

Linear thermal expansivity of Pb:

$$\alpha = \frac{1}{L}\left(\frac{\partial L}{\partial T}\right)_p = 2.9 \times 10^{-5}\,\text{K}^{-1}. \tag{4.135}$$

4.2 A van der Waals gas at temperature T_0 is confined by a partition to half of an insulated rigid cylinder with total volume $2V$ (see Figure 4.1). The unoccupied part, with volume V, is initially evacuated.
The partition spontaneously self-destructs and the gas surges into the formerly unoccupied part, eventually coming to equilibrium by uniformly occupying the entire cylinder.

1. What is the change in temperature of the van der Waals gas?
2. What is the change in entropy of the van der Waals gas?

4.3 The isothermal compressibility of a gas is defined as

$$\kappa_T = -\frac{1}{V}\left(\frac{\partial V}{\partial p}\right)_T, \tag{4.136}$$

while the adiabatic compressibility is defined as

$$\kappa_S = -\frac{1}{V}\left(\frac{\partial V}{\partial p}\right)_S. \tag{4.137}$$

Show that

$$\kappa_T = \frac{C_p}{C_V}\kappa_S. \tag{4.138}$$

4.4 The velocity of sound in a gas is given by

$$c_s^2 = \frac{B_S}{\rho}, \tag{4.139}$$

where ρ is the density of the gas,

$$\rho = \frac{N\mu}{V}, \tag{4.140}$$

and B_S is the adiabatic bulk modulus defined by

$$B_S = -V\left(\frac{\partial p}{\partial V}\right)_S, \tag{4.141}$$

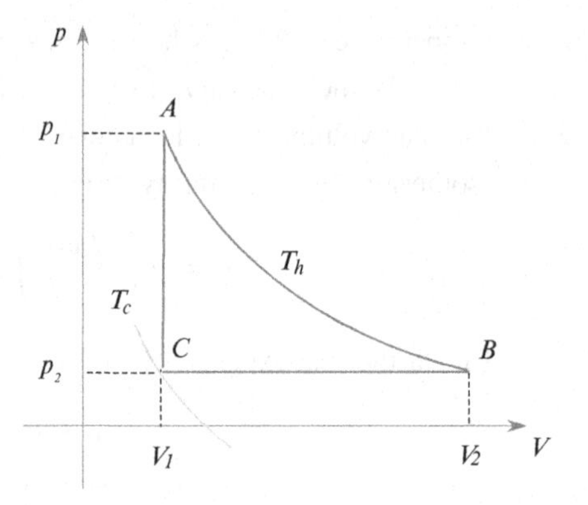

Fig. 4.6 p–V diagram for problem 4.5.

with N the number of gas molecules, μ the mass per molecule and S the entropy.

Show that for an ideal gas at temperature T the sound velocity is given by

$$c_s^2 = \frac{\gamma k_B T}{\mu}, \tag{4.142}$$

where

$$\gamma = \frac{C_p}{C_V}. \tag{4.143}$$

4.5 N molecules of an ideal gas undergo the cycle $A \to B \to C \to A$ as shown in the p–V diagram of Figure 4.6. The curve $A \to B$ is an isotherm at temperature T_h. When the gas is in the state (p_2, V_1) the temperature is T_c.

a. Show that the net heat Q added to the gas during the cycle is

$$Q = Nk_B \left[T_h \ln\left(\frac{T_h}{T_c}\right) - (T_h - T_c) \right]. \tag{4.144}$$

b. Find the engine efficiency.

4.6 A body of mass m with specific heat C_p at temperature 500 K is brought into contact with an identical body at temperature 100 K, with both bodies isolated from their surroundings. After a while the systems come to thermal equilibrium. Show that the change in entropy ΔS of the total system is

$$\Delta S = mC_p \ln(9/5). \tag{4.145}$$

5 Thermodynamic potentials

Rudolph Clausius obtained distinction by mathematical calculations based on the dynamical theory of heat – calculations which, it is claimed, show the necessity of a Creator and the possibility of miracles.

Obituary notice, *New York Times*, 26 Aug. 1888

5.1 Introduction

Thermodynamic processes can by design or by nature take place under conditions in which particular macroscopic parameters – e.g. temperature, pressure, volume, magnetic field, entropy, particle number – are kept constant. For an experimentalist, holding some variables constant, such as entropy or internal energy, is inconvenient or exceptionally difficult, while other variables, such as pressure and temperature, are easier or more natural to fix. Therefore, in addition to an energy potential $\mathcal{U}$ (*internal* energy) which accommodates constant extensive variables $\mathcal{S}$, V and N, other potential functions are introduced which are adapted to fixing other variables.

5.1.1 Internal energy $\mathcal{U}$

The differential

$$d\mathcal{U} = T\, d\mathcal{S} - p\, dV + \mu\, dN, \tag{5.1}$$

a consequence of the First Law of Thermodynamics, is called a *fundamental equation* or *the thermodynamic identity*. With $\mathcal{U} = \mathcal{U}(\mathcal{S}, V, N)$, the total differential is

$$d\mathcal{U} = \left(\frac{\partial \mathcal{U}}{\partial \mathcal{S}}\right)_{V,N} d\mathcal{S} + \left(\frac{\partial \mathcal{U}}{\partial V}\right)_{\mathcal{S},N} dV + \left(\frac{\partial \mathcal{U}}{\partial N}\right)_{\mathcal{S},V} dN. \tag{5.2}$$

Correspondences between Eqs. 5.1 and 5.2 give the following:

$$\left(\frac{\partial \mathcal{U}}{\partial \mathcal{S}}\right)_{V,N} = T, \tag{5.3}$$

$$\left(\frac{\partial \mathcal{U}}{\partial V}\right)_{\mathcal{S},N} = -p \tag{5.4}$$

and

$$\left(\frac{\partial \mathcal{U}}{\partial N}\right)_{\mathcal{S},V} = \mu. \tag{5.5}$$

Because partial derivatives of $\mathcal{U}$ with respect to $\mathcal{S}$, V and N yield accessible thermal properties of the system (in this case T, p and μ) they are called "natural variables" of $\mathcal{U}$. Applying Euler's criterion to Eqs. 5.3 and 5.4 gives the identity

$$\left(\frac{\partial T}{\partial V}\right)_{\mathcal{S},N} = -\left(\frac{\partial p}{\partial \mathcal{S}}\right)_{V,N}, \tag{5.6}$$

which is called a *Maxwell relation.*[1]

Equation 5.1 contains only quasi-static mechanical work, $p\,dV$, and particle work, $\mu\,dN$. But thermodynamics also applies to other conjugate work pairs which can be included in the fundamental equation, Eq. 5.1.

For example, from Table 5.1,

$$d\mathcal{U} = T\,d\mathcal{S} - p\,dV + \boldsymbol{\tau}\cdot d\boldsymbol{\chi} + \boldsymbol{B}_0\cdot d\boldsymbol{M} + \mathcal{E}_0\cdot d\mathcal{P} + \Sigma\cdot d\boldsymbol{A}_s + \boldsymbol{\sigma}\,d\boldsymbol{\epsilon} \\ + \mu\,dN. \tag{5.7}$$

Since $\mathcal{U}$, $\mathcal{S}$, V, χ, $\boldsymbol{M}$, $\mathcal{P}$, $\boldsymbol{A}_s$, $\boldsymbol{\epsilon}$ and N are extensive variables, a corresponding Euler equation can be derived:

$$\mathcal{U} = T\mathcal{S} - pV + \boldsymbol{\tau}\cdot\boldsymbol{\chi} + \boldsymbol{B}_0\cdot\boldsymbol{M} + \mathcal{E}_0\cdot\mathcal{P} + \Sigma\cdot\boldsymbol{A}_s + \boldsymbol{\sigma}\cdot\boldsymbol{\epsilon} + \mu N. \tag{5.8}$$

[1] In standard Maxwell relations N is usually suppressed.

Table 5.1 Here $\mathcal{B}_0$ is the magnetic field prior to insertion of matter, $\boldsymbol{M}$ is the magnetization, $\mathcal{E}_0$ is an electric field prior to insertion of matter, $\mathcal{P}$ is electric polarization, τ is elastic tension, χ is elastic elongation, Σ is surface tension, A_S is surface area, σ is stress, ϵ is strain, μ is the chemical potential, N is the number of particles, m is the mass of a particle, g is the acceleration due to gravity and z is the particle's distance above the Earth's surface.

magnetic	$\mathcal{B}_0 \cdot \mathrm{d}\boldsymbol{M}$
electric	$\mathcal{E}_0 \cdot \mathrm{d}\mathcal{P}$
elastic	$\tau \cdot \mathrm{d}\chi$
surface	$\Sigma\ \mathrm{d}A_S$
strain	$\sigma\ \mathrm{d}\epsilon$
particle number	$\mu\ \mathrm{d}N$
Earth's gravity	$mgz\ \mathrm{d}N$

5.1.2 Enthalpy *H*

Constant pressure is a common thermodynamic constraint, so common that to simplify isobaric analysis the variable enthalpy

$$H = \mathcal{U} + pV \tag{5.9}$$

is introduced to simplify isobaric analysis. Enthalpy is a total energy in the sense of being the sum of:

- energy of the constituent particles $\mathcal{U}$;
- mechanical energy pV needed to provide physical space for these constituents.

Enthalpy may have been in use since the time of Clausius under different names and symbols, but Gibbs' "heat function", $\chi = \epsilon + pV$, seems to be its first formalization. As for the name *enthalpy*, physicists are so conditioned to associate Kamerlingh Onnes with liquefying helium and discovering superconductivity that it might be a surprise to learn that "enthalpy" is also his creation (1901).[2] The symbol H was proposed twenty years later by Alfred W. Porter[3] to represent "heat content".[4]

Taking the total differential of enthalpy we have

$$\mathrm{d}H = \mathrm{d}\mathcal{U} + p\ \mathrm{d}V + V\ \mathrm{d}p. \tag{5.10}$$

[2] From the Greek *enthalpos*, which translates as "to warm within".

[3] A. W. Porter, *Transactions of the Faraday Society* **18**, 139 (1922).

[4] Irmgard K. Howard, "*H* is for Enthalpy", *Journal of Chemical Education* **79**, 697–698 (2002).

When combined with the fundamental equation, Eq. 5.1,

$$dH = T\, d\mathcal{S} + V\, dp + \mu\, dN. \tag{5.11}$$

With $H = H(\mathcal{S}, p, N)$, the total differential is

$$dH = \left(\frac{\partial H}{\partial \mathcal{S}}\right)_{p,N} d\mathcal{S} + \left(\frac{\partial H}{\partial p}\right)_{\mathcal{S},N} dp + \left(\frac{\partial H}{\partial N}\right)_{\mathcal{S},p} dN. \tag{5.12}$$

Comparing this with Eq. 5.11,

$$\left(\frac{\partial H}{\partial \mathcal{S}}\right)_{p,N} = T, \tag{5.13}$$

$$\left(\frac{\partial H}{\partial p}\right)_{\mathcal{S},N} = V \tag{5.14}$$

and

$$\left(\frac{\partial H}{\partial N}\right)_{\mathcal{S},p} = \mu, \tag{5.15}$$

which affirms $\mathcal{S}$, p and N as the natural variables of H.

Finally, applying Euler's criterion to Eqs. 5.13 and 5.14,

$$\left(\frac{\partial V}{\partial \mathcal{S}}\right)_{p,N} = \left(\frac{\partial T}{\partial p}\right)_{\mathcal{S},N}, \tag{5.16}$$

which is another Maxwell relation.

5.1.3 Helmholtz potential *F*

The Helmholtz potential (Helmholtz free energy) F is formed to accommodate constant T, V and N processes[5] and is defined as

$$F = \mathcal{U} - T\mathcal{S}. \tag{5.17}$$

The nature of this quantity is clarified by taking the total differential

$$dF = d\mathcal{U} - T\, d\mathcal{S} - \mathcal{S}\, dT \tag{5.18}$$

and replacing $d\mathcal{U}$ from the fundamental equation, Eq. 5.1, to arrive at

[5] As well as all other extensive variables.

$$dF = -S\,dT - p\,dV + \mu\,dN. \tag{5.19}$$

For a fixed number of particles at constant temperature,

$$-dF = p\,dV, \tag{5.20}$$

where the right-hand side is quasi-static work. Since reversible (quasi-static, friction-free) work is the maximum work that can be done by a system, the decrease in Helmholtz potential at constant temperature is the maximum work a system can perform. Moreover, for a fixed number of particles at constant T and V,

$$dF = 0, \tag{5.21}$$

which is the "extremal" condition for equilibrium of a system under those conditions, i.e. when $dF = 0$, $F = F_{min}$, the equilibrium state.

Taking $F = F(T, V, N)$, the total differential dF is

$$dF = \left(\frac{\partial F}{\partial T}\right)_{V,N} dT + \left(\frac{\partial F}{\partial V}\right)_{T,N} dV + \left(\frac{\partial F}{\partial N}\right)_{T,V} dN. \tag{5.22}$$

Comparing this with Eq. 5.19,

$$\left(\frac{\partial F}{\partial T}\right)_{V,N} = -S, \tag{5.23}$$

$$\left(\frac{\partial F}{\partial V}\right)_{T,N} = -p \tag{5.24}$$

and

$$\left(\frac{\partial F}{\partial N}\right)_{T,V} = \mu, \tag{5.25}$$

affirming T, V and N as natural variables for F.

Applying Euler's criterion to Eqs. 5.23 and 5.24,

$$\left(\frac{\partial S}{\partial V}\right)_{T,N} = \left(\frac{\partial p}{\partial T}\right)_{V,N}, \tag{5.26}$$

a third Maxwell relation.

5.1.4 Gibbs potential G

The Gibbs potential (Gibbs free energy) G, defined as

$$G = \mathcal{U} - T\mathcal{S} + pV, \tag{5.27}$$

accommodates experimental constraints on T, p and N. Differentiating Eq. 5.27,

$$\mathrm{d}G = \mathrm{d}\mathcal{U} - T\,\mathrm{d}\mathcal{S} - \mathcal{S}\,\mathrm{d}T + p\,\mathrm{d}V + V\,\mathrm{d}p, \tag{5.28}$$

and substituting $\mathrm{d}\mathcal{U}$ from Eq. 5.1 its total differential is

$$\mathrm{d}G = -\mathcal{S}\,\mathrm{d}T + V\,\mathrm{d}p + \mu\,\mathrm{d}N. \tag{5.29}$$

Writing $G = G(T, p, N)$, a total differential is

$$\mathrm{d}G = \left(\frac{\partial G}{\partial T}\right)_{p,N} \mathrm{d}T + \left(\frac{\partial G}{\partial p}\right)_{T,N} \mathrm{d}p + \left(\frac{\partial G}{\partial N}\right)_{T,p} \mathrm{d}N, \tag{5.30}$$

which, on comparison with Eq. 5.29, gives

$$\left(\frac{\partial G}{\partial T}\right)_{p,N} = -\mathcal{S}, \tag{5.31}$$

$$\left(\frac{\partial G}{\partial p}\right)_{T,N} = V \tag{5.32}$$

and

$$\left(\frac{\partial G}{\partial N}\right)_{T,p} = \mu, \tag{5.33}$$

confirming T, p and N as natural variables of G. Applying Euler's criterion to Eqs. 5.31 and 5.32,

$$\left(\frac{\partial \mathcal{S}}{\partial p}\right)_{T,N} = -\left(\frac{\partial V}{\partial T}\right)_{p,N}, \tag{5.34}$$

which is a fourth Maxwell relation.

5.2 Enthalpy and throttling

A gas that is slowly forced under high pressure across a very narrow, perfectly insulated porous plug (or valve), emerging on the other side at reduced pressure, is said to be *throttled*. For reasons which will shortly be discussed, the process is applied in commercial refrigeration and liquefaction of gases. To simplify the discussion, the normally open system (gas comes from a high-pressure reservoir and escapes to a low-pressure reservoir) is replaced by a "closed" system consisting of two frictionless pistons, one maintaining the constant high pressure P_0 and the other the constant lower pressure P_f (see Figure 5.1). Fixing attention on a small amount of gas, say N molecules, the high-pressure piston forces the gas through the plug onto the low-pressure piston. At the higher constant pressure P_0, the N molecules have a temperature T_0 and occupy a volume V_0. Therefore the work $\mathcal{W}_0$ done on the gas by the high-pressure piston in moving the N molecules through the plug is

$$\mathcal{W}_0 = P_0\, V_0. \tag{5.35}$$

At the lower pressure P_f the N molecules eventually occupy a volume V_f at temperature T_f so that the work done by the N molecules as they emerge and push the piston outward is

$$\mathcal{W}_f = P_f\, V_f\,. \tag{5.36}$$

Therefore the total work done by the N molecules of gas is

$$\mathcal{W}_f - \mathcal{W}_0 = P_f V_f - P_0 V_0\,. \tag{5.37}$$

Applying the First Law to this adiabatic process,

$$\mathcal{Q} = 0 = (\mathcal{U}_f - \mathcal{U}_0) + (\mathcal{W}_f - \mathcal{W}_0) \tag{5.38}$$

and therefore

$$\mathcal{U}_0 + P_0\, V_0 = \mathcal{U}_f + P_f\, V_f\,, \tag{5.39}$$

which in terms of enthalpy is

$$H_0 = H_f\,. \tag{5.40}$$

The throttling process is isenthalpic.

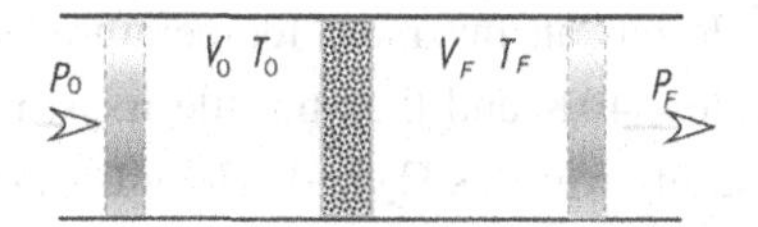

Fig. 5.1 Gas forced across a porous plug.

Of principal interest in throttling is the temperature change it produces in the gas, if any. To further investigate the process write the temperature as $T = T(H, P)$ and take the total differential

$$dT = \left(\frac{\partial T}{\partial H}\right)_P dH + \left(\frac{\partial T}{\partial P}\right)_H dP. \tag{5.41}$$

The good news is that since $dH = 0$ we have, so far,

$$dT = \left(\frac{\partial T}{\partial P}\right)_H dP. \tag{5.42}$$

Unfortunately the "unfriendly" partial derivative $(\partial T/\partial P)_H = \mathcal{J}$ must be evaluated.[6] Typically, there is more than one way to do this. For now, start by writing Eq. 5.11 as

$$d\mathcal{S} = \frac{1}{T} dH - \frac{V}{T} dP. \tag{5.43}$$

Then, since $\mathcal{S}$ is exact, Euler's criterion gives

$$\left(\frac{\partial\left(\frac{1}{T}\right)}{\partial P}\right)_H = -\left(\frac{\partial\left(\frac{V}{T}\right)}{\partial H}\right)_P, \tag{5.44}$$

which becomes, after differentiation,

$$\left(\frac{\partial T}{\partial P}\right)_H = T\left(\frac{\partial V}{\partial H}\right)_P - V\left(\frac{\partial T}{\partial H}\right)_P \tag{5.45}$$

$$= \left[T\left(\frac{\partial V}{\partial T}\right)_P - V\right]\left(\frac{\partial T}{\partial H}\right)_P \tag{5.46}$$

$$= \left[T\left(\frac{\partial V}{\partial T}\right)_P - V\right]\left(\frac{1}{C_P}\right). \tag{5.47}$$

It is easily verified that $\mathcal{J} = 0$ for an ideal gas and, consequently, $\Delta T = 0$. On the other hand, for any real gas, e.g. a van der Waals gas, $\mathcal{J}$ can be positive or negative. If $\mathcal{J} > 0$ the gas will cool whereas if $\mathcal{J} < 0$ the gas will warm. Specifically, below some *inversion temperature* T_{inv} (characteristic of each gas) a gas will cool upon throttling while above T_{inv} it will warm. At atmospheric pressure nitrogen (N_2) has $T_{inv} = 621$ K and oxygen (O_2) has $T_{inv} = 764$ K. On the other hand neon (Ne) has $T_{inv} = 231$ K, and hydrogen (H_2) has $T_{inv} = 202$ K (see Figure 5.2).

^{4}He has an inversion temperature of 40 K. To liquify ^{4}He, one must first cool it below 40 K and then throttle it. At room temperature N_2 and O_2 will cool upon throttling whereas H_2 and ^{4}He will warm.

[6] $\mathcal{J}$ is the Joule–Thomson coefficient.

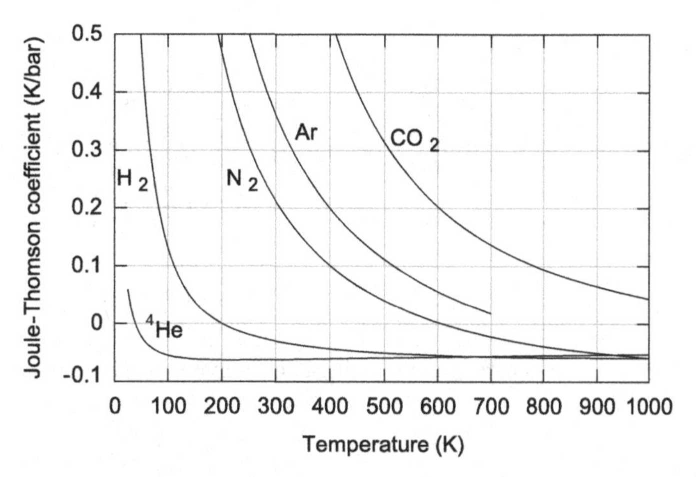

Fig. 5.2 Joule–Thomson coefficients $\mathscr{J}$ for various gases at atmospheric pressure.

5.3 Entropy and heat capacity

Low-temperature heat capacities are among those macroscopic phenomena that require quantum explanations. For that reason they stand high on the list of important thermodynamic experiments. Anomalous low-temperature values in diamond first suggested to Einstein (1905) that nature was not as classical as was generally believed at the time, and he applied quantization – as proposed by Planck (1900) – to develop the first quantum theory of solids 25 years before the arrival of a generally accepted quantum mechanics.

The appearance of entropy in thermodynamics brings a welcome and important uniformity to heat capacity definitions. Whereas in Chapter 2 we derived the result

$$\mathcal{C}_V = \left(\frac{\partial U}{\partial T}\right)_V, \tag{5.48}$$

we can instead, by applying the simple chain rule, write

$$\left(\frac{\partial U}{\partial T}\right)_V = \left(\frac{\partial U}{\partial S}\right)_V \left(\frac{\partial S}{\partial T}\right)_V \tag{5.49}$$

followed by Eq. 5.3, arrive at

$$\left(\frac{\partial U}{\partial T}\right)_V = T\left(\frac{\partial S}{\partial T}\right)_V \tag{5.50}$$

so that

$$\mathcal{C}_V = T\left(\frac{\partial S}{\partial T}\right)_V. \tag{5.51}$$

Similarly

$$\mathcal{C}_p = \left(\frac{\partial H}{\partial T}\right)_p \tag{5.52}$$

$$= \left(\frac{\partial H}{\partial \mathcal{S}}\right)_p \left(\frac{\partial \mathcal{S}}{\partial T}\right)_p \tag{5.53}$$

$$= T\left(\frac{\partial \mathcal{S}}{\partial T}\right)_p. \tag{5.54}$$

In general there is a single, convenient heat capacity definition:

$$\mathcal{C}_\alpha = T\left(\frac{\partial \mathcal{S}}{\partial T}\right)_\alpha. \tag{5.55}$$

5.3.1 Relationship between $\mathcal{C}_p$ and $\mathcal{C}_V$

The entropy-based relations $\mathcal{C}_p = T\left(\partial \mathcal{S}/\partial T\right)_p$ and $\mathcal{C}_V = T\left(\partial \mathcal{S}/\partial T\right)_V$ put this problem on a level footing. Considering the final objective, expressing the entropy as $\mathcal{S} = \mathcal{S}(T, p)$ is one sensible starting point. Then

$$\mathrm{d}\mathcal{S} = \left(\frac{\partial \mathcal{S}}{\partial T}\right)_p \mathrm{d}T + \left(\frac{\partial \mathcal{S}}{\partial p}\right)_T \mathrm{d}p, \tag{5.56}$$

which already contains a piece of the answer, i.e.

$$d\mathcal{S} = \frac{\mathcal{C}_p}{T}\,\mathrm{d}T + \left(\frac{\partial \mathcal{S}}{\partial p}\right)_T \mathrm{d}p. \tag{5.57}$$

An equally sensible choice is $\mathcal{S} = \mathcal{S}(T, V)$, in which case

$$\mathrm{d}\mathcal{S} = \left(\frac{\partial \mathcal{S}}{\partial T}\right)_V \mathrm{d}T + \left(\frac{\partial \mathcal{S}}{\partial V}\right)_T \mathrm{d}V \tag{5.58}$$

$$= \frac{\mathcal{C}_V}{T}\,\mathrm{d}T + \left(\frac{\partial \mathcal{S}}{\partial V}\right)_T \mathrm{d}V. \tag{5.59}$$

Then combining Eqs. 5.57 and 5.59 gives

$$\left(\frac{\mathcal{C}_p - \mathcal{C}_V}{T}\right)\mathrm{d}T = \left[\left(\frac{\partial \mathcal{S}}{\partial V}\right)_T \mathrm{d}V - \left(\frac{\partial \mathcal{S}}{\partial p}\right)_T \mathrm{d}p\right]. \tag{5.60}$$

Next:

1. Apply the pair of Maxwell Relations, Eq. 5.26 and Eq. 5.34:

$$\left(\frac{\mathcal{C}_p - \mathcal{C}_V}{T}\right)\mathrm{d}T = \left[\left(\frac{\partial p}{\partial T}\right)_V \mathrm{d}V + \left(\frac{\partial V}{\partial T}\right)_p \mathrm{d}p\right]. \tag{5.61}$$

2. Apply the cyclic chain rule to the first partial derivative on the right-hand side:

$$\left(\frac{\mathcal{C}_p - \mathcal{C}_V}{T}\right) \mathrm{d}T = -\left(\frac{\partial V}{\partial T}\right)_p \left[\left(\frac{\partial p}{\partial V}\right)_T \mathrm{d}V - \mathrm{d}p\right]. \quad (5.62)$$

3. Noting that

$$\mathrm{d}p = \left(\frac{\partial p}{\partial V}\right)_T \mathrm{d}V + \left(\frac{\partial p}{\partial T}\right)_V \mathrm{d}T, \quad (5.63)$$

we finally arrive with

$$\left(\frac{\mathcal{C}_p - \mathcal{C}_V}{T}\right) = \left(\frac{\partial V}{\partial T}\right)_p \left(\frac{\partial p}{\partial T}\right)_V. \quad (5.64)$$

For an ideal gas

$$\left(\frac{\partial V}{\partial T}\right)_p = \frac{Nk_B}{p} \quad (5.65)$$

and

$$\left(\frac{\partial p}{\partial T}\right)_V = \frac{Nk_B}{V}, \quad (5.66)$$

which gives

$$\mathcal{C}_p - \mathcal{C}_V = Nk_B, \quad (5.67)$$

as stated in Eq. 3.66, without proof.

Eq. 5.64 is a result that can be applied to any gas law – for example, to the van der Waals equation.[7]

This type of result is not limited to gases. We will later derive a relation between $\mathcal{C}_{\mathcal{B}}$ (constant magnetic induction) and $\mathcal{C}_{\mathcal{M}}$ (constant magnetization) for magnetic systems, and between $\mathcal{C}_\tau$ (constant tension) and $\mathcal{C}_\chi$ (constant elongation) for linear elastic systems.[8]

Problems and exercises

5.1 N molecules of an ideal gas at temperature T_0 and volume V_0 undergo an isothermal expansion to a volume V_1. What is the change in:

a. internal energy of the expanding gas?
b. enthalpy of the expanding gas?

[7] Also see problem 5.3.
[8] $\chi = \langle \chi_{op} \rangle$ is the mean elongation.

c. Helmholtz potential of the expanding gas?
d. Gibbs potential of the expanding gas?

5.2 Show that for an ideal gas

a. $\left(\frac{\partial C_V}{\partial p}\right)_T = 0;$
b. $\left(\frac{\partial C_p}{\partial V}\right)_T = 0.$

5.3 If a gas is not ideal but has the equation of state

$$p = \frac{N k_B T}{V}\left[1 + \frac{N}{V} B(T)\right], \tag{5.68}$$

find expressions for:

a. $\left(\frac{\partial C_V}{\partial p}\right)_T;$
b. $\left(\frac{\partial C_p}{\partial V}\right)_T;$
c. $C_p - C_V.$

5.4 It is possible for a specific gas to emerge on the low-pressure side of a throttling plug at either a lower or higher temperature than it had entering on the high-pressure side. Show that for the van der Waals gas, whose equation of state is

$$\left(p + \frac{aN^2}{V^2}\right)(V - Nb) = N k_B T, \tag{5.69}$$

where a and b are gas-specific temperature-independent constants, that at temperature T_0, where

$$k_B T_0 = \frac{2a}{b}\left(\frac{V - Nb}{V}\right)^2, \tag{5.70}$$

throttling produces no temperature change.

6 Knowing the "unknowable"

> Suppose we were asked to arrange the following in two categories – distance, mass, electric force, entropy, beauty, melody. I think there are the strongest grounds for arranging entropy alongside beauty and melody and not with the first three. Entropy is found only when the parts are viewed in association, and it is by hearing or viewing the parts in association that beauty and melody are discerned . . . The reason why this stranger can pass itself off among the aborigines of the physical world is that . . . it has a measure number associated with it.
>
> A. Eddington, *The Nature of the Physical World*, Cambridge University Press (1928)

6.1 Entropy: ticket to the Emerald City

Entropy $\mathcal{S}$ was introduced into thermodynamics by Rudolph Clausius[1] as a cryptic state function without a clear intuitive or physical interpretation. Still, he maintained it to be the basis of a sweeping new, abstract, universal physical principle: "In all spontaneous processes the *entropy* of a closed system (system + environment) tends towards a maximum", i.e.

$$\Delta\mathcal{S} \geq 0. \tag{6.1}$$

It was in his series of memoirs (1856–1865) on irreversible heat loss, i.e. inefficiency of heat engines, that he chose the name *entropy* (Greek for "transformation") because it sounded similar to *energy*, under the mistaken impression that the two were related.

Clausius' principle, in which a particular state variable – entropy – always increases (or remains constant) as a system and its surroundings (i.e. the "universe") evolve from one equilibrium state to the next, is called the *Second Law of Thermodynamics*. The Second Law is *not* a conservation rule. Moreover, entropy is unlike objective observables so familiar in physics, such as mass, length, velocity, energy, acceleration, etc. Instead, emotive analogies or proxies such as disorder, chaos and disorganization – which point in the right direction – are reasonably applied.

[1] R. Clausius, "On the Moving Force of Heat, and the Laws regarding the Nature of Heat itself which are deducible therefrom" (English translation of original), *Phil. Mag.* **2**, 121, 102119 (1851).

The following encounter, said to have taken place in 1948, could serve to acknowledge entropy's subtle character.

Over half a century ago while creating a fledgling Information Theory, Claude Shannon (of Bell Laboratories) sought advice from Princeton's famous mathematician–physicist John von Neumann in naming the function central to his [Shannon's] (information) arguments:[2]

> My greatest concern was what to call it. I thought of calling it information, but the word is overly used. So I decided to call it uncertainty. When I discussed it with John von Neumann, he had a better idea. Von Neumann told me – You should call it entropy, for two reasons. In the first place your uncertainty function has been used in statistical mechanics under that name, so it already has a name. In the second place, and more important, even after nearly one hundred years nobody knows what entropy really is, so in a debate you will always have the advantage.

Although entropy is not a part of quantum mechanics, it plays a key role in postulating a bridge to thermodynamics.

6.2 The bridge

As a consequence of environmental decoherence, a "thermodynamic" density operator, ρ_{op}^{τ},

$$\rho_{op}^{\tau} = \sum_{s} \mathcal{P}(E_s) \left| E_s \right\rangle \left\langle E_s \right| , \tag{6.2}$$

emerges as – potentially – the quantity essential in the evolution to thermodynamics. For example, using the macroscopic internal eigen-energies from

$$\mathbf{h}_{op} | \epsilon_s \rangle = \epsilon_s | \epsilon_s \rangle , \tag{6.3}$$

the average internal energy $\mathcal{U}$ enshrined in the First Law could – in principle – be expressed as

$$\mathcal{U} = Tr \rho_{op}^{\tau} \mathbf{h}_{op} \tag{6.4}$$

or, equivalently, as

$$\mathcal{U} = \sum_{s} \epsilon_s \mathcal{P}(\epsilon_s) . \tag{6.5}$$

This looks promising – but, alas, the $\mathcal{P}(\epsilon_s)$ are *unknown* and, except in trivial cases, *unknowable*. The emerging thermodynamic theory is not complete! Can it ever be?

[2] M. Tribus and E. McIrvine, "Energy and information", *Scientific American* **225**, 179–188 (1971). Reproduced with permission.

An alternative is to replace the unknowable $\mathcal{P}(E_s)$ by inferred "surrogates" $\mathbf{P}(E_s)$ with which to construct a "surrogate" thermal density operator $\hat{\rho}^{\tau}_{op}$,

$$\hat{\rho}^{\tau}_{op} = \sum_s \mathbf{P}(E_s)|E_s\rangle\langle E_s|, \tag{6.6}$$

that describes behavior in accord with the macroscopic physical universe.

6.3 Thermodynamic hamiltonians

A simple dynamical situation would be the just discussed

$$\mathcal{H}_{op} = \mathbf{h}_{op}, \tag{6.7}$$

in which $\mathbf{h}_{op}$ represents the internal energy (kinetic + potential + interaction) of, say, a gas.[3]

If gas particles also have a magnetic moment and are in a local magnetic field $\mathcal{B}_{\mathbf{0}}$ – the field prior to insertion of matter[4] – a macroscopic hamiltonian is

$$\mathcal{H}_{op} = \mathbf{h}_{op} - \boldsymbol{M}_{op} \cdot \mathcal{B}_{\mathbf{0}}, \tag{6.8}$$

where $\boldsymbol{M}_{op}$ represents macroscopic magnetization

$$\boldsymbol{M}_{op} = \sum_i \boldsymbol{m}_{op}(i). \tag{6.9}$$

If the particles are molecular dipoles with electric dipole moments $\mathcal{P}_{op}$ in an electric field $\mathcal{E}_{\mathbf{0}}$, then

$$\mathcal{H}_{op} = \mathbf{h}_{op} - \mathcal{P}_{op} \cdot \mathcal{E}_{\mathbf{0}}, \tag{6.10}$$

where $\mathcal{P}_{op}$ represents the macroscopic electric polarization.

Euler's fundamental equation, Eq. 5.8, shows how other "thermodynamic" hamiltonians can be formulated by adding physically relevant macroscopic operators (see Table 5.1). A more general result is then[5,6]

$$\mathcal{H}_{op} = \mathbf{h}_{op} + \boldsymbol{p}_{op} V - \boldsymbol{\sigma} \cdot \boldsymbol{\epsilon}_{op} - \boldsymbol{\tau} \cdot \boldsymbol{\chi}_{op} - \mathcal{B}_0 \cdot \boldsymbol{M}_{op} - \mathcal{E}_0 \cdot \mathcal{P}_{op} - \boldsymbol{\Sigma} \cdot \mathbf{A}^{s}_{op} - \mu \mathcal{N}_{op}. \tag{6.12}$$

[3] The gas may be atoms or cold molecules (no internal modes) or even electrons.

[4] This subtle but important point will be discussed in detail in Chapter 11.

[5] Schrödinger's non-relativistic wave mechanics is based on closed systems (fixed number of particles). It makes no provision for destroying and creating particles. Therefore a number operator is not accommodated within that theory. Particle creation and destruction is, however, integral to quantum field theories so that number operators play a natural role.

[6] The particle number operator $\mathcal{N}_{op}$ defines the eigenvalue equation

$$\mathcal{N}_{op}\,|N\rangle = n\,|N\rangle \tag{6.11}$$

with eigenvalues $n = 0,\ 1,\ 2,\ 3,\ \ldots$

Since the extensive parameters describe a macroscopic system completely, Eq. 6.12 contains all relevant information to enable a model macroscopic system's average total energy to be expressed as

$$\langle \mathcal{H}_{op} \rangle = \mathcal{T}r\, \hat{\rho}^{\tau}_{op} \mathcal{H}_{op}. \tag{6.13}$$

But the inescapable question remains – what criteria can be devised for constructing the "best" $\mathbf{P}(E_s)$? This question has no a-priori answer apart from a postulate that corresponds to time-tested thermodynamics and describes the measurable universe.[7]

Towards this goal:

1. Minimize

$$\langle \mathcal{H}_{op} \rangle = \mathcal{T}r\, \tilde{\rho}^{\tau}_{op}\, \mathcal{H}_{op} \tag{6.14}$$

$$= \sum_s \tilde{\mathbf{P}}(E_s) E_s \tag{6.15}$$

with respect to the $\tilde{\mathbf{P}}(E_s)$.[8,9,10]

2. With no a-priori basis for choosing the $\tilde{\mathbf{P}}(E_s)$ that minimizes Eq. 6.15, they shall be chosen without "fear or favor", i.e. with minimum bias[11,12,13] (maximum uncertainty).

   Recalling the function(al)[14] $\mathcal{F}$ of Eq. 2.55,

$$\mathcal{F}[\rho] = -\kappa \mathcal{T}r\, \rho^{\tau}_{op} \ln \rho^{\tau}_{op}, \quad \kappa > 0, \tag{6.16}$$

   as a measure of bias in a collection of probabilities,[15] the function(al)

$$\mathcal{F}\left[\tilde{\mathbf{P}}(E)\right] = -\kappa \sum_s \tilde{\mathbf{P}}(E_s) \ln \tilde{\mathbf{P}}(E_s), \quad \kappa > 0, \tag{6.17}$$

   is taken to be a measure of bias in the probabilities $\tilde{\mathbf{P}}(E_s)$ and is maximized. (See Appendix B.)

3. The $\tilde{\mathbf{P}}(E_s)$ are normalized:

$$\sum_s \tilde{\mathbf{P}}(E_s) = 1. \tag{6.18}$$

[7] This is not unlike Ludwig Boltzmann's motivation in postulating $\mathcal{S} = k_B \ln \mathcal{W}$.

[8] L. H. Thomas, "The calculation of atomic fields", *Proc. Cambridge Phil. Soc.* **23**, 542–548 (1927).

[9] E. Fermi,"Un Metodo Statistico per la Determinazione di alcune Priopriet dell'Atomo", *Rend. Accad. Naz. Lincei* **6**, 602–607 (1927).

[10] E. H. Lieb, "Thomas–Fermi and related theories of atoms and molecules", *Rev. Mod. Phys.* **53**, 603–641 (1981); *Errata* **54**, 311 (1982).

[11] J. N. Kapur, *Maximum Entropy Models in Science and Engineering*, John Wiley and Sons, New York (1989).

[12] E. T. Jaynes, "Information theory and statistical mechanics", *Phys. Rev.* **106**, 620 (1957).

[13] R. Balian, "Incomplete descriptions and relevant entropies", *Am. J. Phys.* **67**, 1078 (1999).

[14] $\mathcal{F}$ is a function of functions, which is called a *functional*.

[15] See Appendix B.

To meet these requirements a Lagrangian is formed:

$$\mathcal{L} = -\left\{\kappa \sum_s \tilde{\mathbf{P}}(E_s) \ln \tilde{\mathbf{P}}(E_s) + \sum_s \tilde{\mathbf{P}}(E_s) E_s\right\} - \lambda_0 \sum_s \tilde{\mathbf{P}}(E_s), \tag{6.19}$$

where λ_0 is a Lagrange multiplier and the $\tilde{\mathbf{P}}(E_s)$ are varied until $\mathcal{L}$ is maximized, i.e. $\tilde{\mathbf{P}}(E_s) \to \mathbf{P}(E_s)$.[16] As discussed in Appendix B, with $\kappa \to k_B T$ a *thermal Lagrangian* is

$$\mathcal{L}\left[\tilde{\mathbf{P}}\right] = -\left\{k_B \sum_s \tilde{\mathbf{P}}(E_s) \ln \tilde{\mathbf{P}}(E_s) + T^{-1} \sum_s \tilde{\mathbf{P}}(E_s) E_s\right\} - \lambda_0 \sum_s \tilde{\mathbf{P}}(E_s). \tag{6.20}$$

Then with $\tilde{\mathbf{P}} \to \mathbf{P}$ (least bias), it follows that $\mathcal{F}\left[\tilde{\mathbf{P}}(E)\right] \to \mathcal{F}\left[\mathbf{P}(E)\right] = \mathcal{F}_{max}$, and

$$\mathcal{S} = -k_B \sum_s \mathbf{P}(E_s) \ln \mathbf{P}(E_s), \tag{6.21}$$

the entropy of thermodynamics.

6.4 Microcanonical (Boltzmann) theory

At the time Boltzmann published his ingenious entropy postulate (1864) skepticism about molecular models was continuing almost unabated and classical mechanics was the only operative particle paradigm. Yet his accomplishments were so radical and ahead of his time that it would take another 25 years for them to be reformulated by Max Planck as quantization.

With our 150-year advantage, we illustrate a thermodynamic density operator $\hat{\rho}^{\tau}_{op}$ approach to Boltzmann's fundamental thermodynamic contribution.

Consider a many-particle, closed system with:

- a fixed number of particles N,
- a fixed volume V,
- an isolated g_0-fold degenerate macroscopic eigen-energy $\varepsilon_{0,\alpha}$,

$$\varepsilon_{0,\alpha} \equiv \left\{\varepsilon_{0,1} = \varepsilon_{0,2} = \varepsilon_{0,3} = \cdots = \varepsilon_{0,g_0}\right\}. \tag{6.22}$$

[16] This is equivalent to minimizing the Helmholtz potential, as in Section 5.1.3.

These effectively constitute Boltzmann's "microcanonical" conditions. For this model system the average hamiltonian (internal energy) is

$$\mathcal{U} = \sum_{\substack{s \\ \alpha=1,2,\ldots,g_0}} \mathbf{P}^B\left(\varepsilon_{s,\alpha}\right)\varepsilon_{s,\alpha}\,\delta_{s,0}, \tag{6.23}$$

where $\mathbf{P}^B\left(\varepsilon_{s,\alpha}\right)$ is the probability that the system is in one of the degenerate macroscopic particle states with eigen-energy $\varepsilon_{s,\alpha}$. The Kronecker $\delta_{s,\alpha}$ selects from within this spectrum a g_0-fold degenerate single value $\varepsilon_{0,\alpha}$. Therefore, after summing over the index s, Eq. 6.23 becomes

$$\mathcal{U} = \sum_{\alpha=1}^{g_0} \mathbf{P}^B\left(\varepsilon_{0,\alpha}\right)\varepsilon_{0,\alpha}. \tag{6.24}$$

The remaining sum over α covers the g_0-fold degeneracy.[17]

To determine the surrogate probabilities $\mathbf{P}^B(\varepsilon_{0,\alpha})$ (i.e. $\hat{\rho}^{\tau}_{op}$) a thermal Lagrangian $\mathcal{L}\left[\tilde{\mathbf{P}}^B\right]$ is postulated which, when maximized with respect to the $\tilde{\mathbf{P}}^B(\varepsilon_{0,\alpha})$, minimizes both macroscopic energy $\left[\mathcal{T}r\,\tilde{\rho}^{\tau}_{op}\mathcal{H}_{op}\right]$ and any "bias" (see Eq. 6.20) in the choice of $\tilde{\mathbf{P}}^B(\varepsilon_{0,\alpha})$. In particular,

$$\begin{aligned}\mathcal{L}\left[\tilde{\mathbf{P}}^B\right] &= -k_B \sum_{\alpha=1}^{g_0} \tilde{\mathbf{P}}^B\left(\varepsilon_{0,\alpha}\right)\ln \tilde{\mathbf{P}}^B\left(\varepsilon_{0,\alpha}\right) - T^{-1}\sum_{\alpha=1}^{g_0} \tilde{\mathbf{P}}^B\left(\varepsilon_{0,\alpha}\right)\varepsilon_{0,\alpha} \\ &\quad - \lambda_0 \sum_{\alpha=1}^{g_0} \tilde{\mathbf{P}}^B\left(\varepsilon_{0,\alpha}\right),\end{aligned} \tag{6.25}$$

where λ_0 is a Lagrange multiplier that insures normalized "surrogate" probabilities.[18] The thermal Lagrangian used here, and in the remaining examples of this book, is discussed and justified in Appendix B.

If results from this procedure are in agreement with an observable macroscopic universe and with time-honored thermodynamics – whose validity is not in doubt – the Yellow Brick Road will have guided us, at last, to the Emerald City.

(The procedure resembles a maximum entropy formalism of E. T. Jaynes.[19,20] Jaynes' formulation is, fundamentally, a statistical method in cases of partial information (Bayesian statistics). It adopts Gibbs–Shannon[21] communication (information) theory (see Section 6.1) and is the basis of the "information theory" approach

[17] Boltzmann pictured "microstates" (classical configurations) contributing to a single classical "internal energy".

[18] The Internet is a rich source of tutorials on this powerful mathematical technique that is widely used in physics and other technical fields. See, for example, www.slimy.com/~steuard/teaching/tutorials/Lagrange.html and www.cs.berkeley.edu/~klein/papers/lagrange-multipliers.pdf

[19] E. T. Jaynes, "Information theory and statistical mechanics", *Phys. Rev.* **106**, 620 (1957).

[20] E. T. Jaynes, "Information theory and statistical mechanics II", *Phys. Rev.* **108**, 171 (1957).

[21] Claude E. Shannon, "Prediction and entropy", *Bell System Tech. J.*, **30**, 50 (1951).

to statistical mechanics applied by several authors.[22,23,24] Here, rather than pursuing alternative classical statistics, a reasoned path from quantum mechanics via decoherence to $\hat{\rho}_{op}^{\tau}$ – and hence to thermal physics – is postulated.)

Differentiating[25] the thermal Lagrangian with respect to the $\mathbf{P}^B(\varepsilon_{0,\alpha})$ and setting the resulting expressions to zero we have

$$-k_B\left(\ln \mathbf{P}^B\left(\varepsilon_{0,\alpha}\right)+1\right)-T^{-1}\varepsilon_{0,\alpha}-\lambda_0=0; \qquad \alpha=1,\,2,\,3,\,\ldots,\,g_0, \quad (6.26)$$

where g_0 is the degeneracy of the eigen-energy $\varepsilon_{0,\alpha}$. Solving for each of the probabilities $\mathbf{P}^B(\varepsilon_{0,\alpha})$ we find

$$\mathbf{P}^B\left(\varepsilon_{0,\alpha}\right)=\exp\left\{\frac{\left(-T^{-1}\varepsilon_{0,\alpha}-\lambda_0\right)}{k_B}-1\right\}. \quad (6.27)$$

Applying normalization

$$\sum_{\alpha=1}^{g_0}\mathbf{P}^B\left(\varepsilon_{0,\alpha}\right)=\sum_{\alpha=1}^{g_0}\exp\left\{\frac{\left(-T^{-1}\varepsilon_{0,\alpha}-\lambda_0\right)}{k_B}-1\right\}=1 \quad (6.28)$$

which after summing α gives

$$g_0\exp\left\{\frac{\left(-T^{-1}\varepsilon_0-\lambda_0\right)}{k_B}-1\right\}=1. \quad (6.29)$$

From Eq. 6.27 the Boltzmann probabilities are therefore

$$\mathbf{P}^B\left(\varepsilon_{0,\alpha}\right)=\frac{1}{g_0} \quad (6.30)$$

and from Eqs. 6.24 and 6.30 the (internal) energy is

$$\mathcal{U}=\sum_{\alpha=1}^{g_0}\mathbf{P}^B\left(\varepsilon_{0,\alpha}\right)\varepsilon_{0,\alpha} \quad (6.31)$$

$$=\sum_{\alpha=1}^{g_0}\frac{1}{g_0}\varepsilon_{0,\alpha} \quad (6.32)$$

$$=\varepsilon_0. \quad (6.33)$$

[22] A. Katz, *Principles of Statistical Mechanics*, W.H. Freeman, San Francisco (1967).
[23] R. Baierlein, *Atoms and Information Theory*, W.H. Freeman, San Francisco (1971).
[24] Myron Tribus, *Thermodynamics and Thermostatics*, D. Van Nostrand Company Inc., New York (1961).
[25] The Lagrangian is a functional of the $\mathbf{P}(\varepsilon)$ so the required process is functional differentiation as is done in Lagrangian (classical) mechanics. However, in this uncomplicated case functional differentiation is a subtlety we can pretend to be unaware of.

$\mathcal{U}$ is identified with the g_0-fold degenerate macroscopic Boltzmann eigen-energy ε_0. Finally, from Eqs. 6.30 and 6.21 the entropy is

$$\mathcal{S} = -k_B \sum_{\alpha=1}^{g_0} \mathbf{P}^B\left(\varepsilon_{0,\alpha}\right) \ln \mathbf{P}^B\left(\varepsilon_{0,\alpha}\right) \tag{6.34}$$

$$= -k_B \sum_{\alpha=1}^{g_0} \frac{1}{g_0} \ln \frac{1}{g_0} \tag{6.35}$$

$$= k_B \ln g_0. \tag{6.36}$$

An imaginary trip to Vienna's Zentralfriedhof cemetery finds a tombstone bearing Boltzmann's stony visage. Above the sculpture is carved his enduring formula for entropy $\mathcal{S}$,

$$\mathcal{S} = k \log \mathcal{W}, \tag{6.37}$$

where k was later to become k_B, the Boltzmann constant[26,27] and "log" is the natural logarithm. Comparing Eqs. 6.36 and 6.37, $\mathcal{W}$ (Boltzmann's number of microstates)[28] corresponds to g_0 – the degeneracy of the eigen-energy ε_0.[29,30]

If Boltzmann's equation for entropy, Eq. 6.37, is regarded here as the fundamental "equation of motion" of thermodynamics then, in much the same way that the Lagrangian of classical mechanics generates Newton's Laws, the postulated thermal Lagrangian, Eq. 6.25, generates Boltzmann's "equation of motion". This effectively provides a central organizing principle whose absence in thermal and statistical physics has already been duly noted by Baierlein.[31]

Equations 6.36 and 6.37 provide relatively easily grasped insights into entropy:

- High degeneracy – more statistically equivalent states – means greater uncertainty (i.e. less information) about any particular outcome $\rightarrow$ less "bias" (higher entropy).
- In evolving from configurations of high predictability (large "bias") to ones of low predictability (minimum "bias") the entropy increases – the essence of the Second Law.

26 Boltzmann's constant is universally abbreviated k_B.

27 M. Planck seems to have been the first to write entropy in this form and k was initially called Planck's constant. Yes! Another one!

28 $\mathcal{W}$ represents the German wahrscheinlichkeit (probability), i.e. $\mathcal{W} = \frac{1}{g_0}$.

29 The logarithm in Boltzmann's theory is consistent with the thermal Lagrangian postulate.

30 Of course, Boltzmann knew nothing about quantum eigen-energies. That time was nearly half a century away. His idea of degeneracy was "microstate" counting, i.e. classical "microscopic" configurations which corresponded to a single classical internal energy.

31 R. Baierlein, *Am J. Phys.* **63**, 108 (1995).

6.5 Gibbs' canonical theory

J.W. Gibbs' "canonical" theory[32] may be seen as an extension of the microcanonical case for a closed system (constant particle number) with many internal eigen-energies

$$\mathbf{h}_{op}|\epsilon_s\rangle = \epsilon_s|\epsilon_s\rangle, \tag{6.38}$$

some of which may be degenerate.

A thermal Lagrangian will again be taken as

$$\mathcal{L}\left[\tilde{\mathbf{P}}\right] = -k_B \sum_s \tilde{\mathbf{P}}(\epsilon_s) \ln \tilde{\mathbf{P}}(\epsilon_s) - T^{-1} \sum_s \tilde{\mathbf{P}}(\epsilon_s)\,\epsilon_s - \lambda_0 \sum_s \tilde{\mathbf{P}}(\epsilon_s). \tag{6.39}$$

Maximizing $\mathcal{L}\left[\tilde{\mathbf{P}}\right]$ by (functionally) differentiating with respect to $\tilde{\mathbf{P}}(\epsilon_s)$ and setting the results to zero, we get the set of equations (one for each eigen-index s)

$$-k_B\left[\ln \mathbf{P}(\epsilon_s) + 1\right] - T^{-1}\epsilon_s - \lambda_0 = 0, \qquad s = 1, 2, 3, \ldots, \tag{6.40}$$

whose solutions are

$$\mathbf{P}(\epsilon_s) = \exp\left\{\left(\frac{-T^{-1}\epsilon_s - \lambda_0}{k_B}\right) - 1\right\}. \tag{6.41}$$

Defining $\beta = \frac{1}{k_B T}$ and then imposing normalization to eliminate λ_0 gives the result

$$\mathbf{P}(\epsilon_s) = \frac{\exp(-\beta\epsilon_s)}{\sum_\sigma \exp\{-\beta\epsilon_\sigma\}}. \tag{6.42}$$

Replacing the denominator in Eq. 6.42 with

$$\mathcal{Z} = \sum_\sigma \exp(-\beta\epsilon_\sigma), \tag{6.43}$$

which is *Gibbs' partition function*, we can write

$$\mathbf{P}(\epsilon_s) = \frac{\exp(-\beta\epsilon_s)}{\mathcal{Z}}. \tag{6.44}$$

[32] This is commonly called the "canonical ensemble". But quantum mechanics already brings with it an ensemble interpretation, so an extra use of that term seems redundant.

Note again that the sums are over all states, some of which may be degenerate. Explicitly accounting for degeneracy in carrying out the partition function sum[33]

$$\mathcal{Z} = \sum_{\sigma'} g(\epsilon_{\sigma'}) \exp(-\beta \epsilon_{\sigma'}), \tag{6.45}$$

where $g(\epsilon_{\sigma'})$ is the degeneracy of the state $|\epsilon_{\sigma'}\rangle$.

6.6 Canonical thermodynamics

1. Internal energy:

$$\mathcal{U} = \langle \mathbf{h}_{op} \rangle \tag{6.46}$$

$$= \sum_s \epsilon_s \mathbf{P}(\epsilon_s) \tag{6.47}$$

which is identical with

$$\mathcal{U} = -\frac{\partial \ln \mathcal{Z}}{\partial \beta}. \tag{6.48}$$

2. Entropy:

$$\mathcal{S} = -k_B \sum_s \mathbf{P}(\epsilon_s) \ln \mathbf{P}(\epsilon_s) \tag{6.49}$$

$$= -k_B \sum_s \frac{\exp(-\beta \epsilon_s)}{\mathcal{Z}} [-\beta \epsilon_s - \ln \mathcal{Z}] \tag{6.50}$$

$$= \frac{\mathcal{U}}{T} + k_B \ln \mathcal{Z}. \tag{6.51}$$

Then with Eq. 6.48

$$\mathcal{S} = -k_B \beta \frac{\partial}{\partial \beta} \ln \mathcal{Z} + k_B \ln \mathcal{Z} \tag{6.52}$$

[33] The commonly used symbol $\mathcal{Z}$ arises from the German *Zustandssumme* – the sum over states, a reminder of what has to be summed!

and finally

$$\mathcal{S} = -k_B \beta^2 \frac{\partial}{\partial \beta}\left(\frac{1}{\beta} \ln \mathcal{Z}\right). \tag{6.53}$$

3. Pressure:

   Since the pressure operator is $\mathbf{p}_{op} = -\left(\frac{\partial \mathbf{h}_{op}}{\partial V}\right)_T$ thermodynamic pressure is

$$p = \sum_s -\left(\frac{\partial \epsilon_s}{\partial V}\right)_T \mathbf{P}(\epsilon_s) \tag{6.54}$$

   where $\epsilon_s(V)$ is assumed known. Applying Eq. 6.43 this is identical to

$$p = \frac{1}{\beta}\left(\frac{\partial \ln \mathcal{Z}}{\partial V}\right)_T. \tag{6.55}$$

4. Helmholtz potential F:

   From Eq. 6.51 and the definition of the Helmholtz potential (see Eq. 5.17) it follows that

$$F = -\frac{1}{\beta} \ln \mathcal{Z}, \tag{6.56}$$

   which from Eqs. 6.53 and 6.55 follow the usual thermodynamic results

$$\begin{aligned} \mathcal{S} &= -\left(\frac{\partial F}{\partial T}\right)_V \\ p &= -\left(\frac{\partial F}{\partial V}\right)_T. \end{aligned} \tag{6.57}$$

Although $\mathcal{Z}$ is not a measurable, it is among the most important quantities in statistical thermodynamics.

6.7 Degeneracy and $\mathcal{Z}$

The sum in Eq. 6.45 is over discrete energies. State "multiplicity" (degeneracy) is taken into account by the factor $g(\epsilon_s)$.

There are, generally, two sources of degeneracy:

1. Microscopic degeneracy: Generally arising from solutions to the eigenvalue problem

$$\mathcal{H}_{op}|E_s\rangle = E_s|E_s\rangle, \tag{6.58}$$

   in which several eigenstates, e.g. $|E_1\rangle, |E_2\rangle, |E_3\rangle, \ldots,$ correspond the same eigen-energy. This source of degeneracy will be referred to as *internal degeneracy*. For example the spin independent s-states of the hydrogen atom are 1-fold degenerate (this is called *non-degenerate*). If spin is included, the s-states of the hydrogen atom are 2-fold degenerate.
2. Macroscopic (configurational) degeneracy: Usually associated with systems composed of large numbers of macroscopic, energetically equivalent, configurations, e.g. lattices. These degeneracies can usually be determined by combinatoric arguments.

6.7.1 Examples

Example 1

A simple model of a crystal consists of N sites, each occupied by an atom[34] with spin $\frac{1}{2}$. With no local magnetic field ($\mathcal{B} = 0$) the two spin states, identified with quantum numbers $s_z = \pm\frac{1}{2}$ (or, as cartoons, $\uparrow, \downarrow$), have identical eigen-energies, i.e. they are *internally* 2-fold degenerate.

In this case there is a combination of internal and configurational degeneracy. Since each atom is 2-fold degenerate, for N identical atoms the degeneracy factor is

$$g = 2 \times 2 \times 2 \times 2 \ldots \tag{6.59}$$

$$= 2^N. \tag{6.60}$$

If, for definiteness, this state is taken to have energy $\varepsilon = 0$, then according to Eq. 6.45

$$\mathcal{Z} = 2^N \, e^{-\beta \times 0} \tag{6.61}$$

$$= 2^N. \tag{6.62}$$

[34] Spin–spin interaction is neglected.

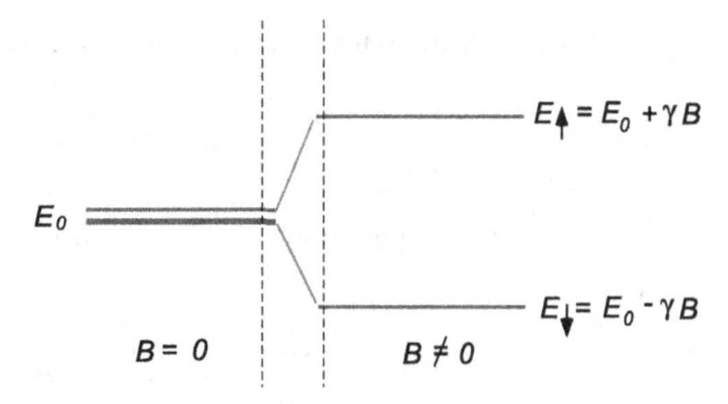

Fig. 6.1 Zeeman effect with spin-$\frac{1}{2}$ atoms in a weak magnetic field.

Since the entropy of a Gibbs system is

$$\mathcal{S} = -k_B \beta^2 \frac{\partial}{\partial \beta}\left(\frac{1}{\beta} \ln \mathcal{Z}\right), \tag{6.63}$$

we have the result

$$\mathcal{S} = N k_B \ln 2, \tag{6.64}$$

which is, of course, identical with Boltzmann's $\mathcal{S} = k_B \ln W$.

If the atoms had higher internal degeneracy the entropy would increase – less statistical knowledge (greater uncertainty). If the atoms were spin $= 0$ (non-degenerate) then $g = 1$ and the entropy would be zero – i.e. statistical certainty – and the smallest possible entropy.

Example 2

Assume now that the $S = \frac{1}{2}$ atoms are immersed in a weak but uniform local[35] magnetic field $\mathcal{B}$ so that the degeneracy is "lifted" (see Figure 6.1) with $E_\uparrow = \gamma \mathcal{B}$ and $E_\downarrow = -\gamma \mathcal{B}$.[36] If $n_\uparrow$ are in the state $E_\uparrow$ and $n_\downarrow$ are in the state $E_\downarrow$, with $N = n_\uparrow + n_\downarrow$, the total macroscopic system energy for any configuration $\{n_\uparrow, n_\downarrow\}$ is, with $E_0 = 0$,

$$E = n_\uparrow \gamma \mathcal{B} - n_\downarrow \gamma \mathcal{B}. \tag{6.65}$$

Although the states $E_\uparrow$ and $E_\downarrow$ are no longer internally degenerate, the N-atom system has configurational degeneracy which is determined by counting the number of unique ways N atoms can be arranged with $n_\uparrow$ in the state $E_\uparrow$ and $n_\downarrow$ in the state $E_\downarrow$.

Among the N total atoms there are $N!$ permutations, no matter what the spin state. But not all the permutations are unique arrangements of spins on sites. In fact, $n_\uparrow! \times n_\downarrow!$ are redundant and must be divided out. Therefore the total configurational

[35] The field at the atom site is taken to be the same as the field external to the sample before matter is inserted.

[36] $\gamma = S g \mu_B$ with S the intrinsic spin of the atom, g the electron g-factor and μ_B the Bohr magneton.

degeneracy for the particular macroscopic energy E in Eq. 6.65 is

$$g(N, n_\uparrow, n_\downarrow) = \frac{N!}{n_\uparrow!\, n_\downarrow!}. \tag{6.66}$$

Consequently the partition function (sum over states) for the system in a magnetic field is

$$\mathcal{Z} = \sum_{\substack{n_\uparrow, n_\downarrow \\ n_\uparrow + n_\downarrow = N}} \frac{N!}{n_\uparrow \times n_\downarrow} \exp\left[-\beta\left(n_\uparrow - n_\downarrow\right)\gamma\mathcal{B}\right], \tag{6.67}$$

which is just the binomial expansion of

$$\mathcal{Z} = \left(e^{-\beta\gamma\mathcal{B}} + e^{\beta\gamma\mathcal{B}}\right)^N. \tag{6.68}$$

The internal energy of the system is[37]

$$\mathcal{U} = -\frac{\partial}{\partial\beta}\ln\mathcal{Z} \tag{6.69}$$

$$= N\gamma\mathcal{B}\frac{e^{\beta\gamma\mathcal{B}} - e^{-\beta\gamma\mathcal{B}}}{e^{\beta\gamma\mathcal{B}} + e^{-\beta\gamma\mathcal{B}}}, \tag{6.70}$$

and from Eq. 6.53 the entropy is

$$\mathcal{S} = Nk_B\left\{\ln\left(e^{\beta\gamma\mathcal{B}} + e^{-\beta\gamma\mathcal{B}}\right) - \beta\gamma\mathcal{B}\left(\frac{e^{\beta\gamma\mathcal{B}} - e^{-\beta\gamma\mathcal{B}}}{e^{\beta\gamma\mathcal{B}} + e^{-\beta\gamma\mathcal{B}}}\right)\right\}. \tag{6.71}$$

Example 3

A simple model of a linear elastomer[38] consists of a chain with N identical links each of length ℓ (see Figure 6.2). The elastomer can be stretched under tension until its maximum length is $L_{max} = N\ell$. (A thin rubber band is an imperfect but helpful illustration.) In this simplified model the linkage energies are very weak so that each molecule (link) can point either to the left or to the right – and *only* to the left or right – with negligible energy difference.[39] Referring to Eq. 6.45, the partition function for this model depends only on configurational degeneracy.

The fully extended polymer with length $L = N\ell$ (see Figure 6.2) has only one possible configuration (actually two, since the links can all point to the left or all to the right). If the tension is relaxed the elastomer will retract to length $L < L_{max}$ as depicted in Figure 6.3.

[37] Strictly speaking, with $E_0 = 0$, the internal energy $\mathcal{U} = 0$. The result given here is a magnetic energy.

[38] An elastic polymer.

[39] Although not without significant deficiencies, the model is useful for understanding unusual elastic and thermodynamic properties of rubber and other elastic polymers.

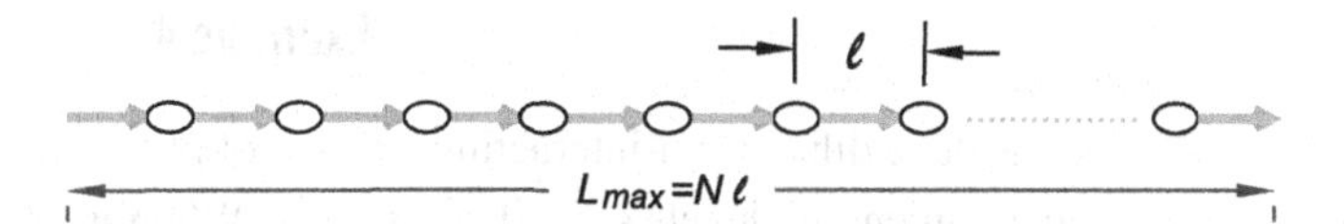

Fig. 6.2 Elastomer completely stretched under tension.

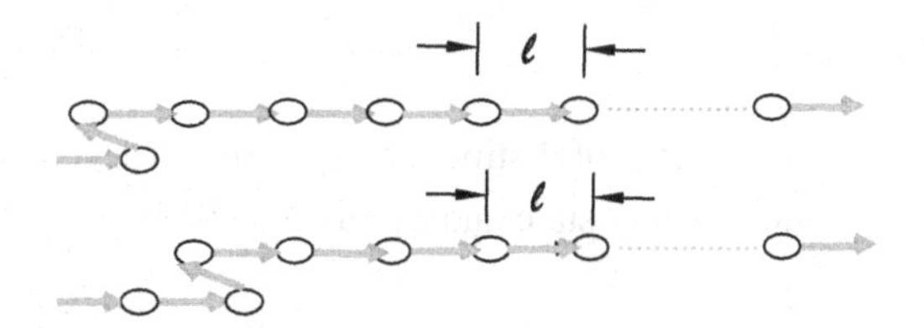

Fig. 6.3 Depiction of "retraction" in an elastic chain.

Let n_L be the number of left-pointing links and n_R be the number of right-pointing links. For this configuration the polymer length is

$$L = \ell \left| n_R - n_L \right|, \tag{6.72}$$

where the magnitude defines the 2-fold left–right symmetry. The total number of links is obviously

$$N = n_R + n_L. \tag{6.73}$$

There are $N!$ permutations among the N links. But not all are unique since any arrangements with n_R right-pointing links and n_L left-pointing links are equivalent. So the number of distinct elastomer configurations is

$$g(n_L, n_R) = 2 \times \frac{N!}{n_L! n_R!}, \tag{6.74}$$

where the factor 2 accounts for the left–right symmetry. Thus the partition function is simply

$$\mathcal{Z} = \sum_{\substack{n_L, n_R \\ n_L + n_R = N}}^{N} g\left(n_L, n_R\right) \tag{6.75}$$

$$= 2 \sum_{\substack{n_L, n_R \\ n_L + n_R = N}}^{N} \frac{N!}{n_L! n_R!} \tag{6.76}$$

$$= 2 \times 2^N, \tag{6.77}$$

and in accord with Eq. 6.63 as well as Boltzmann's fundamental result (applying Stirling's formula, Eq. 7.26, to approximate the factorials in Eq. 6.74)

$$\mathcal{S} = (N + 1)\, k_B \ln 2. \tag{6.78}$$

Example 4

Consider a dilute (non-interacting) system of N two-level atoms with non-degenerate quantum energy levels $\varepsilon_0 = 0$ and $\varepsilon_1 = \varepsilon$. When n_0 atoms are in the ground state ε_0 and n_1 atoms are in the excited state ε_1 the macroscopic eigen-energies are

$$E = n_0 \times 0 + n_1\,\varepsilon. \tag{6.79}$$

The number of distinct configurations (degeneracy) associated with n_0 atoms in the lower energy state and n_1 atoms in the excited energy state is

$$g = \frac{N!}{n_0!n_1!} \tag{6.80}$$

so that the partition function is

$$\mathcal{Z} = \sum_{\substack{n_0,n_1 \\ n_0+n_1=N}}^{N} \frac{N!}{n_0!n_1!} \exp\left(-\beta n_1 \varepsilon\right) \tag{6.81}$$

$$= \left(1 + e^{-\beta\varepsilon}\right)^N. \tag{6.82}$$

Example 5

Ignoring spin, if the excited state of the N impurities in the last example was a hydrogenic 4-fold degenerate $(2s + 2p)$ configuration with energy $E = \epsilon$ and the ground state was a non-degenerate $1s$ state with $E = 0$, what would be the degeneracy g?

Generalizing to the case where n_0 impurity atoms are in an r-fold degenerate ground state and n_1 are in an s-fold degenerate excited state, with $n_0 + n_1 = N$,

$$g = \frac{N!}{n_0!n_1!} r^{n_0} s^{n_1}, \tag{6.83}$$

so that for the hydrogenic impurity,

$$g = \frac{N!}{n_0!n_1!} (1)^{n_0} (4)^{n_1}, \tag{6.84}$$

and the partition function is

$$\mathcal{Z} = \sum_{\substack{n_0,n_1 \\ n_0+n_1=N}}^{N} \frac{N!}{n_0!n_1!} (1)^{n_0} (4)^{n_1} e^{-n_0\beta(0)} e^{-n_1\beta\varepsilon} \tag{6.85}$$

$$= \left(1 + 4e^{-\beta\varepsilon}\right)^N. \tag{6.86}$$

Note: In the case of an arbitrary number of states with energies $\varepsilon_1, \varepsilon_2, \ldots, \varepsilon_m$, and degeneracies $s_1, s_2, \ldots, s_m$ with site occupancies $n_1, n_2, \ldots, n_m$,

$$g = \frac{N!}{n_1!n_2!\ldots n_m!} s_1^{n_1} s_2^{n_2} \ldots s_m^{n_m}, \tag{6.87}$$

which is a generalized multinomial coefficient[40] with a corresponding partition function

$$\mathcal{Z} = \sum_{\substack{n_1, n_2, \ldots, n_m \\ n_1+n_2+\cdots+n_m=N}}^{N} \frac{N!}{n_1!n_2!\ldots n_m!} s_1^{n_1} s_2^{n_2} \ldots s_m^{n_m} e^{-\beta\varepsilon_1 n_1} e^{-\beta\varepsilon_2 n_2} \ldots e^{-\beta\varepsilon_m n_m} \tag{6.88}$$

$$= \left(s_1 e^{-\beta\varepsilon_1} + s_2 e^{-\beta\varepsilon_2} + \cdots + s_m e^{-\beta\varepsilon_m}\right)^N. \tag{6.89}$$

Example 6

Consider a lattice of N identical one-dimensional quantum harmonic oscillators, each oscillator with allowed eigen-energies

$$\mathcal{E}_i = \hbar\omega_0 \left(n_i + \frac{1}{2}\right) \tag{6.90}$$

where ω_0 is the oscillator's natural frequency, $i = 1, 2, 3, \ldots, N$ and $n_i = 1, 2, 3, \ldots, \infty$ are the ith oscillator's allowed quantum numbers. The eigen-energies of the entire oscillator lattice are

$$\mathcal{E}(n) = \sum_{i=1}^{N} \mathcal{E}_i \tag{6.91}$$

$$= \frac{\hbar\omega_0 N}{2} + \hbar\omega_0 \sum_{i=1}^{N} n_i \tag{6.92}$$

$$= \frac{\hbar\omega_0 N}{2} + \hbar\omega_0 n, \tag{6.93}$$

where n is the total lattice quantum number,[41]

$$n = \sum_{i=1}^{N} n_i. \tag{6.94}$$

[40] This result and its widely used generalizations can be found in M. Abramowitz and I. Segun, *Handbook of Mathematical Functions*, Dover (1965), Chapter 24.

[41] Here we have the opportunity to introduce the idea of *quasi-particles*, which are notional particles used to represent the energy excitations of fields. In this case the harmonic oscillator is represented by quasi-particles identified with a "phonon" field, i.e. the quantum numbers n. Clearly these quasi-particles must be "notional" in the sense that, even though there is a fixed number N of vibrating atoms, there is no similar quasi-particle number n conservation.

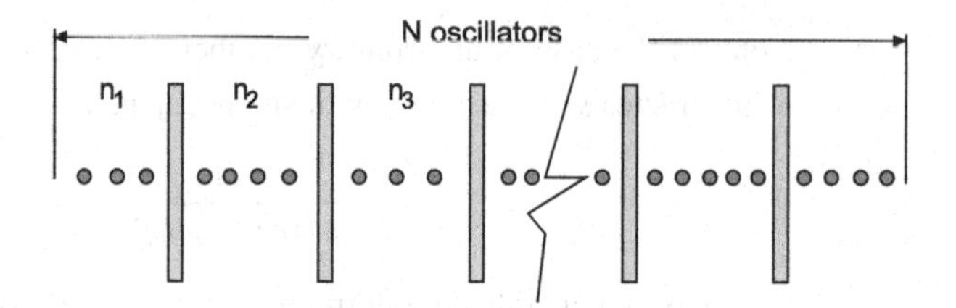

Fig. 6.4 Occupation representation for a lattice of N identical oscillators.

In principle, using Eq. 6.93 the partition function could be written

$$\mathcal{Z} = \sum_{n=0}^{\infty} g(N,n) \exp\left[-\beta\hbar\omega_0\left(\frac{N}{2}+n\right)\right], \tag{6.95}$$

where $g(N,n)$ is the degeneracy of the lattice eigen-energy (see Eq. 6.93).

In Figure 6.4 the N oscillators are represented with compartments created by $N-1$ partitions (shaded bars) while the quantum numbers n_i are represented by n quasi-particles (grey bullets in the boxes). The n quasi-particles are distributed in all possible ways in the N boxes. The figure shows an arrangement of two kinds of objects – partitions and quasi-particles. The number of distinct ways these[42] can be arranged is

$$g(N,n) = \frac{(N-1+n)!}{(N-1)!\,n!}, \tag{6.96}$$

which gives the partition function

$$\mathcal{Z} = \sum_{n=0}^{\infty} \frac{(N-1+n)!}{(N-1)!\,n!} \exp\left[-\beta\hbar\omega_0\left(\frac{N}{2}+n\right)\right] \tag{6.97}$$

$$= e^{-\frac{N\beta\hbar\omega_0}{2}} \sum_{n=0}^{\infty} \frac{(N-1+n)!}{(N-1)!\,n!} \exp(-n\beta\hbar\omega_0). \tag{6.98}$$

This sum is not a binomial expansion but may be approximated using Stirling's asymptotic expansion (see Eq. 7.26). (An alternative exact evaluation using the Euler Γ function

$$\Gamma(n) = (n-1)! = \int_0^{\infty} dt\, t^{n-1} e^{-t} \tag{6.99}$$

is left to Chapter 9 where the harmonic oscillator is discussed.)

[42] There are n quasi-particles, $N-1$ partition walls and $N-1+n$ total objects.

6.8 Closing comments

Boltzmann's pre-quantum insight linked degeneracy (multiplicity) with entropy. But as indicated here, Boltzmann's conjecture follows, in principle, by applying the thermal Lagrangian postulate introduced in this chapter in the special circumstance of a single, isolated, degenerate macroscopic eigen-energy. Applying the same rule to a continuous spectrum of macroscopic eigen-energies leads to Gibbs' "canonical" case. With the thermal Lagrangian (least-biased postulate) thermal physics becomes a direct consequence of macroscopic quantum theory.

Thermodynamics was a serious and practical science long before there was a quantum mechanics. Doubtless, it will continue to be widely applied in its pragmatic, time-honored classical form by engineers and chemists in which the role played by quantum mechanics may be of only academic interest or of no interest at all.

But quantum mechanics lends modern thermodynamics its distinctive and profound character in which microscopic models can be used as a basis for even the most exotic macroscopic systems. This union has fashioned one of the most versatile and widely applied paradigms in modern science.

Problems and exercises

6.1 A hypothetical macroscopic system hamiltonian $\mathbf{h}_{op}$ has eigen-energies and eigenstates given by

$$\mathbf{h}_{op}|\phi_j\rangle = \phi_j|\phi_j\rangle, \tag{6.100}$$

where the eigenvalues are

$$\phi_j = j\gamma, \qquad j = 0, 1, 2, \ldots, \infty. \tag{6.101}$$

Here γ is a positive constant and each eigenstate $|\phi_j\rangle$ is j-fold degenerate.

a. Find $\mathbf{P}_j$, the least biased, normalized $\left(\sum_{j=1}^{\infty} \mathbf{P}_j = 1\right)$ probability of the macroscopic outcome $\phi_j = j\gamma$.
b. What is the entropy of the hypothetical system?
c. Find the average value $\langle \mathbf{h}_{op}\rangle = \mathcal{T}r\rho_{op}^{\tau}\, \mathbf{h}_{op}$.
d. Find the mean uncertainty

$$\left\langle (\Delta\phi_j)^2 \right\rangle = \sum_{j=1}^{\infty} \left(\phi_j - \langle \mathbf{h}_{op}\rangle\right)^2 \mathbf{P}_j. \tag{6.102}$$

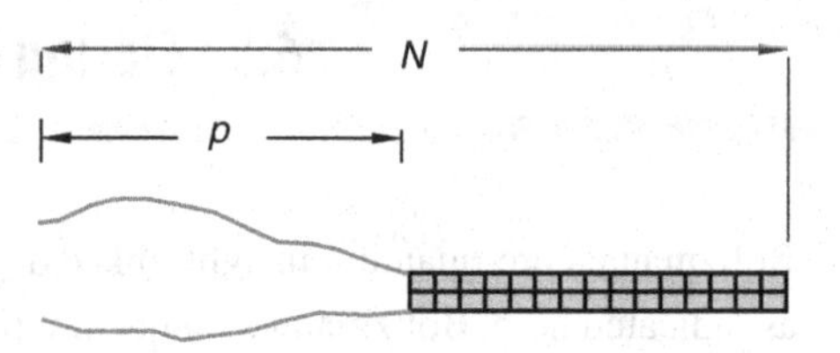

Fig. 6.5 The melting DNA double helix showing p broken links in a chain of N total bonds.

6.2 DNA consists of a pair of molecular strands curled to form a double helix. With increasing temperature the strands dissociate, unraveling the helix – i.e. the DNA "melts" (see Figure 6.5). A simple model of DNA melting[43] consists of two linked strands whose links – starting from only one end of the chain – break up successively. For a link to break, all links to the left of it must already be broken – like a single-ended zipper. The only interior link that can break is one immediately adjacent to one that has already broken. The last link is considered unbreakable.

Assuming the links $1, 2, \ldots, p-1$ are already broken, the energy required to break the pth link is $+\epsilon$. Unbroken links are taken to have energy 0. Each broken link has degeneracy Γ due to the multitude of spatial orientations it can assume.

(a) Show that the DNA partition function is

$$\mathcal{Z} = \frac{1 - x^N}{1 - x}, \tag{6.103}$$

where

$$x = \Gamma \exp(-\beta\epsilon) \tag{6.104}$$

with

$$\beta = \frac{1}{k_B T}. \tag{6.105}$$

(b) In this model an order parameter θ is defined as the average number of broken bonds, $\langle p \rangle$. Show that

$$\theta = \frac{N x^N}{x^N - 1} - \frac{x}{x - 1}. \tag{6.106}$$

(c) Plot θ vs. x for $N = 1000$ to demonstrate the transition and critical point at $x = 1$.

[43] C. Kittel, "Phase transition of a molecular zipper", *Am. J. Phys.* **37**, 917 (1969).

(d) Show that for $N\epsilon \ll 1$ the entropy of "zipper" DNA is

$$\mathcal{S} = \theta \ln \Gamma. \tag{6.107}$$

Hint:

$$S_{N-1} = 1 + x + x^2 + \ldots + x^{N-1} \tag{6.108}$$

$$= \frac{1 - x^N}{1 - x}. \tag{6.109}$$

7 The ideal gas

J. Willard Gibbs is probably the most brilliant person most people have never heard of.

Bill Bryson, *A Short History of Nearly Everything*, Broadway Publishing, NY (2003)

7.1 Introduction

An *ideal gas* is a collection of uncorrelated identical particles with negligible forces of interaction.[1] These particles can be fundamental particles such as electrons, protons or neutrons or even quarks. They can be more complex systems such as atoms, atomic nuclei or molecules. Particle specifics determine, especially at low temperatures, details of their ideal gas behavior even without mutual interactions. For example, in addition to the usual translational motion, molecules can rotate or vibrate, imparting additional properties to these ideal gases. Most interesting however, are low-temperature behaviors associated with quantum mechanical properties.

Even before quantum mechanics had attained a generally accepted form, W. Pauli[2] conjectured that only one electron can occupy a single-particle energy eigen state – a restriction called the *Pauli exclusion principle* (PEP).

The following year, E. Fermi[3] and P. Dirac[4] further showed that quantum mechanics required *all* particles – depending on their intrinsic spin S, which can be integer or half-integer – to belong to one of two possible classes:

1. Particles with *half-integer* spin ($S = \frac{1}{2}, \frac{3}{2}, \frac{5}{2}, \ldots$) obey the Pauli exclusion principle and are thus called Fermi–Dirac (FD) particles, or *fermions*. Members of this class are quarks, electrons, protons, neutrons and neutrinos, as well as their antiparticles, and many composite systems. For example, He^3 is a composite of 2 protons, 2 electrons and 1 neutron, each with spin $\frac{1}{2}$. This odd-number aggregate

[1] Just enough interaction to eventually reach thermal equilibrium.

[2] W. Pauli, "On the connexion between the completion of electron groups in an atom with the complex structure of spectra", *Z. Phys.* **31**, 765 (1925).

[3] E. Fermi, "Sulla quantizzazione del gas perfetto monoatomico," *Rend. Lincei* **3**, 145 (1926).

[4] P. Dirac, "On the theory of quantum mechanics", *Proc. Royal Soc. A* **112**, 661 (1926).

ensures that He^3 has $\frac{1}{2}$-integer spin. Electrons (together with a uniform positive background) constitute a gas of Fermi–Dirac particles.

2. Particles with *integer* spin ($S = 0, 1, 2, \ldots$) have no eigenstate occupation restriction. Integer-spin particles are called Bose–Einstein (BE) particles, or *bosons*. Although most bosons are composite systems, e.g. the H atom (1 proton and 1 electron), He^4 (2 protons, 2 electrons and 2 neutrons) and mesons (2 spin-$\frac{1}{2}$ quarks), there are elementary particle bosons, e.g. $W^{\pm}$ and Z particles, carriers of the weak nuclear force and gluons which are associated with the strong nuclear force. Also included is the γ: particle (photon) and the feverishly sought Higgs boson H^0. But in this era of boson condensate physics, composite boson systems, such as atoms of Li^7, Na^{23}, K^{41}, Rb^{85} and Rb^{87}, are a focus of interest.

In the case of many-particle systems enormous differences exist between low-temperature behavior of He^3 (fermions) and He^4 (bosons). These distinctions will be discussed in Chapters 15 and 16.

7.2 Ideal gas law

In the limit of low particle density and high temperature, ideal FD and BE gasses shed nearly all their quantum properties so that century-and-a-half-old kinetic (classical) gas models successfully describe much of their thermodynamic behavior, in particular the equation of state

$$pV = Nk_B T, \tag{7.1}$$

where p is the gas pressure, N is a fixed number of particles, V the confining volume, T the gas temperature and k_B is Boltzmann's constant.[5] However, in the final analysis the quasi-classical theory is internally inconsistent and must be rescued by quantum corrections.

Taking the opportunity to apply a thermal Lagrangian to a physical problem whose results (and contradictions) are well known, gaseous atoms and molecules are modeled near room temperature and atmospheric pressure as ideal and "quasi-classical". The "ideal gas" is so important in thermodynamic pedagogy as well as for many low-order approximations that we forgo absolute rigor for the present, returning to the subject in Chapter 15 with a more credible approach.

[5] A gas which obeys Eq. 7.1 can be used as an operational thermometer in which T defines the absolute (K) temperature scale.

7.3 Quasi-classical model

The single-particle microscopic hamiltonian operator for non-interacting particles is

$$\mathbf{h}_{op} = -\frac{\hbar^2}{2m}\nabla^2 \tag{7.2}$$

with an eigenvalue equation

$$\mathbf{h}_{op}|\epsilon_s\rangle = \epsilon_s|\epsilon_s\rangle \tag{7.3}$$

with ϵ_s the microscopic eigen-energies.

Consider now a low-density system consisting of N independent particles each with mass m confined to a volume V (no FD or BE occupation restrictions). Each particle can occupy any of the allowed microscopic eigen-energies $\epsilon_1, \epsilon_2, \epsilon_3, \ldots$ with the corresponding macroscopic eigen-energies

$$E(n_1, n_2, n_3, \ldots) = n_1\epsilon_1 + n_2\epsilon_2 + n_3\epsilon_3 + \ldots, \tag{7.4}$$

where n_1 particles are in the state ϵ_1, n_2 particles are in the state ϵ_2, etc., with a fixed total number of particles

$$n_1 + n_2 + n_3 + \ldots = N. \tag{7.5}$$

Guided by discussions in Chapter 6 the thermal Lagrangian is

$$\begin{aligned}\mathcal{L}[\mathbf{P}] = -k_B &\sum_{\substack{n_1,n_2,n_3,\ldots=0,1,2\ldots\\ n_1+n_2+\cdots+n_N=N}}^{N} \mathbf{P}(n_1, n_2, \ldots)\ln\mathbf{P}(n_1, n_2, \ldots)\\ -T^{-1} &\sum_{\substack{n_1,n_2,n_3,\ldots=0,1,2\ldots\\ n_1+n_2+\cdots+n_N=N}}^{N} \mathbf{P}(n_1, n_2, \ldots)[n_1\epsilon_1 + n_2\epsilon_2 + n_3\epsilon_3 + \cdots + n_N\epsilon_N]\\ -\lambda_0 &\sum_{\substack{n_1,n_2,n_3,\ldots=0,1,2\ldots\\ n_1+n_2+\cdots+n_N=N}}^{N} \mathbf{P}(n_1, n_2, \ldots).\end{aligned} \tag{7.6}$$

With $\beta = 1/k_BT$, maximizing $\mathcal{L}[\mathbf{P}]$

$$\mathbf{P}(n_1, n_2, \ldots, n_N) = \frac{\exp\{-\beta(n_1\epsilon_1 + n_2\epsilon_2 + \ldots + n_N\epsilon_N)\}}{\displaystyle\sum_{\substack{n_1,n_2,n_3,\ldots=0,1,2\ldots\\ n_1+n_2+\cdots+n_N=N}}^{N} \exp\{-\beta(n_1\epsilon_1 + n_2\epsilon_2 + \ldots + n_N\epsilon_N)\}}. \tag{7.7}$$

The denominator is the partition function $\mathcal{Z}$ and the sums are over all states. Following the procedure for calculating a partition function,

$$\mathcal{Z} = \sum_{\substack{n_1,n_2,\ldots=0,1,2,\ldots \\ n_1+n_2+\ldots=N}}^{N} g_N(n_1, n_2, \ldots) \exp\{-\beta(n_1\epsilon_1 + n_2\epsilon_2 + \cdots + n_N\epsilon_N)\}, \quad (7.8)$$

where

$$g_N(n_1, n_2, \ldots) = \frac{N!}{n_1! n_2! n_3! \ldots} \quad (7.9)$$

is the configurational degeneracy of an N-particle state[6] with macroscopic eigenenergies given by Eq. 7.4. Therefore

$$\mathcal{Z} = \sum_{\substack{n_1,n_2,\ldots=0,1,2,\ldots \\ n_1+n_2+\ldots=N}}^{N} \frac{N!}{n_1! n_2! \ldots} \exp\{-\beta(n_1\epsilon_1 + n_2\epsilon_2 + \cdots + n_N\epsilon_N)\}, \quad (7.10)$$

which is the expansion of

$$\mathcal{Z} = \left(e^{-\beta\epsilon_1} + e^{-\beta\epsilon_2} + e^{-\beta\epsilon_3} + \ldots\right)^N. \quad (7.11)$$

Although Eq. 7.11 is mathematically correct, the physical result is *not*! For at its heart is a problem in many-particle quantum mechanics with no rigorous semi-classical argument.[7] Indeed, J. W. Gibbs quickly realized that applying Eq. 6.53 to Eq. 7.11 did not result in an extensive entropy. To correct this, the partition function was appended by an ad-hoc $\frac{1}{N!}$ prefactor (known as the Gibbs correction).[8] With this correction, the partition function becomes

$$\mathcal{Z} = \frac{1}{N!}\left(e^{-\beta\epsilon_1} + e^{-\beta\epsilon_2} + e^{-\beta\epsilon_3} + \ldots\right)^N. \quad (7.12)$$

7.4 Ideal gas partition function

Practical ideal gas results can be obtained from Eq. 7.12 by applying the quantum mechanical energy states of free particles confined to a cube[9,10] of side L (volume

6 See Chapter 6.

7 This issue will be addressed in Chapter 15.

8 The correction is a consequence of normalizing quantum many-particle state functions.

9 Periodic boundary conditions are almost universally applied for free particles and fields. But in this problem we choose the equivalent and probably more familiar "large 3-D box" boundary conditions.

10 Lower dimensional systems, which are of considerable importance in modern materials science, might be dealt with in an analogous way.

$V = L^3$),

$$\epsilon_{\nu} = \frac{\hbar^2}{2m}\left(\frac{\pi}{L}\right)^2\left(\nu_x^2 + \nu_y^2 + \nu_z^2\right), \tag{7.13}$$

where $\nu_x,\ \nu_y,\ \nu_z = 1, 2, 3, \ldots$ The eigen-energies in Eq. 7.13 correspond to particles with purely translation modes, e.g. a monatomic gas or a polyatomic gas without vibrational, rotational or electronic excitations.[11]

Using the energies in Eq. 7.13 and the partition function, Eq. 7.12,

$$\mathcal{Z} = \frac{1}{N!}\left[\sum_{\nu_x,\nu_y,\nu_z=1}^{\infty} e^{-\beta\frac{\hbar^2}{2m}\left(\frac{\pi}{L}\right)\left(\nu_x^2+\nu_y^2+\nu_z^2\right)}\right]^N, \tag{7.14}$$

which is

$$\mathcal{Z} = \frac{1}{N!}\left[\sum_{\nu_x=1}^{\infty} e^{-\beta\frac{\hbar^2}{2m}\left(\frac{\nu_x\pi}{L}\right)^2}\right]^{3N}. \tag{7.15}$$

Substituting the new counting variable

$$k_x = \frac{\pi\nu_x}{L} \tag{7.16}$$

implies

$$\frac{L}{\pi}\Delta k_x = \Delta\nu_x. \tag{7.17}$$

Since the quantities $\nu_j, \ldots$ are the integers $1, 2, \ldots$ it follows that all $\Delta\nu_j \equiv 1$ and the partition function is

$$\mathcal{Z} = \frac{1}{N!}\left[\sum_{\nu_x=1}^{\infty}\Delta\nu_x\, e^{-\beta\frac{\hbar^2}{2m}\left(\frac{\nu_x\pi}{L}\right)^2}\right]^{3N} \rightarrow \frac{1}{N!}\left[\frac{L}{\pi}\sum_{k_x}^{\infty}\Delta k_x\, e^{-\beta\frac{\hbar^2}{2m}k_x^2}\right]^{3N}. \tag{7.18}$$

When the cube becomes large, i.e. $L \to \infty$, it follows that $\Delta k_x \to dk_x$, and the sum over Δk_x in Eq. 7.18 can be replaced by the equivalent integral

$$\mathcal{Z} = \frac{1}{N!}\left[\frac{L}{\pi}\int_0^{\infty} \mathrm{d}k_x\, e^{-\beta\frac{\hbar^2}{2m}k_x^2}\right]^{3N}. \tag{7.19}$$

[11] These eigen-energies are also appropriate for ideal FD and BE quantum gases so long as FD and BE occupation number restrictions are observed. FD and BE quantum gases have strikingly different partition functions, even with the same allowed eigen-energies. (See Chapters 15 and 16.)

Using the standard Gaussian integral result

$$\int_0^\infty dx\, e^{-\alpha x^2} = \frac{1}{2}\sqrt{\frac{\pi}{\alpha}}, \tag{7.20}$$

the partition function is

$$\mathcal{Z} = \frac{1}{N!}\left[\frac{L}{2\pi}\sqrt{\frac{2m\pi}{\beta\hbar^2}}\right]^{3N} \tag{7.21}$$

$$= \frac{1}{N!}\left(n_Q V\right)^N, \tag{7.22}$$

where

$$n_Q = \left(\frac{m}{2\pi\hbar^2\beta}\right)^{\frac{3}{2}} \tag{7.23}$$

is the *quantum concentration* – a particle density parameter that measures the gas's degree of dilution. In particular, if $n/n_Q \ll 1$, where $n = N/V$ is the particle concentration, then the gas is sufficiently dilute for the quasi-classical ideal gas result Eq. 7.21 to apply. If, on the other hand, $n/n_Q \gg 1$ quantum considerations are inescapable and the arguments that brought us this far are inadequate.[12,13]

7.5 Thermodynamics of the ideal gas

This ideal gas model is replete with familiar thermodynamic results:

1. Internal energy $\mathcal{U}$:
   With $\mathbf{P}(n_1, n_2, \ldots)$ from Eq. 7.7, the internal energy is

$$\mathcal{U} = \sum_{\substack{n_1,n_2,n_3,\ldots=0,1,2,\ldots \\ n_1+n_2+\cdots+n_N=N}}^{N} \mathbf{P}(n_1, n_2, \ldots)\left[n_1\epsilon_1 + n_2\epsilon_2 + n_3\epsilon_3 + \cdots + n_N\epsilon_N\right], \tag{7.24}$$

   which is identical with

$$\mathcal{U} = -\frac{\partial \ln \mathcal{Z}}{\partial \beta}. \tag{7.25}$$

   Using Eq. 7.21 together with Stirling's approximation

[12] See Chapters 15 and 16

[13] In addition to the quantum mechanical $1/N!$, Planck's constant $\hbar$ also appears in the partition function – yet another intrusion of quantum mechanics. Equation 7.21 shows that even the classical ideal gas – often imagined as tiny billiard balls banging about in a box – has quantum imprinting.

$$n! \sim n^n e^{-n}, \qquad n \gg 1,$$
$$\ln n! \sim n \ln n - n, \qquad n \gg 1, \tag{7.26}$$

we have

$$\mathcal{U} = \frac{3N}{2\beta} \tag{7.27}$$

$$= \frac{3}{2} N k_B T. \tag{7.28}$$

2. Heat capacity $\mathcal{C}_V$:

   With only translational (kinetic) energy

$$\mathcal{C}_V = \left(\frac{\partial \mathcal{U}}{\partial T}\right)_V \tag{7.29}$$

$$= \frac{3}{2} N k_B. \tag{7.30}$$

3. Entropy $\mathcal{S}$:

$$\mathcal{S} = k_B \beta^2 \left(\frac{\partial}{\partial \beta}\right)\left(-\frac{1}{\beta} \ln \mathcal{Z}\right) \tag{7.31}$$

$$= N k_B \left[\frac{5}{2} + \frac{3}{2} \ln\left(\frac{m}{2\pi\hbar^2\beta}\right) - \ln\left(\frac{N}{V}\right)\right] \tag{7.32}$$

$$= N k_B \left[\frac{5}{2} + \ln\left(\frac{n_Q}{n}\right)\right], \tag{7.33}$$

   which is called the Sackur–Tetrode equation.[14,15]

4. Helmholtz potential F:

   Applying Stirling's approximation to Eq. 6.56, where it was shown that

$$F = -\frac{1}{\beta} \ln \mathcal{Z}, \tag{7.34}$$

   we find

$$F = -\frac{N}{\beta}\left(\ln \frac{n_Q}{n} + 1\right). \tag{7.35}$$

[14] The inclusion of $1/N!$ – the Gibbs correction – is a fundamental quantum mechanical result derived from the requirement that under permutation of identical particles the many-particle state vector changes, at most, by a phase factor.

[15] The Gibbs $1/N!$ correction makes an essential contribution to the entropy. Without it $\mathcal{S} \stackrel{?}{=} N k_B \left[\frac{3}{2} + \ln\left(V n_Q\right)\right]$, which does not satisfy Euler's homogeneity condition $\mathcal{S}(\lambda V) = \lambda \mathcal{S}(V)$. $\mathcal{S}$ would not be extensive.

5. Equation of state:
   The expression for pressure is

$$p = \frac{1}{\beta}\left(\frac{\partial \ln \mathcal{Z}}{\partial V}\right)_T \tag{7.36}$$

$$= -\left(\frac{\partial F}{\partial V}\right)_T, \tag{7.37}$$

giving the "ideal gas law"

$$p = \frac{Nk_BT}{V}. \tag{7.38}$$

7.6 Gibbs' entropy paradox

Consider a cylinder that is partitioned into two sections with volumes V_1 and V_2 so that the total volume is $V = V_1 + V_2$. N_1 and N_2 absolutely identical ideal gas molecules occupy the volumes V_1 and V_2, respectively – same quantum concentration, same temperature T, same concentration n. If the partition magically dissolves the identical gases will spontaneously mix and occupy the total volume V with the equilibrium temperature and concentration unchanged.

Objectively, nothing has happened. There is no way to distinguish between the "before" and "after". There is no change in the state of knowledge. For this process mixing has, apparently, no objective meaning! Is there a change in entropy?

The Sackur–Tetrode equation gives the entropy of the original system (before the partition dissolves) as

$$\mathcal{S} = \mathcal{S}_1 + \mathcal{S}_2 \tag{7.39}$$

$$= N_1 k_B\left[\ln\left(n_Q/n\right) + \frac{5}{2}\right] + N_2 k_B\left[\ln\left(n_Q/n\right) + \frac{5}{2}\right] \tag{7.40}$$

$$= N k_B\left[\ln\left(n_Q/n\right) + \frac{5}{2}\right], \tag{7.41}$$

which is identical to the entropy of the system after the partition disintegrates – entirely in accord with physical intuition and our interpretation of entropy's informational meaning.[16]

If, however, the partition function of Eq. 7.22 had not been "quantum" corrected by the Gibbs $1/N!$ there would be an increase in the entropy after the partition disintegrates, which, based on our understanding of entropy, is nonsense!

[16] Note that when calculating thermodynamic averages such as $\mathcal{U}$ or p the Gibbs $1/N!$ correction cancels. The same cancellation does not occur in entropy calculations.

7.7 Entropy of mixing

Consider the situation just described, but now with two different ideal gases, a and b, with unique quantum concentrations n_Q^a and n_Q^b. Initially, species a occupies a volume V_a and species b occupies a volume V_b. The partition function for gas a is

$$\mathcal{Z}_a = \frac{\left(V_a n_Q^a\right)^{N_a}}{N_a!} \tag{7.42}$$

and the partition function for gas b is

$$\mathcal{Z}_b = \frac{\left(V_b n_Q^b\right)^{N_b}}{N_b!}. \tag{7.43}$$

Applying the Sackur–Tetrode equation to each species, the total initial entropy is

$$\mathcal{S} = \mathcal{S}_a + \mathcal{S}_b \tag{7.44}$$

$$= N_a k_B \left[\frac{5}{2} + \ln\left(n_Q^a / n_a\right)\right] + N_b k_B \left[\frac{5}{2} + \ln\left(n_Q^b / n_b\right)\right], \tag{7.45}$$

where $n_a = N_a / V_a$ and $n_b = N_b / V_b$.

After the chamber partition magically dissolves and the gases mix, the total entropy of the mixed system is

$$\mathcal{S}_{a+b} = N_a k_B \left\{\ln\left[(V_a + V_b)\, n_Q^a / N_a\right] + \frac{5}{2}\right\}$$

$$+ N_b k_B \left\{\ln\left[(V_a + V_b)\, n_Q^b / N_b\right] + \frac{5}{2}\right\}. \tag{7.46}$$

Defining an entropy of mixing as $\mathcal{S}_{mixing} = \mathcal{S}_{a+b} - (\mathcal{S}_a + \mathcal{S}_b)$ then

$$\mathcal{S}_{mixing} = N_a k_B \ln\left(\frac{V_a + V_b}{V_a}\right) + N_b k_B \ln\left(\frac{V_a + V_b}{V_b}\right) > 0. \tag{7.47}$$

The mixing entropy $\mathcal{S}_{mixing}$ has the same value even if the partition functions had not been Gibbs (quantum) corrected by $1/N!$. Moreover, Eq. 7.47 now says – nonsensically – that a finite mixing entropy remains even after the limit

$$n_Q^a \rightarrow n_Q^b \tag{7.48}$$

has been taken! Mixing entropy is obviously not a continuous function of n_Q. This, physically, may be attributable to the "quantization" of mass units.

7.8 The non-ideal gas

The idea of a "classical" ideal gas is clearly a fiction since $\hbar$ and the Gibbs correction factor $1/N!$, both from quantum mechanics, must appear in the ideal gas entropy. Moreover, atoms and molecules in a real gas do interact and with significant consequences. An approximate equation of state that improves upon the ideal gas law is due to van der Waals (1873) who took account of long-range attraction and short-range repulsion between particles. Van der Waals' equation not only improves descriptions of gas phase thermodynamics, it also suggests gas–liquid phase transitions and gas critical points as consequences of interactions. This eponymous equation of state is:[17]

$$\left[p + (N/V)^2 a\right](V - Nb) = Nk_B T, \tag{7.49}$$

where a and b are coefficients characteristic of specific gases.[18]

The van der Waals equation, by quantitatively providing two parameters that characterize real molecules, would seem to have settled a bitter ongoing debate of the times: Did gases, and matter in general, have microscopic structure? But it didn't.

Problems and exercises

7.1 Assume an N-particle ideal gas in two-dimensions with eigen-energies

$$E_n = \frac{\hbar^2}{2m}\left(\frac{\pi}{L}\right)^2\left(n_x^2 + n_y^2\right), \tag{7.50}$$

where

$$n_x, n_y = 1, 2, 3, \ldots \tag{7.51}$$

The confining surface area is

$$A = L^2. \tag{7.52}$$

(a) Write an expression for the partition function $\mathcal{Z}_{2D}$ of the two-dimensional ideal gas.

(b) Find the internal energy $\mathcal{U}_{2D}$ of the gas.

(c) Find the heat capacity $\mathcal{C}_A$ for the two-dimensional gas.

(d) Find the entropy $\mathcal{S}_{2D}$ of the gas.

[17] A derivation of the van der Waals equation is given in Chapter 13.

[18] R. C. Weast (ed.), *Handbook of Chemistry and Physics* (53rd edn.), Chemical Rubber Co., Cleveland (1972).

7.2 The quantity

$$\pi_T = \left(\frac{\partial \mathcal{U}}{\partial V}\right)_T \tag{7.53}$$

is called internal pressure and expresses the role of particle–particle interactions in the behavior of a gas. Find π_T for

(a) an ideal gas;

(b) a van der Waals gas;

(c) a gas with the equation of state

$$p = \frac{Nk_BT}{V}\left[1 + \frac{N}{V}B(T)\right]. \tag{7.54}$$

8 The two-level system

A theory is the more impressive the greater the simplicity of its premises, the more different kinds of things it relates and the more extended is its area of applicability.

A. Einstein. Reproduced by permission of Open Court Publishing Company, a division of Carus Publishing Company, from *Albert Einstein: Philosopher-Scientist*, edited by Paul Arthur Schilpp (The Library of Living Philosophers Volume VII), 1949, 1951 and 1970 by The Library of Living Philosophers

8.1 Anomalous heat capacity

Insulators or semiconductors containing impurities with closely spaced electronic or nuclear ground states[1] show an anomalous low-temperature heat capacity peak – the so-called *Schottky anomaly*.[2] It is an example of microscopic quantum effects having macroscopic (thermodynamic) signatures.

8.2 Schottky model

A simple model for the "Schottky" anomalous heat capacity consists of a crystal containing a dilute[3] concentration of N identical atomic impurities, each having the same pair of closely spaced, non-degenerate energy levels – a lower state ϵ_1 and a higher state ϵ_2. If, at some temperature T, n_1 impurity atoms are in the low energy state and n_2 are in the higher energy state, the macroscopic eigen-energies are

$$E(n_1, n_2) = n_1\epsilon_1 + n_2\epsilon_2, \tag{8.1}$$

1 Usually degenerate low-energy states whose degeneracy is lifted by crystalline electric fields, magnetic fields, spin-orbit coupling or hyperfine interactions.

2 The phenomenon is named for Walter Hermann Schottky (1886–1976).

3 Diluteness insures negligible inter-atomic effects.

with the thermal Lagrangian

$$\mathcal{L} = -k_B \sum_{\substack{n_1,n_2 \\ n_1+n_2=N}} \mathbf{P}(n_1, n_2) \ln \mathbf{P}(n_1, n_2)$$
$$- T^{-1} \sum_{\substack{n_1,n_2 \\ n_1+n_2=N}} \mathbf{P}(n_1, n_2)\,(n_1\epsilon_1 + n_2\epsilon_2) - \lambda_0 \sum_{\substack{n_1,n_2 \\ n_1+n_2=N}} \mathbf{P}(n_1, n_2), \tag{8.2}$$

where $\mathbf{P}(n_1, n_2)$ is the probability that n_1 impurity atoms are in the state ϵ_1 and n_2 are in the state ϵ_2. The sums are over all states. The method of Lagrange multipliers gives the probabilities

$$\mathbf{P}(n_1, n_2) = \frac{e^{\ \beta(n_1\epsilon_1 \mid n_2\epsilon_2)}}{\mathcal{Z}} \tag{8.3}$$

with the partition function

$$\mathcal{Z} = \sum_{\substack{n_1,n_2 \\ n_1+n_2=N}} e^{-\beta(n_1\epsilon_1+n_2\epsilon_2)}, \tag{8.4}$$

which again represents a sum over all states. One way to calculate this sum is by explicitly including a configurational degeneracy (see Chapter 6)

$$g(n_1, n_2) = \frac{N!}{n_1!\,n_2!} \tag{8.5}$$

to give

$$\mathcal{Z} = \sum_{\substack{n_1,n_2 \\ n_1+n_2=N}}^{N} \frac{N!}{n_1!\,n_2!} e^{-\beta(n_1\epsilon_1+n_2\epsilon_2)}, \tag{8.6}$$

which is the binomial expansion of[4]

$$\mathcal{Z} = \left(e^{-\beta\epsilon_1} + e^{-\beta\epsilon_2}\right)^N. \tag{8.7}$$

The thermodynamic properties that follow from Eq. 8.7 are:

1. Internal energy:

$$\mathcal{U} = -\frac{\partial}{\partial\beta} \ln \mathcal{Z} \tag{8.8}$$

$$= N \frac{\epsilon_1 e^{-\beta\epsilon_1} + \epsilon_2 e^{-\beta\epsilon_2}}{e^{-\beta\epsilon_1} + e^{-\beta\epsilon_2}}. \tag{8.9}$$

[4] Some authors prefer to calculate the partition function – one atom at a time – and then take the product of N single-atom partition functions. This avoids the mathematics of combinatoric analysis in obtaining g. But it also evades the macroscopic character of the physics.

2. Entropy:

$$\mathcal{S} = -k_B \sum_n \mathbf{P}(n_1, n_2) \ln \mathbf{P}(n_1, n_2) \tag{8.10}$$

$$= -k_B \beta^2 \frac{\partial}{\partial \beta} \left(\frac{1}{\beta} \ln \mathcal{Z} \right) \tag{8.11}$$

$$= N k_B \left[\ln \left(e^{-\beta \epsilon_1} + e^{-\beta \epsilon_2} \right) + \beta \frac{\left(\epsilon_1 e^{-\beta \epsilon_1} + \epsilon_2 e^{-\beta \epsilon_2} \right)}{\left(e^{-\beta \epsilon_1} + e^{-\beta \epsilon_2} \right)} \right]. \tag{8.12}$$

3. Heat capacity:
   The heat capacity $\mathcal{C}_V$ can be calculated from Eq. 8.12 with

$$\mathcal{C}_V = T \left(\frac{\partial \mathcal{S}}{\partial T} \right)_V. \tag{8.13}$$

In solids this is usually assumed not to differ significantly from $\mathcal{C}_p$, which is a measured heat capacity. Then with $\epsilon_1 = 0$, $\epsilon_2 - \epsilon_1 = \Delta\epsilon$

$$\mathcal{C}_V = N k_B (\beta \Delta\epsilon)^2 \frac{e^{\beta \Delta\epsilon}}{\left(e^{\beta \Delta\epsilon} + 1 \right)^2}, \tag{8.14}$$

which in terms of a dimensionless plotting coordinate $X = \beta\, \Delta\epsilon$ becomes

$$\frac{\mathcal{C}_V}{N k_B} = X^2 \frac{e^X}{\left(e^X + 1 \right)^2}. \tag{8.15}$$

Plotting this in Figure 8.1, a Schottky peak appears at $X^{-1} = \frac{k_B T}{\Delta\epsilon} = 0.417$. Although the low-temperature heat capacity of insulating crystals usually has a dominant $\sim T^3$ contribution from lattice vibrations, the sharp anomalous peak can be distinguished above this lattice baseline.

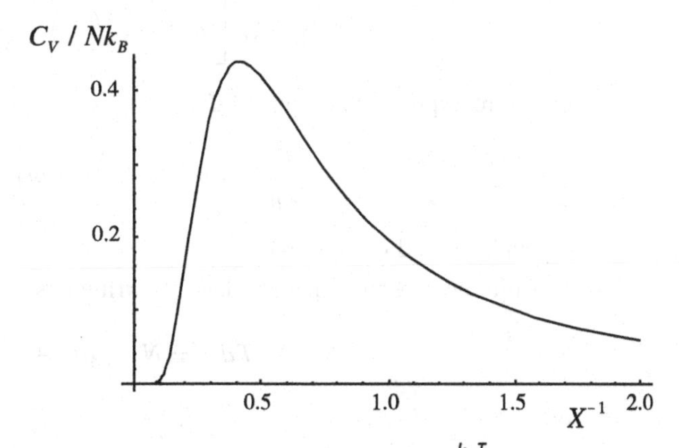

Fig. 8.1 Schottky heat capacity anomaly. The abscissa is taken to be $X^{-1} = \frac{k_B T}{\Delta\epsilon}$ to emphasize the characteristic low-temperature Schottky peak.

In the high-temperature limit, $X \ll 1$, Eq. 8.15 reduces to

$$\frac{\mathcal{C}_V}{Nk_B} \rightarrow \frac{X^2}{4} \tag{8.16}$$

$$= \left(\frac{\Delta\epsilon}{2k_BT}\right)^2. \tag{8.17}$$

At low temperature, $X \gg 1$, Eq. 8.15 becomes

$$\frac{\mathcal{C}_V}{Nk_B} \rightarrow X^2 e^{-X} \tag{8.18}$$

$$= \left(\frac{\Delta\epsilon}{k_BT}\right)^2 e^{-\frac{\Delta\epsilon}{k_BT}}, \tag{8.19}$$

which has the limiting Schottky behavior

$$\lim_{T\rightarrow 0} \mathcal{C}_V = 0. \tag{8.20}$$

8.3 Two-level systems and negative temperature

Writing internal energy, Eq. 8.9, as

$$\mathcal{U} = N\frac{\Delta\epsilon}{1+e^{\beta\Delta\epsilon}} \tag{8.21}$$

a dimensionless internal energy $\Phi = \frac{\mathcal{U}}{N\Delta\epsilon}$ can be defined. Then the entropy of Eq. 8.12,

$$\mathcal{S} = Nk_B\left[\ln\left(1+e^{-\beta\Delta\epsilon}\right) + \frac{\beta\Delta\epsilon}{e^{\beta\Delta\epsilon}+1}\right], \tag{8.22}$$

can be re-expressed in terms of Φ as

$$\frac{\mathcal{S}}{Nk_B} = (\Phi - 1)\ln(1-\Phi) - \Phi\ln\Phi, \tag{8.23}$$

which is plotted in Figure 8.2.

From the "thermodynamic identity" written as

$$Td\mathcal{S} = N\Delta\epsilon\; d\Phi + pdV \tag{8.24}$$

or

$$d\mathcal{S} = \frac{N\Delta\epsilon}{T}d\Phi + \frac{p}{T}dV, \tag{8.25}$$

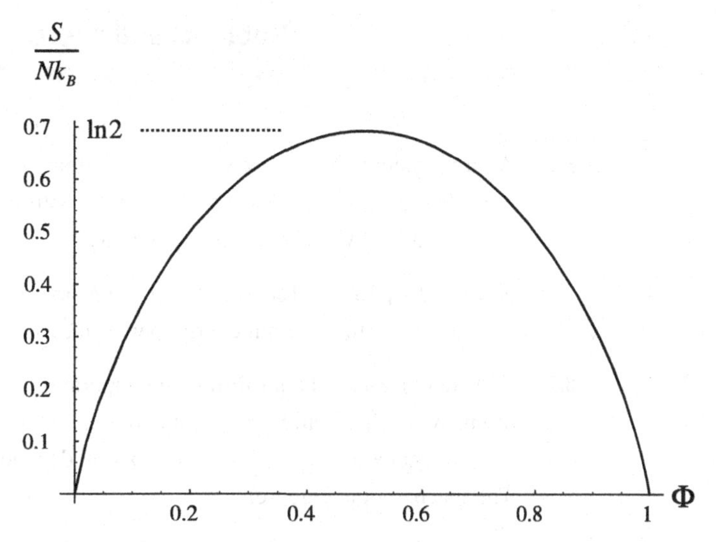

Fig. 8.2 $\mathcal{S}$ vs. Φ. The region $0.5 < \Phi \leq 1$ corresponds to negative temperature.

the slope at any point is (see Figure 8.2)

$$\frac{1}{N}\left(\frac{\partial \mathcal{S}}{\partial \Phi}\right)_N = \frac{\Delta \varepsilon}{T} \tag{8.26}$$

which displays the properties:

- for $\Phi < 0.5$, slope > 0 – temperature is positive;
- at $\Phi = 0.5$, slope $= 0$ – temperature is infinite;
- for $\Phi > 0.5$, slope < 0 – temperature is negative.

Negative temperature means that a population of excited atoms is unstably inverted, i.e. the system is not in a state of thermodynamic equilibrium and is, at the smallest perturbation, ready to dump its excess energy into the ground state. In this sense negative temperatures are "hotter" than positive temperatures.[5] To achieve negative temperatures the excitation energies of the system must have an upper bound, a condition satisfied by the isolated two-level Schottky model.

The typical two-level system is, however, rarely isolated – i.e. phonons, radiation, magnetic excitations, etc. are available for interaction. Therefore quasi-stable negative temperature conditions are unusual. There are, however, special circumstances in which strongly prohibitive optical selection rules prevent "triggers" from acting except on long time scales. Population inversion in lasers is one example in which a negative temperature is maintained until stray photons "stimulate" the inverted population into a downward avalanche.

[5] Nor do they belong to equilibrium thermodynamics.

Problems and exercises

8.1 A large system of N weakly interacting atoms is in thermal equilibrium. Each atom has only three possible eigenstates with energies $E = 0$, $E = \epsilon$ and $E = 3\epsilon$, all of which are non-degenerate.

a. Find the partition function $\mathcal{Z}$ for the N-atom system.
b. Find $\mathcal{U}/N$, the internal energy per atom, when $\beta\epsilon \ll 1$.

8.2 A system in thermal equilibrium at temperature T consists of a large number of atoms N_0 each of which can exist in only two quantum energy states: E_1 which is non-degenerate, and E_2 which is p-fold degenerate, with $E_2 - E_1 = \epsilon > 0$. The macroscopic eigen-energies are

$$E\left(n_1, n_2\right) = n_1 E_1 + n_2 E_2 \tag{8.27}$$

with n_1 atoms in state E_1 and n_2 atoms in state E_2.

a. Find the partition function $\mathcal{Z}$ for the N_0 atom system.
b. Find an expression for the internal energy $\mathcal{U}$.
c. Show that in the high-temperature limit the internal energy is

$$\mathcal{U} = N_0 \frac{p\epsilon}{1+p} \tag{8.28}$$

and the entropy is

$$\mathcal{S} = N_0 k_B \ln\left(1+p\right). \tag{8.29}$$

d. Find an expression for the high-temperature heat capacity $\mathcal{C}_V$.

8.3 Suppose each atom in a lattice of N atoms has three states with energies $-\epsilon$, 0 and $+\epsilon$. The states with energies $\pm\epsilon$ are non-degenerate whereas the state with energy 0 is two-fold degenerate.

a. Find the partition function $\mathcal{Z}$ of the system.
b. Calculate the internal energy of the system as a function of temperature.
c. Show that the heat capacity per atom, $\mathcal{C}_V/N$, is

$$\mathcal{C}_V/N = \frac{\beta^2\epsilon^2}{1+\cosh\left(\beta\epsilon\right)}. \tag{8.30}$$

9 Lattice heat capacity

I believe that no one who has won, through long years of experience, a reasonably reliable sense for the not always easy experimental evaluation of a theory, will be able to contemplate these results without immediately becoming convinced of the huge logical power of the quantum theory.

W. Nernst, *Z. fur Elektrochem.* **17**, 265–275 (1911) [trans. A. Wasserman]

9.1 Heat capacity of solids

The nearly 200-year-old Dulong–Petit "rule" for molar heat capacities of crystalline matter, c_v, predicts the constant value

$$c_v = \frac{3}{2} N_A k_B \tag{9.1}$$

$$= 24.94\,\mathrm{J\,mole^{-1}}, \tag{9.2}$$

where N_A is Avagadro's number. Although Dulong–Petit, which assumes solids to be dense, classical, ideal gases (see Eq. 7.29), is in amazingly good agreement with the high-temperature ($\sim$300 K) molar heat capacities of many solids, it fails to account for the observed rapid fall in c_v at low temperature. An especially large effect in diamond caught Einstein's (1907) attention and with extraordinary insight he applied Planck's "quanta" to an oscillator model of an atomic lattice to predict a universal decline in c_v as $T \to 0$ K. Several years later, when low-temperature molar heat capacities could be accurately measured, they were indeed found to behave in approximate agreement with Einstein's prediction.[1] It was this result that ultimately succeeded in making the case for quantum theory and the need to radically reform physics to accommodate it.

[1] The c_v in metals has an additional very-low-temperature contribution $\sim T$ from conduction electrons, which, of course, Einstein could not account for.

9.2 Einstein's model

Einstein's model assumes a solid composed of N atoms, each of mass M, bound to equilibrium sites within a unit cell by simple harmonic forces. The potential energy of each atom is

$$V(\mathbf{R}_\alpha) = \frac{1}{2} M\omega_0^2 \, \delta\mathbf{R}_\alpha \cdot \delta\mathbf{R}_\alpha, \tag{9.3}$$

with a classical equation of motion

$$\delta\ddot{\mathbf{R}}_\alpha + \omega_0^2 \delta\mathbf{R}_\alpha = 0, \tag{9.4}$$

where $\delta\,\mathbf{R}_\alpha = \mathbf{R}_\alpha - \mathbf{R}_{\alpha,0}$ is the displacement vector of the αth ion from its origin $\mathbf{R}_{\alpha,0}$. Einstein's independent oscillator model ignores any interactions between ions so there is only a single mode with frequency ω_0. From a modern perspective Einstein's intuitive harmonic assumption is correct, since atoms in a solid are bound by a total potential energy $V(R_\alpha)$ consisting of:

- a short-range repulsive component arising from the screened coulomb interaction between positively charged ion cores

$$V_{\alpha,\beta} \approx \frac{e^2}{2} \sum_{\substack{\alpha,\beta \\ \alpha \neq \beta}} \frac{Z_\alpha Z_\beta \exp^{-\gamma |R_\alpha - R_\beta|}}{|R_\alpha - R_\beta|}; \tag{9.5}$$

- a long-range attractive component arising from quantum mechanical electron–electron correlations and ion–electron interactions.

The two potential energy components are shown in Figure 9.1 together with their sum which has a nearly harmonic[2] minimum near R_0.

9.3 Einstein model in one dimension

Einstein's groundbreaking calculation of the heat capacity of a crystal lattice – the first application of a quantum theory to solids – is based on an independent oscillator model. Its simplest form considers N independent oscillators in a one-dimensional lattice with the αth oscillator potential obtained by expanding $V(r)$ about the solid curve's minimum $R_{0,\alpha}$ (see Figure 9.1).

$$V(R_\alpha) = V(R_{0,\alpha}) + \frac{1}{2} M\omega_0^2 \delta R_\alpha{}^2, \tag{9.6}$$

[2] With increasing displacement from R_0 anharmonicity (departures from harmonicity) has significant physical consequences.

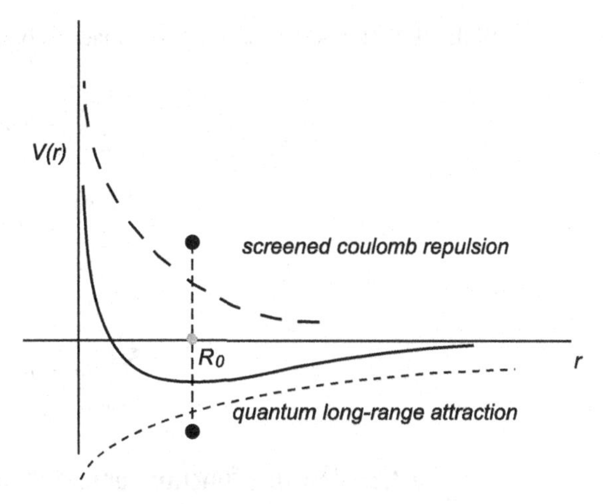

Fig. 9.1 The long dashed line is the screened (short-range) coulomb repulsion between ion cores. The short dashed line is the effective ion–ion attraction due to quantum mechanical electron correlations and ion–electron interactions. The solid line is a nearly harmonic sum of the two contributions, with a potential minimum at R_0.

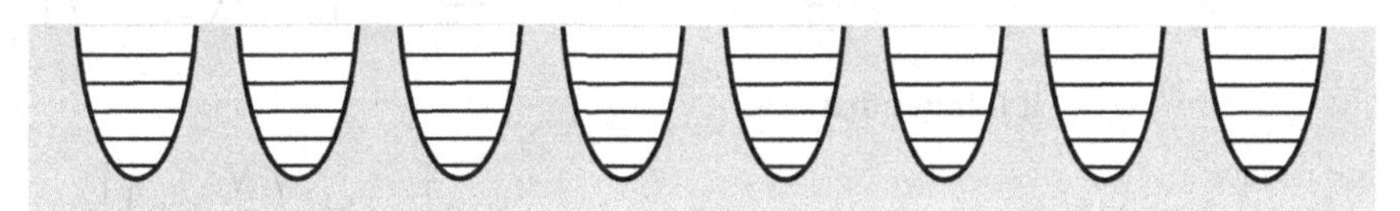

Fig. 9.2 Cartoon array of one-dimensional Einstein harmonic oscillator potentials showing the equally spaced energy levels of Eq. 9.8.

with

$$\delta R_\alpha = R_\alpha - R_{0,\alpha}. \tag{9.7}$$

Quantizing the model results in corresponding quantum energy levels

$$E(n_\alpha) = \hbar\omega_0 \left(n_\alpha + \frac{1}{2}\right), \tag{9.8}$$

where $\alpha = 1, 2, \ldots, N$ and ω_0 is the natural oscillator frequency. The constant term $V\left(R_{0,\alpha}\right)$ is ignored. For the one-dimensional oscillator the quantum number has the integer values $n_\alpha = 0, 1, 2, \ldots, \infty$. The integer lattice energy levels represent *quasi-particles* called "phonons". Although they are not real particles they bear all relevant kinetic attributes of real particles (energy, momentum, etc.) except that they are *not* number-conserved. The lack of number conservation has the consequence that they always have chemical potential $\mu = 0$.

With all N atoms contributing, the macroscopic eigen-energies are

$$E(n) = \frac{N\hbar\omega_0}{2} + \hbar\omega_0 \sum_{\alpha=1}^{N} n_\alpha \tag{9.9}$$

$$= \frac{N\hbar\omega_0}{2} + \hbar\omega_0 n \tag{9.10}$$

with

$$\sum_{\alpha=1}^{N} n_\alpha = n. \tag{9.11}$$

9.3.1 Partition function for one-dimensional Einstein model

From the thermal Lagrangian

$$\mathcal{L} = -k_B \sum_{n=0}^{\infty} \mathbf{P}(n) \ln \mathbf{P}(n) - \frac{1}{T} \sum_{n=0}^{N} \mathbf{P}(n) \left[\beta\hbar\omega_0\left(\frac{N}{2} + n\right)\right] - \lambda_0 \sum_{n=0}^{N} \mathbf{P}(n) \tag{9.12}$$

it follows that

$$\mathbf{P}(n) = \frac{\exp\left[-\beta\hbar\omega_0\left(\frac{N}{2} + n\right)\right]}{\mathcal{Z}}. \tag{9.13}$$

Emphasizing the role of degeneracy in performing the partition function's sum over all states, Eqs. 6.96 and 6.97 are used. The partition function is therefore written

$$\mathcal{Z} = \sum_{n=0}^{\infty} \frac{(N-1+n)!}{(N-1)!\,n!} \exp\left[-\beta\hbar\omega_0\left(\frac{N}{2} + n\right)\right] \tag{9.14}$$

$$= e^{-\frac{N\beta\hbar\omega_0}{2}} \sum_{n=0}^{\infty} \frac{(N-1+n)!}{(N-1)!\,n!} \exp\left(-n\beta\hbar\omega_0\right). \tag{9.15}$$

The resulting sum is not entirely obvious, although it can be approximated using the Stirling formula, Eq. 7.26. But with an integral representation of the Γ function,

$$\Gamma(n) = (n-1)! = \int_0^{\infty} dt\, t^{n-1} e^{-t}, \tag{9.16}$$

we can make the replacement

$$(N+n-1)! = \int_0^{\infty} dt\, t^{(N+n-1)} e^{-t} \tag{9.17}$$

so the partition function becomes

$$\mathcal{Z} = e^{-\frac{N\beta\hbar\omega_0}{2}} \int_0^{\infty} \mathrm{d}t \, \frac{t^{N-1}e^{-t}}{(N-1)!} \sum_{n=0}^{\infty} \frac{1}{n!} t^n \, e^{-n\beta\hbar\omega_0}. \tag{9.18}$$

Summing over n (to get an exponential) and then integrating over t (again using the Γ function)

$$\mathcal{Z} = \frac{e^{\frac{N\beta\hbar\omega_0}{2}}}{\left(e^{\beta\hbar\omega_0} - 1\right)^N}. \tag{9.19}$$

Evaluation of the one-dimensional oscillator partition function in this way is not especially difficult, but it is not obviously extensible to higher dimensionality or to other physically interesting models.

9.3.2 Phonons and the one-dimensional oscillator

A preferable route, using Eq. 9.10 for the eigen-energies, has the one-dimensional oscillator partition function

$$\mathcal{Z} = \sum_{n_1=0}^{\infty} \sum_{n_2=0}^{\infty} \cdots \sum_{n_N=0}^{\infty} \exp\left[-\beta\left(\frac{N\hbar\omega_0}{2} + \hbar\omega_0 \sum_{\alpha=1}^{N} n_\alpha\right)\right] \tag{9.20}$$

$$= e^{-\beta N\hbar\omega_0/2} \sum_{n_1=0}^{\infty} \sum_{n_2=0}^{\infty} \cdots \sum_{n_N=0}^{\infty} \exp\left[-\beta\hbar\omega_0 \sum_{\alpha=1}^{N} n_\alpha\right], \tag{9.21}$$

where the sum over states is equivalent to the sum over all numbers of phonons n_α. Explicitly summing over α gives a product of N identical geometrical series

$$\mathcal{Z} = e^{-\frac{\beta N\hbar\omega_0}{2}} \left(\sum_{n_1=0}^{\infty} \exp\left[-\beta\hbar\omega_0 n_1\right]\right) \left(\sum_{n_2=0}^{\infty} \exp\left[-\beta\hbar\omega_0 n_2\right]\right) \cdots \times \left(\sum_{n_N=0}^{\infty} \exp\left[-\beta\hbar\omega_0 n_N\right]\right) \tag{9.22}$$

and hence

$$\mathcal{Z} = e^{-\beta N\hbar\omega_0/2} \left(\sum_{n=0}^{\infty} e^{-\beta\hbar\omega_0 n}\right)^N, \tag{9.23}$$

which is summed to give the result

$$\mathcal{Z} = \frac{e^{\beta N \hbar\omega_0/2}}{\left(e^{\beta\hbar\omega_0} - 1\right)^N}, \tag{9.24}$$

identical to Eq. 9.19.

9.4 The three-dimensional Einstein model

The three-dimensional Einstein model, although suggesting a better approximation, still falls short of what is physically observed. (See the Debye model discussion later in the chapter.) Here the coordinate components x, y, z, of the αth independent oscillator displacements are taken into account with the eigen-energies

$$E\left(n_{\alpha,x}, n_{\alpha,y}, n_{\alpha,z}\right) = \hbar\omega_0\left(n_{\alpha,x} + n_{\alpha,y} + n_{\alpha,z} + 3/2\right), \tag{9.25}$$

where again $\alpha = 1, 2, 3, \ldots, N$ and $n_{\alpha,x}, n_{\alpha,y}, n_{\alpha,z} = 0, 1, 2, \ldots, \infty$ with ω_0 the oscillator frequency. The N-oscillator lattice has the macroscopic eigen-energies

$$E_{x,y,z} = \frac{3N\hbar\omega_0}{2} + \hbar\omega_0 \sum_{\alpha=1}^{N} \left(n_{\alpha,x} + n_{\alpha,y} + n_{\alpha,z}\right). \tag{9.26}$$

9.4.1 Partition function for the three-dimensional Einstein model

Using the result of Eq. 9.26, the thermal Lagrangian is as Eq. 9.12 and the partition function written

$$\mathcal{Z} = e^{-\frac{3N\beta\hbar\omega_0}{2}} \sum_{\substack{n_{1,x}=0\\ n_{1,y}=0\\ n_{1,z}=0}}^{\infty} \sum_{\substack{n_{2,x}=0\\ n_{2,y}=0\\ n_{3,z}=0}}^{\infty} \cdots$$

$$\times \sum_{\substack{n_{N,x}=0\\ n_{N,y}=0\\ n_{N,z}=0}}^{\infty} \exp\left\{-\beta\left[\hbar\omega_0 \sum_{\alpha=1}^{N}\left(n_{\alpha,x} + n_{\alpha,y} + n_{\alpha,z}\right)\right]\right\}. \tag{9.27}$$

The sum is managed, as in Section 9.3.2, by first explicitly summing over α in the exponential. Then, because the three coordinate sums $\left(n_{\alpha,x}, n_{\alpha,y}, n_{\alpha,z}\right)$ are identical, what remains is

$$\mathcal{Z} = e^{-\frac{3\beta N\hbar\omega_0}{2}} \left(\sum_{n_1=0}^{\infty} \exp\left[-\beta\hbar\omega_0 n_1\right] \right)^3 \left(\sum_{n_2=0}^{\infty} \exp\left[-\beta\hbar\omega_0 n_2\right] \right)^3$$

$$\cdots \left(\sum_{n_N=0}^{\infty} \exp\left[-\beta\hbar\omega_0 n_N\right] \right)^3 \tag{9.28}$$

and finally

$$\mathcal{Z} = e^{-\frac{3N\beta\hbar\omega_0}{2}} \left(\sum_{n=0}^{\infty} e^{-\beta\hbar\omega_0 n} \right)^{3N}, \tag{9.29}$$

where the remaining sum gives

$$\mathcal{Z} = \left[e^{-\frac{\beta\hbar\omega_0}{2}} \left(\frac{1}{1 - e^{-\beta\hbar\omega_0}} \right) \right]^{3N}. \tag{9.30}$$

9.4.2 Thermodynamics of the three-dimensional Einstein model

Following steps from previous chapters, the internal energy is

$$\mathcal{U} = -\frac{\partial}{\partial\beta} \ln \mathcal{Z} \tag{9.31}$$

$$= 3N\hbar\omega_0 \left[\frac{1}{2} + \langle n \rangle \right], \tag{9.32}$$

where[3]

$$\langle n \rangle = \frac{1}{\left(e^{\beta\hbar\omega_0} - 1 \right)}. \tag{9.33}$$

Therefore for the three-dimensional model

$$\mathcal{C}_V = \left(\frac{\partial U}{\partial T} \right)_V$$

$$= -k_B \beta^2 \left(\frac{\partial U}{\partial \beta} \right)_V$$

$$= 3Nk_B \frac{(\beta\hbar\omega_0)^2 \, e^{\beta\hbar\omega_0}}{\left(e^{\beta\hbar\omega_0} - 1 \right)^2}. \tag{9.34}$$

[3] $\langle n \rangle$ is the average phonon (quasi-particle) occupation number. It is also loosely referred to as a Bose–Einstein "function".

It is conventional to replace the oscillator frequency ω_0 with an Einstein temperature θ_E

$$k_B\theta_E = \hbar\omega_0, \tag{9.35}$$

so that

$$\mathcal{C}_V = 3Nk_B\frac{(\theta_E/T)^2\, e^{\theta_E/T}}{\left(e^{\theta_E/T}-1\right)^2}, \tag{9.36}$$

with which specific materials may be characterized by fitting to experimental data.

In the low-temperature limit, Eq. 9.36 becomes

$$\lim_{T\to 0}\mathcal{C}_V \to 3Nk_B\,(\theta_E/T)^2\, e^{-\theta_E/T} \tag{9.37}$$

as shown in Figure 9.3. This steep exponential decline in $\mathcal{C}_V$ is never observed. Universally observed is $\mathcal{C}_V \sim T^3$.

Einstein was aware that a single oscillator frequency model was bound to be inadequate and he tried to improve it, without success. But he achieved his primary objective – to apply Planck's quantum theory and show that it explained poorly understood low-temperature heat capacities.

At high temperature the Einstein result $\lim_{T\to\infty}\mathcal{C}_V = 3Nk_B$ is in accord with Dulong–Petit.

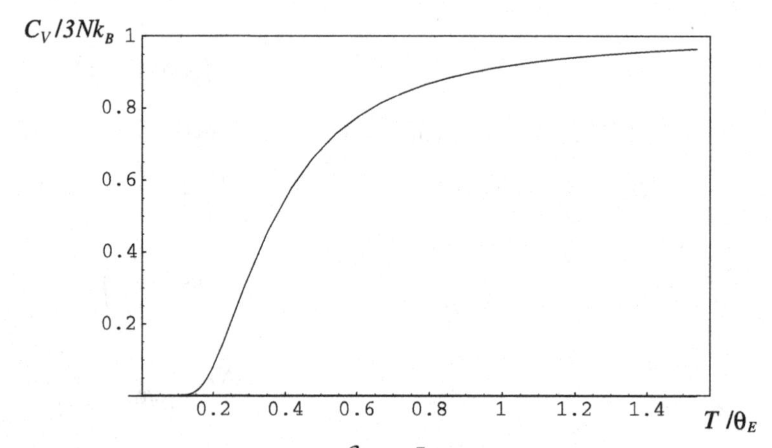

Fig. 9.3 Three-dimensional Einstein model heat capacity $\frac{\mathcal{C}_V}{3Nk_B}$ vs. $\frac{T}{\theta_E}$. Note the sharp exponential drop as $T \to 0$.

9.5 Debye model

Atom–atom interactions were added to Einstein's theory by Debye.[4] Their effect is to introduce dispersion into the oscillator frequencies, which is precisely the correction Einstein sought but never achieved.

As a result of atom–atom interactions:

1. Translational (crystal) symmetry is introduced, with a new (wave vector) quantum number $\mathbf{k}$, sometimes called *crystal momentum*, with values

$$k_j = \frac{2\pi}{N_j a_j}\nu_j, \quad j = x, y, z, \tag{9.38}$$

where a_j is the lattice spacing in the jth crystal direction, N_j the number of atoms in the jth periodic cell direction and where $\nu_j = 1, 2, 3, \ldots, N_j$.[5,6]

2. As shown in Figure 9.4, rather than a single Einstein lattice frequency ω_0 there is now a range of oscillator frequencies[7] which Debye assumed varied linearly with $|\mathbf{k}|$

$$\omega = \omega(\mathbf{k}) \tag{9.39}$$

$$= \langle c_s \rangle |\mathbf{k}|, \tag{9.40}$$

where $\langle c_s \rangle$ is an average speed of sound in the crystal.[8]

3. The infinitely sharp Einstein "phonon" density of states implied in Figure 9.4

$$\mathscr{D}_E(\omega) = N\delta(\omega - \omega_0) \tag{9.41}$$

is replaced in the Debye model by[9]

$$\mathscr{D}_D(\omega) = \frac{V\omega^2}{2\pi^2 \langle c_s \rangle^3}. \tag{9.42}$$

4. Since in a finite crystal the quantum number $\mathbf{k}$ is bounded, the range of Debye's oscillator frequencies is also bounded, i.e.

$$0 \le \omega < \Omega_D. \tag{9.43}$$

[4] P. Debye, "Zur Theorie der spezifischen Waerme", *Annalen der Physik (Leipzig)* **39**, 789 (1912).

[5] In a three-dimensional crystal the lattice atom displacements $\delta\mathbf{R}$ are in three mutually perpendicular directions (polarizations) such that $\mathbf{k} \cdot \delta\mathbf{R} = 0$.

[6] In solid state physics it is conventional to choose $-\frac{N}{2} < \nu \le \frac{N}{2}$ which, in this example, would define a single Brillouin zone.

[7] One might say that the atom–atom interactions have lifted the degeneracy among single atom oscillator frequencies ω_0.

[8] There are three different sound velocities corresponding to the two independent transverse phonons and single longitudinal phonon. This turns out to be an approximation that accurately replicates the small-$|\mathbf{k}|$ behavior of lattice vibrations in many 3-D crystals.

[9] See Appendix E.

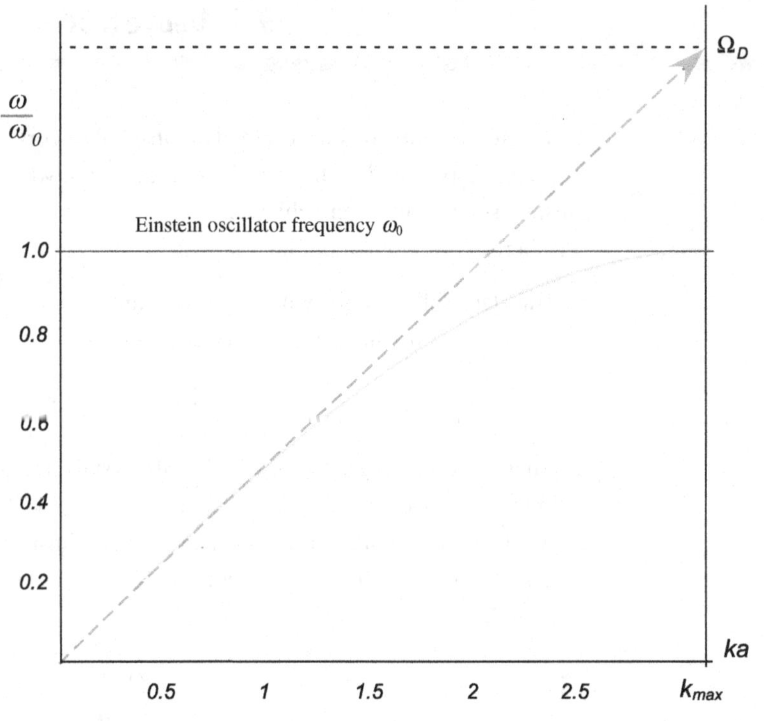

Fig. 9.4 Mode-dependent frequencies. The solid curve represents an approximate result for a real lattice. k_{max} is the Brillouin zone boundary for the crystal ($-k_{max} < k \leq k_{max}$). The dashed line represents Debye's linear approximation. The slope of the dashed line is the average speed of sound in the crystal. Ω_D is the Debye approximation's highest attained frequency. The horizontal fine line at $\omega = \omega_0$ represents the dispersion of an Einstein lattice.

Within the Debye model Ω_D is found from

$$\int_0^\infty \mathrm{d}\omega \mathscr{D}_E(\omega) = \int_0^{\Omega_D} \mathrm{d}\omega \mathscr{D}_D(\omega), \tag{9.44}$$

i.e. the total number of modes, $3N$, is the same in both models, so that

$$\Omega_D = \left(6\pi^2 \langle c_s \rangle^3 N/V\right)^{1/3}. \tag{9.45}$$

Whereas the Einstein internal energy $\mathcal{U}_E$ is (see Eq. 9.32)

$$\mathcal{U}_E = 3\hbar \int_0^\infty \omega \, \mathrm{d}\omega \overbrace{\{N\delta(\omega - \omega_0)\}}^{\text{Einstein density of states}} \left[\frac{1}{2} + \frac{1}{e^{\beta\hbar\omega} - 1}\right] \tag{9.46}$$

$$= 3N\omega_0 \left[\frac{1}{2} + \frac{1}{e^{\beta\hbar\omega_0} - 1}\right], \tag{9.47}$$

the changes introduced by Debye give instead the internal energy $\mathcal{U}_D$ (see Appendix E),

$$\mathcal{U}_D = \hbar \int_0^{\Omega_D} \omega \; d\omega \; \overbrace{\left\{ \frac{3V \; \omega^2}{2\pi^2 \langle c_s \rangle^3} \right\}}^{\text{Debye density of states}} \left[\frac{1}{2} + \frac{1}{e^{\beta\hbar\omega} - 1} \right] \tag{9.48}$$

$$= \frac{3V}{2\pi^3 \hbar^3 \langle c_s \rangle^3 \beta^4} \int_0^{\beta\hbar\Omega_D} dx \, x^3 \left(\frac{1}{2} + \frac{1}{e^x - 1} \right). \tag{9.49}$$

9.5.1 Thermodynamics of the Debye lattice

At high temperature, $\beta\hbar\Omega_D \ll 1$, the integral in Eq. 9.49 can be approximated by expanding $e^x \cong 1 + x$. Then using Eq. 9.44 the Debye internal energy is

$$\lim_{T\to\infty} \mathcal{U}_D = \frac{V}{2\pi^2} \left(\frac{\Omega_D}{\langle c_s \rangle} \right)^3 k_B T \tag{9.50}$$

$$= 3Nk_B T, \tag{9.51}$$

consistent with the Dulong–Petit rule. At low temperature, $\beta\hbar\Omega_D \gg 1$, and defining

$$\hbar\Omega_D = k_B \Theta_D, \tag{9.52}$$

with Θ_D the Debye temperature, the internal energy integral can be approximated as

$$\lim_{T\to 0} \mathcal{U}_D \to \frac{3V}{2\pi^3 \hbar^3 \langle c_s \rangle^3 \beta^4} \int_0^{\infty} dx x^3 \left(\frac{1}{2} + \frac{1}{e^x - 1} \right), \tag{9.53}$$

from which, with Eq. 9.45, follows the low-temperature heat capacity

$$\lim_{T\to 0} \mathcal{C}_V * \to \frac{6k_B^4}{\pi^2 \hbar^3 \langle c_s \rangle^3} \times \frac{\pi^4}{15} T^3 \tag{9.54}$$

$$= Nk_B \left(\frac{12\pi^4}{5} \right) \times \left(\frac{T}{\Theta_D} \right)^3. \tag{9.55}$$

The $\mathcal{C}_V \cong T^3$ behavior is almost universally observed in three-dimensional solids. Examples of Debye temperatures are given in Table 9.1.

Problems and exercises

9.1 Two-dimensional materials are now widely fabricated and investigated. Within the framework of Debye's model for lattice vibrations find the low-temperature

Table 9.1 Debye temperatures for some typical solids.

	θ_D
aluminum	428 K
cadmium	209 K
chromium	630 K
copper	343.5 K
gold	165 K
iron	470 K
lead	105 K
nickel	450 K
platinum	240 K
silicon	645 K
titanium	420 K
zinc	327 K
carbon	2230 K

lattice contribution to the constant area heat capacity C_A for a two-dimensional crystalline system.

9.2 Ferromagnetic systems have very-low-temperature collective spin excitations which are called *magnons*.[10] Their quasi-particle behavior is similar to collective lattice, phonon, excitations except that their energy spectrum in zero magnetic field is

$$\hbar\omega_{\mathbf{k}} = D\,|\mathbf{k}|^2\,, \tag{9.56}$$

where

$$D = 2SJa^2 \tag{9.57}$$

with J a magnetic moment coupling constant, S the spin of a magnetic moment and a is a lattice constant. Show that the low-temperature magnon heat capacity C_V is proportional to $T^{3/2}$.

9.3 In microscopic lattice models that include anharmonic contributions (phonon–phonon interactions), phonon frequencies ω change with volume V. Using

$$p = \frac{1}{\beta}\left(\frac{\partial \ln \mathcal{Z}}{\partial V}\right)_T, \tag{9.58}$$

where $\mathcal{Z}$ is the Debye partition function, find the pressure of a Debye solid.

[10] T. Holstein and H. Primakoff, "Field dependence of the intrinsic domain magnetization of a ferromagnet", *Phys. Rev.* **58**, 1098–1113 (1940).

Therefore in taking the volume derivative, Eq. 9.58, $\langle c_s \rangle$ and hence Θ_D depend on volume.

Express the pressure in terms of the dimensionless Grüneisen parameter γ,

$$\gamma \equiv -\frac{V}{\Theta_D}\left(\frac{d\Theta_D}{dV}\right). \tag{9.59}$$

10 Elastomers: entropy springs

Mr. McGuire: "I want to say one word to you. Just one word."
Benjamin: "Yes, sir."
Mr. McGuire: "Are you listening?"
Benjamin: "Yes, I am."
Mr. McGuire: "Plastics."

The Graduate (1967)[1]

10.1 Naive one-dimensional elastomer

Crystalline matter and polymeric materials exhibit considerable difference in elastic behaviors. Under applied stress σ all solids experience strain (elongation) ϵ. But crystalline materials also exhibit a property of "stiffness" due to accompanying large increases in bond energy. These materials also completely lose their elastic properties at small values of strain, $\epsilon \sim 0.001$. Polymeric materials, such as a common rubber band, are neither "stiff" nor, generally, do they lose elasticity even under large strain.

To see that something quite distinctive must account for elasticity differences between polymers and crystalline materials, one need only compare values of isothermal Young's moduli

$$E_T = \left(\frac{\partial \sigma}{\partial \epsilon}\right)_T . \tag{10.1}$$

Values for crystalline materials are given in Table 10.1 and for polymeric materials in Table 10.2. Young's moduli in crystalline matter can be as much as four orders of magnitude larger than for polymeric materials. What makes them so different is that elastic polymers are "softly" linked chains of chemical monomers whose "rubbery" elasticity is associated with freedom for monomer units to bend, rotate and randomly stack, with negligible energy cost. Figure 10.1 illustrates the structure of the polymer polyethylene in which ethylene molecules (monomers) are joined in chains. Elastomer flexibility is enabled by unhindered unit rotation about a carbon "backbone". Truly "rubbery" materials retain their elastic behavior even when subject to strains of $\epsilon \sim$ 1–3, which is much larger than for metals.

[1] Courtesy of The Internet Movie Database (http://www.imdb.com). Used with permission.

Table 10.1 Isothermal Young's moduli for some typical crystalline materials.

	E_T
aluminum	$7 \times 10^{10}\,\mathrm{N\,m^{-2}}$
steel	$20.7 \times 10^{10}\,\mathrm{N\,m^{-2}}$
chromium	$29 \times 10^{10}\,\mathrm{N\,m^{-2}}$
tungsten	$40 \times 10^{10}\,\mathrm{N\,m^{-2}}$
titanium	$40 \times 10^{10}\,\mathrm{N\,m^{-2}}$
copper	$12 \times 10^{10}\,\mathrm{N\,m^{-2}}$
glass	$7.5 \times 10^{10}\,\mathrm{N\,m^{-2}}$
diamond	$110 \times 10^{10}\,\mathrm{N\,m^{-2}}$

Table 10.2 Isothermal Young's moduli for some elastic polymers.

	E_T
rubber	$0.001 \times 10^{10}\,\mathrm{N\,m^{-2}}$
Teflon	$0.05 \times 10^{10}\,\mathrm{N\,m^{-2}}$
polypropylene	$0.1 \times 10^{10}\,\mathrm{N\,m^{-2}}$
nylon	$0.3 \times 10^{10}\,\mathrm{N\,m^{-2}}$
low density polyethylene	$0.02 \times 10^{10}\,\mathrm{N\,m^{-2}}$
polystyrene	$0.3 \times 10^{10}\,\mathrm{N\,m^{-2}}$

Another polymer with a simple carbon backbone is polyvinyl chloride (see Figure 10.2).

Other polymers can be made which restrict the monomers from swinging so freely. An example of a less flexible polymer is polystyrene (Figure 10.3), which is typically used in a rigid "glassy" state.

Elasticity in polymers is primarily a result of stacking arrangement multiplicity for the constituent monomers. Less tangled arrangements (elongated chains together with large elastic tension) imply fewer stacking possibilities, i.e. less uncertainty in describing those arrangements. More tangled arrangements (shorter chains and

Fig. 10.1 Polyethylene segment. Polyethylene tends to crystallize at room temperature and is not the best example of an elastomer.

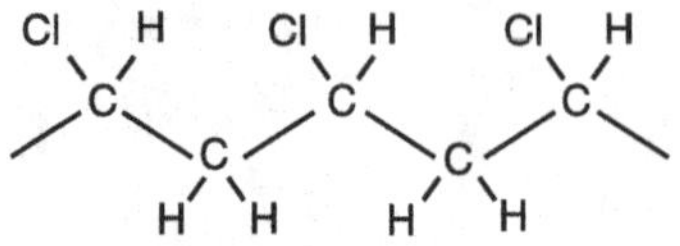

Fig. 10.2 Polyvinyl chloride polymer has a carbon backbone similar to polyethylene, but with a chlorine atom in the side-chains.

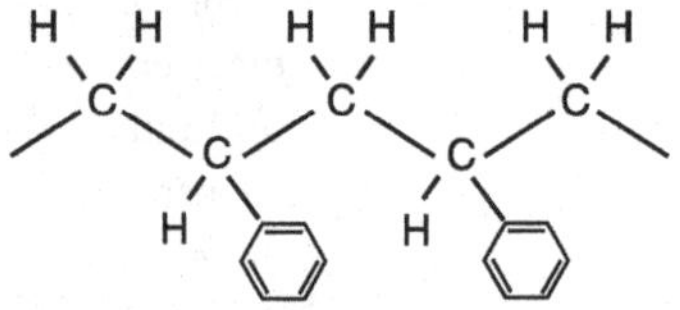

Fig. 10.3 Polystyrene has large aromatic rings in the side chains which severely restrict rotation about the carbon backbone.

smaller elastic tension) correspond to greater stacking possibilities, i.e. more uncertainty. Entropy being a measure of (configurational) uncertainty, an arrangement with less uncertainty has smaller thermodynamic entropy while one with greater uncertainty is a higher entropy configuration. Elastomers tend to resist elongation (spring shut) in order to achieve greater entropy. For this reason elastomers are sometimes called "entropy springs".

In this chapter we first hypothesize a microscopic one-dimensional model of elastomer elasticity and, using thermal Lagrangians, derive an elastomer equation of state. Then, using standard thermodynamic methods as introduced in Chapters 2, 3 and 4, together with the derived equation of state, we discuss the elastomer model's macroscopic physical behaviors.

10.2 Elongation as an extensive observable

Real elastomers generally function in three dimensions. When they are unidirectionally stretched, their cross-sectional areas shrink proportionately (necking) to maintain approximately constant volume. The change in cross-sectional area corresponds to the action of internal tensions that tend to squash the elastomer in directions transverse to its elongation. Metallic rods, although also elastic, due to much higher Young's moduli hardly elongate. Therefore in metal rods there is little change in the rod's cross-sectional area and transverse tensions can be safely ignored.

Even though real elastomers should be treated as three-dimensional to better account for elastic properties, the naive one-dimensional model discussed here does – satisfactorily – illustrate the overwhelming effect of entropy in its elastic properties as well as other unusual empirical features of a "rubber band".

We consider a naive one-dimensional model of polymer elasticity similar to that introduced in the discussion of degeneracy in Chapter 6. We assume here, as well, that any conformation of links is only weakly energy dependent, i.e. any elongation assumed by the polymer has the same (internal) bond energy. The Lagrangian used in implementing least bias for this elastic model includes in the hamiltonian a macroscopic, observable elastomer elongation, the average difference between the polymer's end coordinates. The "thermodynamic" hamiltonian is therefore

$$\mathcal{H}_{op} = \mathbf{h}_0 - \boldsymbol{\tau} \cdot \boldsymbol{\chi}_{op}, \tag{10.2}$$

where $\boldsymbol{\tau}$ is elastic tension – the variable conjugate to the elongation operator, $\boldsymbol{\chi}_{op}$. Note that $\mathbf{h}_0$ is an internal "bond energy" hamiltonian which is neglected in the model.

10.2.1 Naive elastomer model

The naive one-dimensional polymer is assumed to be an assemblage of N links, each of length a, that can point to the left or to the right with no energy difference for either orientation (see Figure 10.4).

In this naive model $\langle \mathbf{h}_0 \rangle = 0$ so that the hamiltonian becomes

$$\mathcal{H}_{op} = -\tau \, \chi_{op} \tag{10.3}$$

and the average macroscopic polymer elongation $\langle \chi_{op} \rangle$ is

$$\langle \chi_{op} \rangle = \mathcal{T}r \, \tilde{\rho}^{\tau}_{op} \chi_{op}. \tag{10.4}$$

The elongation operator's macroscopic eigenvalues are $\chi = a(n_R - n_L)$, where n_L is the number of left-directed monomers, n_R is the number of right-directed monomers and $n_R + n_L = N$ is the total number of monomers. Therefore macroscopic eigen-energies of the system (eigen-energies of the hamiltonian in Eq. 10.3) are

$$E\,(n_R, n_L) = -\tau \, \chi \tag{10.5}$$

$$= -\tau \, a \, (n_R - n_L)\,. \tag{10.6}$$

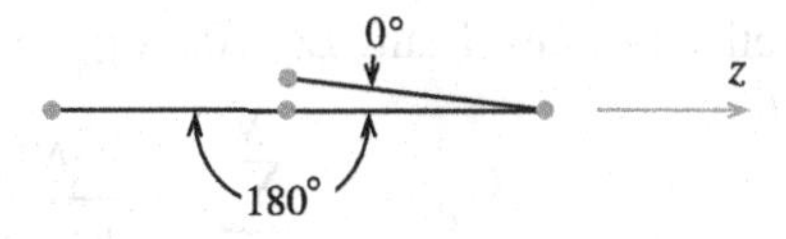

Fig. 10.4 One-dimensional model of monomer folding.

Forming the thermal Lagrangian

$$\mathcal{L} = -k_B \sum_{\substack{n_R,n_L \\ n_R+n_L=N}}^{N} \mathbf{P}(n_L, n_R) \ln \mathbf{P}(n_L, n_R) - \lambda_0 \sum_{\substack{n_R,n_L \\ n_R+n_L=N}}^{N} \mathbf{P}(n_L, n_R) - T^{-1} \sum_{\substack{n_R,n_L \\ n_R+n_L=N}}^{N} \mathbf{P}(n_L, n_R) [-\tau a (n_R - n_L)]], \tag{10.7}$$

we find, with $\beta = \frac{1}{k_B T}$,

$$\mathbf{P}(n_L, n_R) = \frac{e^{\beta \tau a (n_R - n_L)}}{\mathcal{Z}_\chi}, \tag{10.8}$$

where $\mathcal{Z}_\chi$, the partition function, is

$$\mathcal{Z}_\chi = \sum_{\substack{n_L,n_R \\ n_L+n_R=N}}^{N} g(n_L, n_R) e^{\beta \tau a (n_R - n_L)}. \tag{10.9}$$

The degeneracy $g(n_L, n_R)$ is inserted to remind us to sum over all states. Applying the degeneracy for this model from Chapter 6,

$$g(n_L, n_R) = 2\frac{N!}{n_L! n_R!}, \tag{10.10}$$

the partition function is

$$\mathcal{Z}_\chi = \sum_{\substack{n_L,n_R \\ n_L+n_R=N}}^{N} 2\frac{N!}{n_L! n_R!} e^{\beta \tau a (n_R - n_L)} \tag{10.11}$$

$$= 2\left(e^{\beta a \tau} + e^{-\beta a \tau}\right)^N \tag{10.12}$$

$$= 2\left[2\cosh(\beta \tau a)\right]^N. \tag{10.13}$$

10.3 Properties of the naive one-dimensional "rubber band"

10.3.1 Elongation

The average elongation of "rubbery" materials, $\langle \chi_{op} \rangle$, shows some empirically distinctive features. Using Eq. 10.8 $\langle \chi_{op} \rangle$ is

$$\langle \chi_{op} \rangle = \sum_{\substack{n_L,n_R \\ n_L+n_R=N}}^{N} 2\frac{N!}{n_L! n_R!} [a(n_R - n_L)] \frac{e^{\beta \tau a (n_R - n_L)}}{\mathcal{Z}_\chi} \tag{10.14}$$

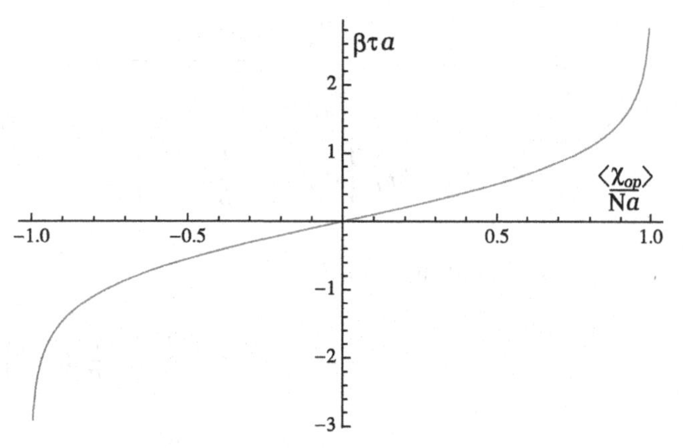

Fig. 10.5 Average extension, $\langle \chi_{op} \rangle / Na$, of "rubber" as a function of $\frac{\tau a}{k_B T}$. At small tension, $\frac{\tau a}{k_B T} \ll 1$, the behavior is linear.

or

$$\langle \chi_{op} \rangle = -\left(\frac{\partial}{\partial \tau}\right)\left[-\frac{1}{\beta} \ln \mathcal{Z}_\chi\right], \tag{10.15}$$

which gives the elastomer equation of state

$$\langle \chi_{op} \rangle = N a \tanh(\beta \tau a). \tag{10.16}$$

Examining this result in Figure 10.5, note the highly non-linear response to large tension, diverging as $\frac{\langle \chi_{op} \rangle}{N a} \to 1$. The elastic limit, where elasticity is permanently lost, is in that region.

At small tension, $\beta \tau a \ll 1$, the elongation is linear, i.e. Hooke's law,

$$\langle \chi_{op} \rangle = \frac{N a^2 \tau}{k_B T} \tag{10.17}$$

or

$$\tau = \frac{k_B T}{N a^2} \langle \chi_{op} \rangle, \tag{10.18}$$

which displays the distinctive feature that the elastic constant of a "rubber band" increases with increasing temperature.[2]

10.3.2 Entropy

"Rubber band" entropy can be found from

$$\mathcal{S} = -k_B \sum_{\substack{n_L, n_R \\ n_L + n_R = N}}^{N} \mathbf{P}(n_L, n_R) \ln \mathbf{P}(n_L, n_R) \tag{10.19}$$

[2] Finding the thermal "expansivity" of the one-dimensional rubber band is assigned as a problem.

$$= -k_B \sum_{\substack{n_L, n_R \\ n_L + n_R = N}}^{N} \left\{ \frac{e^{\beta\tau a(n_R - n_L)}}{\mathcal{Z}_\chi} \left[\beta\tau a \left(n_R - n_L\right)\right] - \ln \mathcal{Z}_\chi \right\} \tag{10.20}$$

$$= -k_B\beta \left\{ \tau \langle\chi_{op}\rangle - \frac{1}{\beta} \ln \mathcal{Z}_\chi \right\}, \tag{10.21}$$

which is identical to

$$\mathcal{S} = -\frac{\partial}{\partial T}\left(-\frac{1}{\beta} \ln \mathcal{Z}_\chi\right) \tag{10.22}$$

$$= k_B\beta^2 \frac{\partial}{\partial \beta}\left(-\frac{1}{\beta} \ln \mathcal{Z}_\chi\right), \tag{10.23}$$

with the result ($N \gg 1$)

$$\mathcal{S} = Nk_B \left[\ln 2 + \left\{\ln\left[2\cosh\left(\beta\tau a\right)\right] - \beta\tau a \tanh\left(\beta\tau a\right)\right\}\right]. \tag{10.24}$$

Recasting the entropy in terms of average elastomer extension $X = \frac{\langle\chi_{op}\rangle}{Na}$ (see Eq. 10.16) we find

$$\frac{\mathcal{S}}{Na} = \ln\left[\frac{2}{\sqrt{1 - X^2}}\right] - X \tanh^{-1} X. \tag{10.25}$$

Extension vs. entropy is shown in Figure 10.6. In this model energy plays no role in elastic behavior. It is evident, then, that rubber's restorative "springiness" is due to favoring an entropy increase in the more compact configuration.

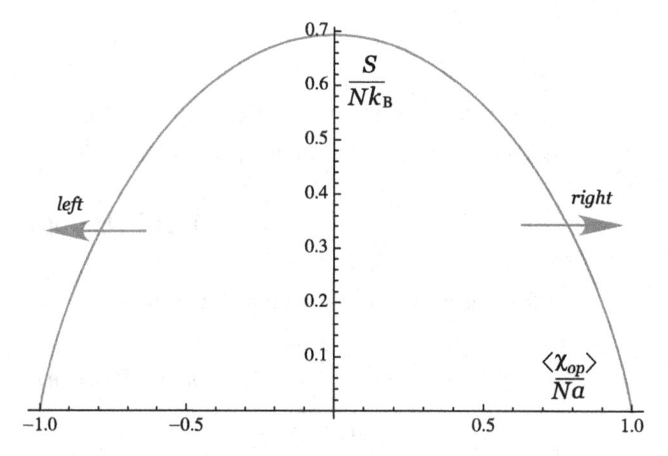

Fig. 10.6 Average extension per monomer, $\langle\chi_{op}\rangle/Na$, of an "entropy spring" as a function of $\mathcal{S}$.

10.4 "Hot rubber bands": a thermodynamic analysis

If a thick *real* rubber band is *quickly* stretched and then *abruptly* relaxed several times in rapid succession, it cannot fail to be noticed that the rubber band warms. Even without sensitive thermometry the temperature rise can be detected by placing the rubber band against the center of your forehead or against your upper lip, both of which are sensitive to temperature. The thermodynamics of this successive pair of processes can be qualitatively studied using the one-dimensional polymer model.[3,4]

If the elastomer is stretched so quickly that there is insufficient time for it to exchange heat with the surroundings, then the process can be described as adiabatic ($đQ = 0$). In that case,

$$0 = d\mathcal{U} + đ\mathcal{W}. \tag{10.26}$$

Since $\mathcal{U}$ is a state variable[5] the integral

$$\int_A^B d\mathcal{U} = \mathcal{U}(B) - \mathcal{U}(A) \tag{10.27}$$

only depends on end-points and not path. Then $\mathcal{W}$ also depends on end-points, and a quasi-static path can be chosen, i.e.

$$đ\mathcal{W}_{QS} = -\tau \; dL. \tag{10.28}$$

The remaining problem-solving steps are typical of those in Chapters 3–5 but adapted to the elastic model.

Starting with the First Law

$$đQ = d\mathcal{U} - \tau \; dL \tag{10.29}$$

and applying the adiabatic (fast stretching) condition, we have[6]

$$0 = d\mathcal{U} - \tau \; dL. \tag{10.31}$$

Taking the differential of $\mathcal{U} = \mathcal{U}(T, L)$

$$d\mathcal{U} = \left(\frac{\partial \mathcal{U}}{\partial T}\right)_L dT + \left(\frac{\partial \mathcal{U}}{\partial L}\right)_T dL, \tag{10.32}$$

[3] The elastic properties of a typical elastomer have, in principle, weak (but not zero) internal energy dependence. In the interest of clarity this energy dependence is, in the one-dimensional model, entirely ignored. To continue the present discussion a dose of reality must be restored. This comes in the form of a non-zero internal energy. But as will be seen, its microscopic origin and its magnitude are irrelevant.

[4] For simplicity, the average extension $\langle \chi_{op} \rangle$ will now be replaced by L.

[5] Note that now $\mathbf{h}_{op} \geq 0$ so that $\mathcal{U} \neq 0$.

[6] Alternatively it might be better to evaluate

$$dT = \left(\frac{\partial T}{\partial L}\right)_{\mathcal{S}} dL. \tag{10.30}$$

Eq. 10.31 becomes

$$0 = \left(\frac{\partial \mathcal{U}}{\partial T}\right)_L dT + \left[\left(\frac{\partial \mathcal{U}}{\partial L}\right)_T - \tau\right] dL. \tag{10.33}$$

However, the constant length heat capacity is

$$\mathcal{C}_L = \left(\frac{\partial \mathcal{U}}{\partial T}\right)_L, \tag{10.34}$$

which leaves only the partial derivative

$$\left(\frac{\partial \mathcal{U}}{\partial L}\right)_T \tag{10.35}$$

to be evaluated. A way forward[7] is to use the Helmholtz potential $F = \mathcal{U} - T\mathcal{S}$ and take the constant temperature partial derivative

$$\left(\frac{\partial \mathcal{U}}{\partial L}\right)_T = \left(\frac{\partial F}{\partial L}\right)_T + T\left(\frac{\partial \mathcal{S}}{\partial L}\right)_T. \tag{10.36}$$

This is handled by noting that combining the (elastic) thermodynamic identity

$$T\, d\mathcal{S} = d\mathcal{U} - \tau\, dL \tag{10.37}$$

with the differential of Helmholtz potential

$$dF = d\mathcal{U} - T\, d\mathcal{S} - \mathcal{S}\, dT \tag{10.38}$$

gives

$$dF = -\mathcal{S}\, dT + \tau\, dL \tag{10.39}$$

and the identification

$$\left(\frac{\partial F}{\partial L}\right)_T = \tau. \tag{10.40}$$

To simplify the last term in Eq. 10.36, apply the Euler criterion to the exact differential dF, i.e. Eq. 10.39,

$$-\left(\frac{\partial \mathcal{S}}{\partial L}\right)_T = \left(\frac{\partial \tau}{\partial T}\right)_L, \tag{10.41}$$

which is a Maxwell relation. Finally Eq. 10.36 becomes

$$\left(\frac{\partial \mathcal{U}}{\partial L}\right)_T = \tau - T\left(\frac{\partial \tau}{\partial T}\right)_L \tag{10.42}$$

and Eq. 10.33 reduces to

$$\frac{\mathcal{C}_L}{T} dT = \left(\frac{\partial \tau}{\partial T}\right)_L dL. \tag{10.43}$$

[7] A different but equivalent method was used on a similar partial derivative in Chapter 2.

Using the approximate (linearized) elastomer equation of state, Eq. 10.17, we find

$$\left(\frac{\partial \tau}{\partial T}\right)_L = \frac{k_B L}{Na^2}, \tag{10.44}$$

arriving at the differential equation

$$\frac{\mathcal{C}_L}{T}\,\mathrm{d}T = \frac{k_B}{Na^2} L\,\mathrm{d}L, \tag{10.45}$$

which is easily integrated to give

$$\frac{T_f}{T_0} = \exp\left(\frac{L_f^2 - L_0^2}{\kappa^2}\right), \tag{10.46}$$

where[8]

$$\kappa^{-2} = \frac{1}{2\langle \mathcal{C}_L \rangle}\left(\frac{k_B}{Na^2}\right). \tag{10.47}$$

Thus, as a consequence of adiabatic stretching, i.e. $L_f > L_0$, the elastomer warms!

10.4.1 Abruptly relaxed rubber band

When the rubber band is suddenly allowed to relax under no effective restraint (snaps shut) we still have a fast process with $đ\mathcal{Q} = 0$. But relaxing without restraint means that no work is done by the rubber band, i.e. $đ\mathcal{W} = 0$. This process is thermodynamically identical to the free expansion of a confined gas studied in Chapter 3. Therefore, from the First Law, there is no internal energy change, $d\mathcal{U} = 0$.

This suggests that to find the temperature change in the rapidly relaxed "rubber band", we write, as in Chapter 3, the infinitesimal

$$\mathrm{d}T = \left(\frac{\partial T}{\partial L}\right)_{\mathcal{U}} \mathrm{d}L. \tag{10.48}$$

We are again temporarily blocked by an "unfriendly" partial derivative. But with the skill gained in Chapters 3–5 we start by applying the cyclic chain rule

$$\left(\frac{\partial T}{\partial L}\right)_{\mathcal{U}} = -\left(\frac{\partial \mathcal{U}}{\partial L}\right)_T \left(\frac{\partial \mathcal{U}}{\partial T}\right)_L^{-1} \tag{10.49}$$

and using Eqs. 10.34 and 10.42 from above

$$\mathrm{d}T = \frac{1}{\mathcal{C}_L}\left[T\left(\frac{\partial \tau}{\partial T}\right)_L - \tau\right] \mathrm{d}L, \tag{10.50}$$

which, together with the linearized polymer equation of state, shows that the temperature of an unrestrained, rapidly relaxed elastomer does not change! Therefore a perceptible temperature increase can accumulate by repeated adiabatic stretching and abrupt (unrestrained) relaxation of the rubber band. Try it!

[8] Assuming an average value of $\mathcal{C}_L$.

10.4.2 Entropy change in the relaxing elastomer

A rubber band snaps shut when the elongating constraint (tension) is suddenly released. Of course it does. What else would you expect from a rubber band? What governs this spontaneous process? As alluded to above, it is an expected increase in the entropy of the universe (i.e. the rubber band).[9] To calculate the change in entropy in this constant $\mathcal{U}$ process write

$$d\mathcal{S} = \left(\frac{\partial \mathcal{S}}{\partial L}\right)_{\mathcal{U}} dL. \tag{10.51}$$

What seems to be an unusual partial derivative has been considered previously in Chapter 3 where we first applied the cyclic chain rule to get

$$\left(\frac{\partial \mathcal{S}}{\partial L}\right)_{\mathcal{U}} = -\left(\frac{\partial \mathcal{U}}{\partial L}\right)_{\mathcal{S}} \left(\frac{\partial \mathcal{U}}{\partial \mathcal{S}}\right)_{L}^{-1}. \tag{10.52}$$

Then from the thermodynamic identity, Eq. 10.37, we find both

$$\left(\frac{\partial \mathcal{U}}{\partial L}\right)_{\mathcal{S}} = \tau \tag{10.53}$$

and

$$\left(\frac{\partial \mathcal{U}}{\partial \mathcal{S}}\right)_{L} = T, \tag{10.54}$$

so that Eq. 10.51 reduces to

$$d\mathcal{S} = -\frac{\tau}{T}\, dL. \tag{10.55}$$

Using the linearized equation of state this further reduces to the differential expression

$$d\mathcal{S} = -\frac{k_B}{Na^2} L\, dL, \tag{10.56}$$

which is integrated to give

$$\Delta\mathcal{S} = \frac{k_B}{2Na^2}\left(L_0^2 - L_f^2\right) > 0. \tag{10.57}$$

Spontaneous, unrestrained contraction of a stretched rubber band is driven by increased entropy of the "universe".

10.5 A non-ideal elastomer

Instead of the one-dimensional elastomer consider a model elastomer as a three-dimensional bundle of roughly parallel but weakly interacting chains (directed, say,

[9] A pretty modest-sized universe, to be sure.

in the z-direction). The general picture is that each chain is a system of N links ($N \gg 1$), each of length a, which can swivel around the bond at any polar or azimuthal angle θ, ϕ with respect to the primary z-axis. However, now an interchain energy, assumed to be proportional to $\sin\theta$, is introduced. But to simplify the model we assume that these links can only point in transverse $\{\pm x, \pm y\}$ directions as well as the longitudinal z-direction.

All links pointing parallel to the primary z-axis are taken to have identical energy, $E=0$, while rotationally hindered links bending in $\{\pm x, \pm y\}$ directions (subject to zero mean displacements in the x- and y-directions) acquire an average energy $E=\epsilon$. Such an interaction introduces non-ideality in a simple but plausible way – reminiscent of cross-linking that takes place in vulcanization of rubber. In the limit $\epsilon \to 0$ transverse links are as easily created as longitudinal links so that links meander randomly in three dimensions. In this limit the elastomer may be thought of as ideal in an analogous sense to a three-dimensional non-interacting gas being ideal. (More will be said about this comparison below.) For $\epsilon \neq 0$ the transverse link interaction energy causes meandering to become biased and the elastomer is no longer ideal.

As before, we formulate a thermal Lagrangian tailored to the model described above, and obtain "surrogate" probabilities and a partition function for the system. The partition function yields all necessary thermodynamic detail from which the microscopic interaction parameter ϵ can, presumably, be found by a suitable experiment.

The macroscopic hamiltonian is

$$\mathcal{H}_{op} = \mathbf{h}_0 - \boldsymbol{\tau} \cdot \boldsymbol{X}_{op}, \tag{10.58}$$

where $\mathbf{h}_0$ accounts for the macroscopic "cross-linking" bond energy, $\boldsymbol{X}_{op}$ is the length vector operator whose average $\langle \boldsymbol{X}_{op} \rangle$ has components in any of three mutually perpendicular directions and $\boldsymbol{\tau}$ is the elastic tension vector conjugate to the displacement operator.

Within the framework of the three-dimensional elastomer model the macroscopic eigen-energies of $\mathcal{H}_{op}$ are

$$\begin{aligned}\hat{E}\left(n_{\pm x}, n_{\pm y}, n_{\pm z}\right) &= \left(n_{+x} + n_{-x} + n_{+y} + n_{-y}\right)\epsilon \\ &\quad - \tau_x a\left(n_{+x} - n_{-x}\right) - \tau_y a\left(n_{+y} - n_{-y}\right) - \tau_z a\left(n_{+z} - n_{-z}\right),\end{aligned} \tag{10.59}$$

where $n_{\pm q}$ are the discrete number of links in the $\pm q$ directions, with $q = x, y, z$ the mutually perpendicular coordinate directions and a the length per link. The thermal Lagrangian $\mathcal{L}$ for this model is therefore

$$\mathcal{L} = -k_B \sum_n \mathbf{P}(\mathbf{n}) \ln \mathbf{P}(\mathbf{n})$$
$$- T^{-1} \left[\left\{ \sum_n \mathbf{P}(\mathbf{n}) \left[\left(n_{+x} + n_{-x} + n_{+y} + n_{-y} \right) \epsilon \right] \right\} \right.$$
$$\left. - \left\{ \sum_n \mathbf{P}(\mathbf{n}) \left[\tau_x a \left(n_{+x} - n_{-x} \right) + \tau_y a \left(n_{+y} - n_{-y} \right) + \tau_z a \left(n_{+z} - n_{-z} \right) \right] \right\} \right]$$
$$- \lambda_0 \sum_n \mathbf{P}(\mathbf{n}), \tag{10.60}$$

with T the absolute temperature, and where

$$\mathbf{P}(\mathbf{n}) \equiv \mathbf{P}\left(n_{\pm x}, n_{\pm y}, n_{\pm z} \right) \tag{10.61}$$

is the probability there are $n_{\pm x}, n_{\pm y}, n_{\pm z}$ links of each type. The symbolic sum

$$\sum_n \equiv \sum_{\substack{n_{+x}, n_{-x} \\ n_{+y}, n_{-y} \\ n_{+z}, n_{-z}}} \tag{10.62}$$

is over the integers $\left\{ n_{\pm x}, n_{\pm y}, n_{\pm z} \right\}$ so that, for example, the average value of the longitudinal component of the elastomer length vector $\langle L_z \rangle$ is

$$\langle L_z \rangle = a \sum_n \mathbf{P}(\mathbf{n}) \left(n_{+z} - n_{-z} \right) \tag{10.63}$$

and the internal energy $\mathcal{U}$ is

$$\mathcal{U} = \sum_n \mathbf{P}(\mathbf{n}) \left(n_{+x} + n_{-x} + n_{+y} + n_{-y} \right) \epsilon \tag{10.64}$$

with the "ideal" elastomer characterized by $\lim_{\epsilon \to 0} \mathcal{U} = 0$. Maximizing the Lagrangian $\mathcal{L}$ with respect to the $\mathbf{P}(\mathbf{n})$ we obtain

$$\mathbf{P}(\mathbf{n}) = \frac{e^{-\beta \epsilon \left(n_{+y} + n_{-y} + n_{+x} + n_{-x} \right)} e^{-\beta a \tau_z \left(n_{+z} - n_{-z} \right)}}{\mathcal{Z}_X}, \tag{10.65}$$

where $\tau_x = \tau_y = 0$ (required by $\langle L_x \rangle = \langle L_y \rangle = 0$) has been imposed. The normalizing denominator

$$\mathcal{Z}_X = \sum_n e^{-\beta \epsilon \left(n_{+y} + n_{-y} + n_{+x} + n_{-x} \right)} e^{-\beta a \tau_z \left(n_{+z} - n_{-z} \right)} \tag{10.66}$$

is the partition function. Equation 10.66 is summed by using the degeneracy $g(\mathbf{n})$ of the macroscopic eigen-energies

$$g\left(n_{\pm x}, n_{\pm y}, n_{\pm z} \right) = \frac{N!}{n_{+x}! n_{-x}! n_{+y}! n_{-y}! n_{+z}! n_{-z}!} \tag{10.67}$$

to give the partition function

$$\mathcal{Z}_X = \sum_n \frac{N!}{n_{+x}!n_{-x}!n_{+y}!n_{-y}!n_{+z}!n_{-z}!} e^{-\beta\epsilon(n_{+y}+n_{-y}+n_{+x}+n_{-x})} e^{-\beta a\tau_z(n_{+z}-n_{-z})}, \tag{10.68}$$

the sum now being over *distinct* integer values restricted by the fixed total

$$n_{+x} + n_{-x} + n_{+y} + n_{-y} + n_{+z} + n_{-z} = N. \tag{10.69}$$

Equation 10.68 is just the multinomial expansion of

$$\mathcal{Z}_X = \left[e^{-\beta\tau_z a} + e^{+\beta\tau_z a} + 4e^{-\beta\epsilon}\right]^N. \tag{10.70}$$

10.6 Three-dimensional elastomer thermodynamics

Substituting probabilities of Eq. 10.65 into

$$\mathcal{S} = -k_B \sum_{\mathbf{n}}^{N} \mathbf{P}(\mathbf{n}) \ln \mathbf{P}(\mathbf{n}) \tag{10.71}$$

gives the thermodynamic entropy

$$\mathcal{S} = \beta k_B \left(\mathcal{U} - \tau_z \langle L_z \rangle + \frac{1}{\beta} \ln \mathcal{Z}_X\right). \tag{10.72}$$

Defining $\mathcal{G}_X$, an elastic Gibbs potential, as

$$\mathcal{G}_X = -\frac{1}{\beta} \ln \mathcal{Z}_X, \tag{10.73}$$

Eq. 10.72 is rewritten as

$$\mathcal{G}_X = \mathcal{U} - T\mathcal{S} - \tau_z \langle L_z \rangle. \tag{10.74}$$

Taking the total differential of $\mathcal{G}_X$ and using the fundamental law expression

$$T\,\mathrm{d}\mathcal{S} = \mathrm{d}\mathcal{U} - \tau_z\,\mathrm{d}\langle L_z \rangle, \tag{10.75}$$

where $d\langle L_z \rangle$ is an infinitesimal change in elastomer length, we also find

$$\mathrm{d}\mathcal{G}_X = -\mathcal{S}\,\mathrm{d}T - \langle L_z \rangle\,\mathrm{d}\tau_z \tag{10.76}$$

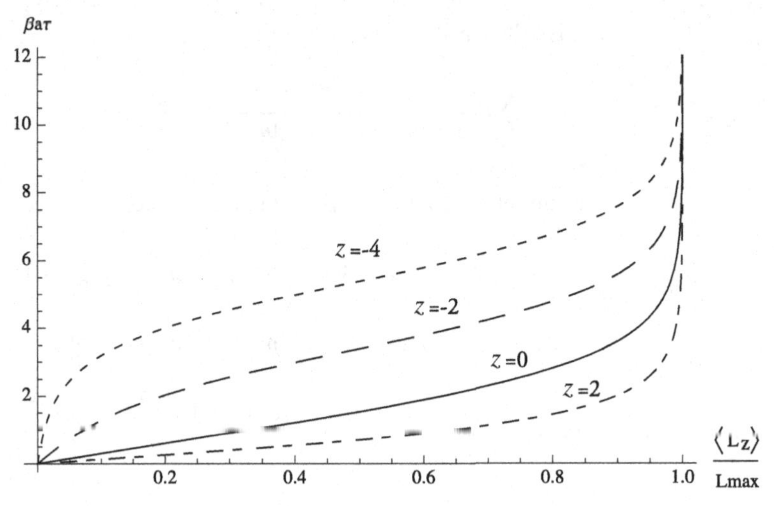

Fig. 10.7 Model elastomer stress–strain relationships as described by Eq. 10.78. Curves with $z = \beta\epsilon < 0$ are typically observed in real rubber-like elastomers.

so that from the partition function we find the elastomer length, i.e. the equation of state,

$$\langle L_z \rangle = \left[\frac{\partial}{\partial(\beta\tau_z)} \ln \mathcal{Z}_X \right]_T \tag{10.77}$$

$$= \frac{Na \sinh(\beta a \tau_z)}{\cosh(\beta a \tau_z) + 2e^{-\beta\varepsilon}}. \tag{10.78}$$

Plotting $\beta a \tau$ vs. $\langle L_z \rangle / L_{max}$ in Figure 10.7, various elastic polymer behaviors obtained from the equation of state, Eq. 10.78, are shown. The curve for $z = \beta\epsilon > 0$ is characteristic of an "elasticity" dominated by coiling and uncoiling of individual chains. On the other hand, the curve for $z = \beta\epsilon < 0$ corresponds to a tendency to form cross-linkages,[10] in which case rubber initially strongly resists stretching followed by a region of easier deformation (possibly associated with physically observed "necking") until, as $\langle L_z \rangle / L_{\max} \to 1$, stretched rubber suddenly ruptures. The case $z = \beta\epsilon < 0$ parallels observed characteristics of "rubber".

Using Eq. 10.76 or, equivalently, Eq. 10.71, the elastomer entropy is

$$\mathcal{S} = -\left(\frac{\partial \mathcal{G}_X}{\partial T} \right)_{\tau_z}$$
$$= -k_B \beta^2 \left[\frac{\partial}{\partial \beta} \left(\frac{1}{\beta} \ln \mathcal{Z}_X \right) \right]_{\tau_z}, \tag{10.79}$$

[10] Such as happens in vulcanization of rubber.

which evaluates to

$$\mathcal{S} = Nk_B \left\{ \ln 2 + \ln \left[2e^{-\beta\epsilon} + \cosh(\beta\tau_z a) \right] + \frac{\beta \left[2\epsilon - a\tau_z e^{\beta\epsilon} \sinh(\beta\tau_z a) \right]}{2 + e^{\beta\epsilon} \cosh(\beta\tau_z a)} \right\}. \tag{10.80}$$

Finally, from Eqs. 10.73, 10.74, 10.78 and 10.80, the internal energy is

$$\mathcal{U} = \frac{2N\varepsilon}{2 + e^{\beta\varepsilon} \cosh(\beta a\tau_z)}. \tag{10.81}$$

10.6.1 An experiment

The typical undergraduate thermal physics course seems to lack realistic hands-on laboratory exercises, especially ones that might explore the synergy between statistical theory, microscopic parameters and experiment. Perhaps this is a result of historical precedence in which gases – with easy kinetic theory visualizations – enjoy perceived pedagogical simplicity. Although having a disproportionate role in thermodynamic instruction they are not especially convenient subjects for non-trivial laboratory studies in a classroom setting nor are statistical theories of non-ideal gases – which should be the basis for exploring any experimental synergy – easily accessible at the undergraduate level.

In Appendix F macroscopic thermodynamic (lab) measurements are used to find a microscopic interaction energy ϵ as modeled in the three-dimensional elastomer of Section 10.6

Problems and exercises

10.1 Find the linear thermal expansivity

$$\alpha = \frac{1}{L} \left(\frac{\partial L}{\partial T} \right)_{\tau} \tag{10.82}$$

for the "naive" one-dimensional rubber band.

10.2 Find the mean fluctuations in the average one-dimensional polymer elongation $\langle \chi_{op} \rangle$

$$\langle\langle (\Delta\chi)^2 \rangle\rangle = \langle\langle (\chi_{op} - \langle \chi_{op} \rangle) \rangle\rangle^2 \tag{10.83}$$

for the naive polymer model.

10.3 A macromolecule is composed of a chain of N chemically identical molecular monomers for which two distinct states are accessible. At equilibrium N_0 units are in the molecular ground state with energy E_0 and N_1 units are in an excited state with energy E_1. Both states are non-degenerate. The ground state monomer length is $\ell = \lambda$ while in the excited state it has length $\ell = (1+\delta)\,\lambda$, $0 < \delta \ll 1$.

a. Write an expression for the polymer partition function $\mathscr{Z}$.
b. Find an equation of state describing the total length L as a function of chain tension τ and temperature T, $L = L(\tau, T)$.

10.4 A one-dimensional polymer that emulates piezoelectric behaviors is modeled as $N = (n_R + n_L)$ molecular links of which n_R are right-pointing $(+x)$ and n_L are left-pointing $(-x)$. Each link has length λ and fixed electric dipole moment p. When placed in a uniform electric field $\mathscr{E}_{0,x}$, each right pointing link has an energy

$$\epsilon_R = -p \cdot \mathscr{E}_0 \tag{10.84}$$

while each left-pointing link has an energy

$$\epsilon_L = +p \cdot \mathscr{E}_0. \tag{10.85}$$

The implied "piezoelectric" behaviors are:

i. a mechanical tension τ applied along the longitudinal $+x$-direction, resulting in an electric polarization $\mathcal{P}$ in that direction;
ii. an electric field $\mathscr{E}_0$ applied in, say, the $+x$-direction, resulting in a mechanical deformation $\Delta\ell$ in that direction.

The thermodynamic hamiltonian for this model is taken to be

$$\hat{\mathbf{h}} = \mathbf{h_0} - \tau \boldsymbol{\chi}_{op} - \mathscr{E}_0 \cdot \mathcal{P}_{op} \tag{10.86}$$

so that the corresponding macroscopic eigen-energies are

$$\hat{E} = N\epsilon_0 - \tau\lambda\,(n_R - n_L) - \mathscr{E}_0 \cdot p\,(n_R - n_L) \tag{10.87}$$

with the average length

$$\langle \boldsymbol{\chi}_{op} \rangle = \boldsymbol{\lambda}\,\langle n_R - n_L \rangle \tag{10.88}$$

and the average polarization

$$\langle \mathcal{P}_{op} \rangle = p\,\langle n_R - n_L \rangle. \tag{10.89}$$

It is assumed that $\mathbf{h}_0$ is field independent and $\epsilon_0 \approx 0$.

a. Write a thermal Lagrangian corresponding to the model.
b. Write a partition function for this piezoelectric polymer in the form

$$\mathcal{Z} = \sum_E g(E) e^{-\beta E} \tag{10.90}$$

and evaluate the expression.
c. Find the entropy of the model in terms of β and $p \cdot \mathcal{E}_0$.
d. Find the Helmholtz potential F in terms of β and $p \cdot \mathcal{E}_0$.
e. Find the average polymer length $\langle \chi_{op} \rangle$ in terms of β and $\mathcal{P} \cdot \mathcal{E}_0$.
A characterizing piezoelectric property is the coefficient

$$\xi = \left(\frac{\partial \mathcal{P}}{\partial \tau} \right)_{\beta, \mathcal{E}} . \tag{10.91}$$

f. Describe ξ in words.
g. Find ξ in terms of β and $p \cdot \mathcal{E}_0$.

11 Magnetic thermodynamics

The nation that controls magnetism controls the universe.

"Diet Smith", from Chester Gould's *Dick Tracy* comic strip (1960)

11.1 Magnetism in solids

Lodestones – fragments of magnetic[1] $FeO + Fe_2O_3$ (Fe_3O_4) – although known to the ancients were, according to Pliny the Elder, first formally described in Greek 6th-century BCE writings. By that time they were already the stuff of myth, superstition and amazing curative claims, some of which survive to this day. The Chinese used lodestones in navigation as early as 200 BCE and are credited with inventing the magnetic compass in the 12th century CE.

Only in the modern era has magnetism become well understood, inspiring countless papers, books[2] and more than a dozen Nobel prizes in both fundamental and applied research.

11.1.1 Forms of macroscopic magnetism

Paramagnetism

In an external magnetic field $\mathcal{B}$ the spin-state degeneracy of local (atomic) or itinerant (conduction) electronic states is lifted (Zeeman effect). At low temperature this results in an induced macroscopic magnetic moment whose vector direction lies parallel to the external field. This is referred to as *paramagnetism*.[3]

For most materials removing the external field restores spin-state degeneracy, returning the net moment to zero.

[1] Named, as one story goes, for Magnus, the Greek shepherd who reported a field of stones that drew the nails from his sandals.

[2] See e.g. Stephen Blundell, *Magnetism in Condensed Matter*, Oxford Maser Series in Condensed Matter Physics (2002); Daniel C. Mattis, *Theory of Magnetism Made Simple*, World Scientific, London (2006); Robert M. White, *Quantum Theory of Magnetism: Magnetic Properties of Materials*, 3rd rev. edn, Springer-Verlag, Berlin (2007).

[3] Itinerant (conduction) electron paramagnetism is referred to as *spin* or *Pauli paramagnetism*.

Diamagnetism

Macroscopic magnetization may also be induced with a magnetization vector antiparallel to the external field, an effect called *diamagnetism.* In conductors diamagnetism arises from the highly degenerate quantum eigen-energies and eigenstates (referred to as *Landau levels*)[4] formed by interaction between mobile electrons and magnetic fields. Diamagnetism is also found in insulators, but largely from surface quantum orbitals rather than interior bulk states.[5] Both cases are purely quantum phenomena, leaving macroscopic diamagnetism without an elementary explanatory model.[6,7] All solids show some diamagnetic response, but it is usually dominated by any paramagnetism that may be present.

At high magnetic fields and low temperatures very pure metals exhibit an oscillatory diamagnetism called the *de Haas–van Alphen effect* whose source is exclusively the Landau levels.[8]

"Permanent" magnetism

"Permanent" magnetism refers to macroscopic magnetization taking place without any external magnetic fields. Several examples of this phenomenon are:

- *Ferromagnetism:* this is an ordered state of matter in which local paramagnetic moments interact to produce an effective internal magnetic field resulting in collective alignment of moments throughout distinct regions called domains. Due to these internal interactions, domains can remain aligned even after the external field is removed.

  Ferromagnetic alignment abruptly disappears at a material-specific temperature called the *Curie temperature* T_c, at which point ordinary local paramagnetism returns.
- *Antiferromagnetism:* at low temperatures, interactions between adjacent identical paramagnetic atoms, ions or sub-lattices can induce collective "anti-alignment" of adjacent paramagnets, resulting in a net zero magnetic moment.
- *Ferrimagnetism:* at low temperatures, interactions between unequivalent paramagnetic atoms, ions or sub-lattices can produce collective "anti-alignment" of moments, resulting in a small residual magnetization.

  In both ferrimagnetism and antiferromagnetism, increasing temperature weakens "anti-alignment" with the collective induced moments approaching a

[4] L. Landau, "Diamagnetism of metals", *Z. Phys.* **64**, 629 (1930).

[5] D. Ceresoli *et al.*, "Orbital magnetization in crystalline solids", *Phys. Rev. B* **74**, 24408 (2006).

[6] Niels Bohr, *Studier over Metallernes Elektrontheori*, Kbenhavns Universitet (1911).

[7] Hendrika Johanna van Leeuwen, "Problèmes de la théorie électronique du magnetisme", *Journal de Physique et le Radium* **2**, 361 (1921).

[8] D. Shoenberg, *Magnetic Oscillations in Metals*, Cambridge University Press, Cambridge (1984).

maximum. Then, at a material-specific temperature called the Néel temperature T_N, anti-alignment disappears and the material becomes paramagnetic.

In this chapter general concepts in the thermodynamics of magnetism and magnetic fields are discussed as well as models of local paramagnetism and ferromagnetism.

11.2 Magnetic work

Central to integrating magnetic fields and magnetizable systems into the First Law of Thermodynamics is formulating magnetic work. Using Maxwell fields,[9] the energy generated within a volume V in a time δt, by an electric field $\mathcal{E}$ acting on true charge currents $\mathcal{J}$ – *Joule heat* – is[10]

$$\delta \mathcal{W}^M = -\delta t \int_V \mathcal{J} \cdot \mathcal{E}\, dV. \tag{11.1}$$

Therefore the quasi-static and reversible[11] magnetic work done *by* the system is

$$\delta \mathcal{W}^M_{QS} = \delta t \int_V \mathcal{J} \cdot \mathcal{E}\, dV. \tag{11.2}$$

Using the Maxwell equation (in cgs–Gaussian units)[12]

$$\nabla \times \mathcal{H} = \frac{4\pi}{c}\mathcal{J} + \frac{1}{c}\frac{\delta \mathcal{D}}{\delta t}, \tag{11.3}$$

the work done by the system is

$$\delta \mathcal{W}^M_{QS} = \delta t \left\{ \frac{c}{4\pi} \int_V (\nabla \times \mathcal{H}) \cdot \mathcal{E}\ dV - \frac{1}{4\pi} \int_V \frac{\delta \mathcal{D}}{\delta t} \cdot \mathcal{E}\ dV \right\}. \tag{11.4}$$

Using the vector identity

$$\boldsymbol{U} \cdot \nabla \times \boldsymbol{V} = \nabla \cdot (\boldsymbol{V} \times \boldsymbol{U}) + \boldsymbol{V} \cdot \nabla \times \boldsymbol{U}, \tag{11.5}$$

[9] Maxwell fields in matter and free space are the local averages that appear in his equations of electromagnetism.

[10] Since heat and work are not state functions they do not have true differentials, so the wiggly δs are used instead to represent incremental work in an interval of time δt.

[11] In specifying reversibility non-reversible hysteresis effects are excluded.

[12] Even though they have fallen out of pedagogical favor in E&M textbooks, cgs units offer unrivaled clarity in presenting the subtle issues involved in thermodynamics of magnetic and electric fields.

this becomes

$$\delta \mathcal{W}_{QS}^{M} = \delta t \left\{ \frac{c}{4\pi} \left[\int_V \nabla \cdot (\mathcal{H} \times \mathcal{E})\ \mathrm{d}V + \int_V \mathcal{H} \cdot \nabla \times \mathcal{E}\ \mathrm{d}V \right] - \frac{1}{4\pi} \int_V \frac{\delta \mathcal{D}}{\delta t} \cdot \mathcal{E}\ \mathrm{d}V \right\}. \tag{11.6}$$

The first integral on the right can be transformed by Gauss' theorem into a surface integral. But since the fields are static (non-radiative), they fall off faster than $\frac{1}{r^2}$ so that for a very distant surface the surface integral can be neglected. Then, with the Maxwell equation (Faraday's Law)

$$\nabla \times \mathcal{E} = -\frac{1}{c} \left(\frac{\delta \mathcal{B}}{\delta t} \right), \tag{11.7}$$

incremental work done by the system is

$$\delta \mathcal{W}_{QS}^{M} = -\delta t \left\{ \frac{1}{4\pi} \int_V \mathcal{H} \cdot \frac{\delta \mathcal{B}}{\delta t}\ \mathrm{d}V + \frac{1}{4\pi} \int_V \frac{\delta \mathcal{D}}{\delta t} \cdot \mathcal{E}\ \mathrm{d}V \right\} \tag{11.8}$$

$$= -\left\{ \frac{1}{4\pi} \int_V \mathcal{H} \cdot \delta \mathcal{B}\ \mathrm{d}V + \frac{1}{4\pi} \int_V \delta \mathcal{D} \cdot \mathcal{E}\ \mathrm{d}V \right\}, \tag{11.9}$$

where the integrals are over the volume of the sample and surrounding free space. (**Note:** The "wiggly" deltas in $\delta\mathcal{B}(\mathbf{x})$, $\delta\mathcal{H}(\mathbf{x})$ and $\delta\mathcal{D}(\mathbf{x})$ represent functional changes, i.e. changes in the fields (states) not the coordinates. The fields themselves are functions of coordinates $\mathbf{x}$ and are not just simple variables.)

Limiting the discussion to magnetic phenomena, the magnetic contribution to quasi-static work done by the system is therefore

$$\delta \mathcal{W}_{QS}^{\mathcal{M}} = -\left\{ \frac{1}{4\pi} \int_V \mathcal{H} \cdot \delta \mathcal{B}\ \mathrm{d}V \right\}, \tag{11.10}$$

so that a fundamental magnetic thermodynamic equation for $\delta\mathcal{U}$ becomes

$$\delta \mathcal{U} = T \delta \mathcal{S} - p\ \mathrm{d}V + \frac{1}{4\pi} \int_V \mathcal{H} \cdot \delta \mathcal{B}\ \mathrm{d}V. \tag{11.11}$$

From the Helmholtz potential, $F = \mathcal{U} - T\mathcal{S}$,

$$\delta F = \delta \mathcal{U} - T \delta \mathcal{S} - \mathcal{S}\ \mathrm{d}T, \tag{11.12}$$

which when combined with Eq. 11.11 gives the change δF:

$$\delta F = -\mathcal{S}\, \mathrm{d}T - p dV + \frac{1}{4\pi} \int_V \mathcal{H} \cdot \delta\mathcal{B}\, \mathrm{d}V. \tag{11.13}$$

Defining magnetic enthalpy H as

$$H = \mathcal{U} + pV - \frac{1}{4\pi} \int_V \mathcal{B} \cdot \mathcal{H}\, dV \tag{11.14}$$

gives, with Eq. 11.11, an enthalpy change δH:

$$\delta H = T\delta\mathcal{S} + V\, \mathrm{d}p - \frac{1}{4\pi} \int_V \mathcal{B} \cdot \delta\mathcal{H}\, \mathrm{d}V. \tag{11.15}$$

Finally, a magnetic Gibbs potential G is defined as

$$G = F + pV - \frac{1}{4\pi} \int_V \mathcal{B} \cdot \mathcal{H}\, \mathrm{d}V, \tag{11.16}$$

which with Eq. 11.13 gives the Gibbs potential change δG:

$$\delta G = -\mathcal{S}\, \mathrm{d}T + V\, \mathrm{d}p - \frac{1}{4\pi} \int_V \mathcal{B} \cdot \delta\mathcal{H}\, \mathrm{d}V. \tag{11.17}$$

Magnetization density[13] $\mathcal{M}$ and polarization density[14] $\boldsymbol{P}$ are inserted by the linear constitutive relations

$$\mathcal{H} = \mathcal{B} - 4\pi\mathcal{M} \tag{11.18}$$

and

$$\mathcal{D} = \mathcal{E} + 4\pi\, \boldsymbol{P} \tag{11.19}$$

allowing quasi-static magnetic work to be written

$$\delta\mathcal{W}_{QS}^{\mathcal{M}^1} = \left\{ \frac{1}{4\pi} \int_V \mathcal{B} \cdot \delta\mathcal{B}\, \mathrm{d}V - \int_{V'} \mathcal{M} \cdot \delta\mathcal{B}\, \mathrm{d}V \right\} \tag{11.20}$$

[13] Total magnetic moment per unit volume.

[14] Total electric dipole moment per unit volume.

or

$$\delta \mathcal{W}_{QS}^{\mathcal{M}^2} = \left\{ \frac{1}{4\pi} \int_V \mathcal{H} \cdot \delta \mathcal{H} \, dV + \int_{V'} \mathcal{H} \cdot \delta \mathcal{M} \, dV \right\}, \tag{11.21}$$

and quasi-static electric work to be written

$$\delta \mathcal{W}_{QS}^{\boldsymbol{P}^1} = \left\{ \frac{-1}{4\pi} \int_V \mathcal{D} \cdot \delta \mathcal{D} \, dV + \int_{V'} \boldsymbol{P} \cdot \delta \mathcal{D} \, dV \right\} \tag{11.22}$$

or

$$\delta \mathcal{W}_{QS}^{\boldsymbol{P}^2} = \left\{ \frac{-1}{4\pi} \int_V \mathcal{E} \cdot \delta \mathcal{E} \, dV - \int_{V'} \mathcal{E} \cdot \delta \boldsymbol{P} \, dV \right\}. \tag{11.23}$$

The first terms in all alternatives are total field energies – integrals over all space, both inside and outside matter. The second terms are integrals over V' which include only the volume of magnetized (polarized) matter. Since magnetic (electric) thermodynamics is primarily concerned with matter that is magnetized (polarized), one practice is to bravely ignore the total field energies. Another is to absorb the field energies into the internal energy $\mathcal{U}$. Since neither option is entirely satisfactory a third way is discussed in Section 11.3 below.

Nevertheless, these results – in terms of local average fields – are general and thermodynamically correct.[15] But they are not convenient to apply. Nor are they the fields that appear in microscopic magnetic (electric) quantum hamiltonians. In quantum magnetic (electric) models the hamiltonians for individual magnetic (electric) moments depend only on the field before the sample is introduced. After the sample is introduced internal fields can additionally result from:

a. interactions between internal moments as additional terms in the hamiltonian – these interactions may be approximately treated as "effective fields" acting in addition to the external field (see, for example, Section 11.6, below);
b. internal "demagnetizing" fields arising from fictitious surface "poles" induced by $\mathcal{B}_0$;[16]
c. internal currents induced by the applied field (especially in conductors).[17]

[15] $\delta \mathcal{W}_{QS}^{\mathcal{M}^2}$ and $\delta \mathcal{W}_{QS}^{\mathcal{P}^2}$ have the correct form for work – intensive × extensive.

[16] "Demagnetizing" fields introduce sample shape dependence into the magnetic properties.

[17] T. Holstein, R.E. Norton and P. Pincus, "de Haas–van Alphen effect and the specific heat of an electron gas", *Physical Review B* **8**, 2649 (1973).

11.3 Microscopic models and uniform fields

Microscopic models of magnetic and electric hamiltonians are expressed in terms of uniform applied fields $(\mathcal{B}_0, \mathcal{E}_0)$ present before matter is introduced. This emphasis on applied fields (rather than average Maxwell fields within matter) results in thermodynamic relations somewhat different from Eqs. 11.11 to 11.17 above. Focusing on magnetic effects in the absence of internal magnetic interactions, quasi-static magnetic work done by the system is (see Eq. G.23 in Appendix G)

$$đ\mathcal{W}_{QS}^{\boldsymbol{M}} = -\mathcal{B}_0 \cdot \mathrm{d}\boldsymbol{M}. \tag{11.24}$$

Since $\mathcal{B}_0$ is uniform, a total macroscopic magnetization vector $\boldsymbol{M}$ has been defined as:

$$\boldsymbol{M} = \int_{V'} \langle \mathcal{M} \rangle \, \mathrm{d}V, \tag{11.25}$$

with $\langle \mathcal{M} \rangle$ the average magnetization per unit volume.[18]

Therefore thermodynamic differential relations become (see Eqs. G.24, G.26, G.25 and G.27)

$$\begin{aligned}
T\,\mathrm{d}\mathcal{S} &= \mathrm{d}\mathcal{U}^* + p\,\mathrm{d}V - \mathcal{B}_0 \cdot \mathrm{d}\boldsymbol{M}, \\
T\,\mathrm{d}\mathcal{S} &= \mathrm{d}H^* - V\,\mathrm{d}p + \boldsymbol{M} \cdot \mathrm{d}\mathcal{B}_0, \\
\mathrm{d}F^* &= -\mathcal{S}dT - p\,\mathrm{d}V + \mathcal{B}_0 \cdot \mathrm{d}\boldsymbol{M}, \\
\mathrm{d}G^* &= -\mathcal{S}\,\mathrm{d}T + V\,\mathrm{d}p - \boldsymbol{M} \cdot \mathrm{d}\mathcal{B}_0,
\end{aligned} \tag{11.27}$$

where the starred potentials are

$$\mathcal{U}^* = \mathcal{U} + \frac{1}{8\pi} \int_V \mathcal{B}_0^2 \, \mathrm{d}V, \tag{11.28}$$

$$H^* = H + \frac{1}{8\pi} \int_V \mathcal{B}_0^2 \, \mathrm{d}V, \tag{11.29}$$

$$F^* = F + \frac{1}{8\pi} \int_V \mathcal{B}_0^2 \, \mathrm{d}V, \tag{11.30}$$

$$G^* = G + \frac{1}{8\pi} \int_V \mathcal{B}_0^2 \, \mathrm{d}V. \tag{11.31}$$

[18] Similarly

$$đ\mathcal{W}_{QS}^{\boldsymbol{P}} = \mathcal{E}_0 \cdot \mathrm{d}\boldsymbol{P}. \tag{11.26}$$

11.4 Local paramagnetism

The classical energy of a magnetic moment **m** in an average local (Maxwell) magnetic field $\mathcal{B}$ is

$$E = -\mathbf{m} \cdot \mathcal{B}. \tag{11.32}$$

For fundamental magnetic moments (electrons, protons, neutrons, etc.) quantum mechanics postulates an operator replacement $\mathbf{m} \to \boldsymbol{m}_{op}$, and a quantum paramagnetic hamiltonian

$$\mathcal{H}_{\mathcal{M}} = -\boldsymbol{m}_{op} \cdot \mathcal{B}_0, \tag{11.33}$$

where $\mathcal{B}_0$ is the uniform field present before matter is introduced.[19]

The paramagnetic hamiltonian for a solid consisting of N identical moments fixed at crystalline sites i is

$$\mathcal{H}_{op} = \mathbf{h}_0 - \mathcal{B}_0 \cdot \sum_{i=1}^{N} \boldsymbol{m}_{op}(i), \tag{11.34}$$

where $\boldsymbol{m}_{op}(i)$ is a magnetic moment operator[20] and $\mathbf{h}_0$ is any non-magnetic part of the hamiltonian.[21,22] The "magnetization" operator (total magnetic moment per unit volume) is

$$\mathcal{M}_{op} = \frac{1}{V} \sum_{i=1}^{N} \boldsymbol{m}_{op}(i). \tag{11.37}$$

One objective is to find the macroscopic equation of state $\langle M \rangle = M(T, \mathcal{B}_0)$, where $\langle M \rangle$ is the average magnetization per unit volume,

$$\langle M \rangle = \mathcal{T}r\, \rho_{op}^{\tau}\, \mathcal{M}_{op}. \tag{11.38}$$

[19] Appendix G includes a discussion of the implications and limitations of using $\mathcal{B}_0$ in the thermodynamics.

[20] $\boldsymbol{m}_{op}$ is proportional to an angular momentum (spin) operator $\boldsymbol{S}_{op}$, with

$$\boldsymbol{m}_{op} = \frac{g\gamma_B}{\hbar} \boldsymbol{S}_{op}. \tag{11.35}$$

Here g is the particle g-factor and $\gamma_B = \frac{e\hbar}{2mc}$ is the Bohr magneton (cgs–Gaussian units.)

For a $spin = \frac{1}{2}$ atom the quantum mechanical z-component spin operator S_z is taken with two eigenstates and two eigenvalues

$$S_z \left| \pm\frac{1}{2} \right\rangle = \pm\frac{\hbar}{2} \left| \pm\frac{1}{2} \right\rangle \tag{11.36}$$

[21] It is assumed that there are no interactions corresponding to internal fields, $\mathcal{B}_{int}$.

[22] Field–particle current terms $\mathcal{H}^{\mathscr{A}\cdot\mathscr{J}} = \frac{1}{2m}\left(p_{op} - \frac{q}{c}\mathscr{A}_{op}\right)^2$, where $\mathscr{A}_{op}$ is the vector potential operator, are ignored.

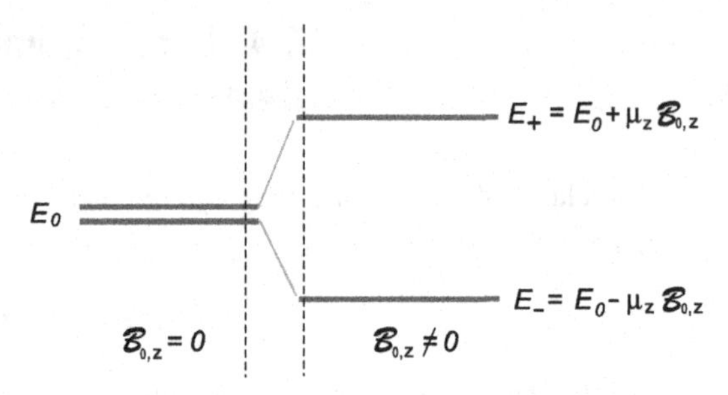

Fig. 11.1 Lifting the $s = \frac{1}{2}$ spin degeneracy with a magnetic field $\mathcal{B}_{0,z}$.

11.5 Simple paramagnetism

Consider a spin $= \frac{1}{2}$ atom, which in the absence of a magnetic field has the pair of degenerate states with energy E_0[23,24,25]

$$\boldsymbol{m}_{z,op}|\bar{\mu}_{\pm\frac{1}{2}}\rangle = \pm\frac{g\gamma_B}{2}|\bar{\mu}_{\pm\frac{1}{2}}\rangle. \tag{11.39}$$

In a uniform magnetic field, $\mathcal{B}_{0,z}$, the degeneracy of each atom state is lifted, creating a pair of non-degenerate states of energy

$$E_- = E_0 - \mu_{\frac{1}{2}}\mathcal{B}_{0,z} \quad \text{and} \quad E_+ = E_0 + \mu_{\frac{1}{2}}\mathcal{B}_{0,z} \tag{11.40}$$

with $\mu_{\frac{1}{2}} = \frac{g\gamma_B}{2}$ (see Figure 11.1).

The macroscopic N-moment eigen-energies are

$$E\,(n_+, n_-) = NE_0 + (n_+ - n_-)\;\mu_{\frac{1}{2}}\;\mathcal{B}_{0,z}, \tag{11.41}$$

where n_+ is the number of atoms in a state with eigen-energy

$$E_+ = E_0 + \mu_{\frac{1}{2}}\;\mathcal{B}_{0,z} \tag{11.42}$$

and n_- is the number of atoms in a state with eigen-energy

$$E_- = E_0 - \mu_{\frac{1}{2}}\;\mathcal{B}_{0,z}, \tag{11.43}$$

with $n_+ + n_- = N$.

[23] $\mathbf{h}_0$ of Eq. 11.34, and hence E_0, is assumed to make no magnetic contribution, from interacting moments, internal currents or other internal fields.

[24] An eigenvalue equation for atomic spin J, $\boldsymbol{m}_{z,op}$ is

$$\boldsymbol{m}_{z,op}\left|\bar{\mu}_{m_J}\right\rangle = g\gamma_B\, m_J\left|\bar{\mu}_{m_J}\right\rangle,$$

where $|\bar{\mu}_{m_J}\rangle$ are the eigenstates and $g\gamma_B\, m_J$ the eigenvalues, with $-J \le m_J \le J$.

[25] To preserve simplicity, discussion is confined to spin-$\frac{1}{2}$ magnetic moments.

The probabilities $\mathbf{P}(\epsilon_s)$ required for the thermodynamic density operator

$$\hat{\rho}^{\tau}_{op} = \sum_{s} \mathbf{P}(\epsilon_s)|E_s\rangle\langle E_s| \tag{11.44}$$

are found by constructing the thermal Lagrangian

$$\begin{aligned}\mathcal{L} = -k_B \sum_{n_+,n_-=0}^{N} \mathbf{P}(n_+, n_-) \ln \mathbf{P}(n_+, n_-) - \lambda_0 \sum_{n_+,n_-=0}^{N} \mathbf{P}(n_+, n_-) \\ - T^{-1} \sum_{n_+,n_-=0}^{N} \mathbf{P}(n_+, n_-) \left[N E_0 + \mu_{\frac{1}{2}} (n_+ - n_-) \mathcal{B}_{0,z} \right]\end{aligned} \tag{11.45}$$

using the N-atom macroscopic eigen-energies of Eq. 11.41. The resulting probabilities are

$$\mathbf{P}(n_+, n_-) = \frac{e^{-\beta \left[N E_0 + \mu_{\frac{1}{2}} (n_+ - n_-) \mathcal{B}_{0,z} \right]}}{\mathcal{Z}_M}, \tag{11.46}$$

where $\beta = 1/k_B T$, and the denominator (the paramagnetic partition function) is

$$\mathcal{Z}_M = \sum_{\substack{n_+,n_- \\ n_+ + n_- = N}}^{N} g(n_+, n_-) e^{-\beta \left[N E_0 + \mu_{\frac{1}{2}} (n_+ - n_-) \mathcal{B}_{0,z} \right]}. \tag{11.47}$$

The sum over all states is accounted for by the configurational degeneracy

$$g(n_+, n_-) = \frac{N!}{n_+! n_-!} \tag{11.48}$$

so that

$$\mathcal{Z}_M = \sum_{\substack{n_+,n_- \\ n_+ + n_- = N}}^{N} \frac{N!}{n_+! n_-!} e^{-\beta \left[N E_0 + \mu_{\frac{1}{2}} (n_+ - n_-) \mathcal{B}_{0,z} \right]}, \tag{11.49}$$

which is the binomial expansion of

$$\mathcal{Z}_M = e^{-\beta N E_0} \left(e^{\beta \mu_{\frac{1}{2}} \mathcal{B}_{0,z}} + e^{-\beta \mu_{\frac{1}{2}} \mathcal{B}_{0,z}} \right)^N \tag{11.50}$$

$$= \left[2 e^{-\beta E_0} \cosh \left(\beta \mu_{\frac{1}{2}} \mathcal{B}_{0,z} \right) \right]^N. \tag{11.51}$$

11.6 Local paramagnet thermodynamics

Using Eq. 11.46, simple paramagnet thermal properties for the spin $= \frac{1}{2}$ system are found.

The average *total magnetization* is

$$\langle \boldsymbol{M} \rangle = -\frac{\displaystyle\sum_{\substack{n_-,n_+ \\ n_+ + n_- = N}}^{N} \frac{N!}{n_+! n_-!} \left[\mu_{\frac{1}{2}} (n_+ - n_-) \right] e^{-\beta \left[N E_0 + \mu_{\frac{1}{2}} (n_+ - n_-) \mathcal{B}_{0,z} \right]}}{\mathcal{Z}_M} \tag{11.52}$$

or

$$\langle \boldsymbol{M} \rangle = -\frac{\partial}{\partial \mathcal{B}_{0,z}} \left(-\frac{1}{\beta} \ln \mathcal{Z}_M \right) \tag{11.53}$$

$$= N \mu_{\frac{1}{2}} \tanh \left(\beta \mu_{\frac{1}{2}} \mathcal{B}_{0,z} \right). \tag{11.54}$$

Equation 11.54 is called the *Langevin paramagnetic equation.* Note in Figure 11.2 that the magnetization saturates as $\beta \mu_{1/2} \mathcal{B}_0 \to \infty$, where

$$\tanh \left(\beta \mu_{\frac{1}{2}} \mathcal{B}_{0,z} \right) \to 1, \tag{11.55}$$

with a saturation value

$$\langle \boldsymbol{M} \rangle \approx N \mu_{\frac{1}{2}}. \tag{11.56}$$

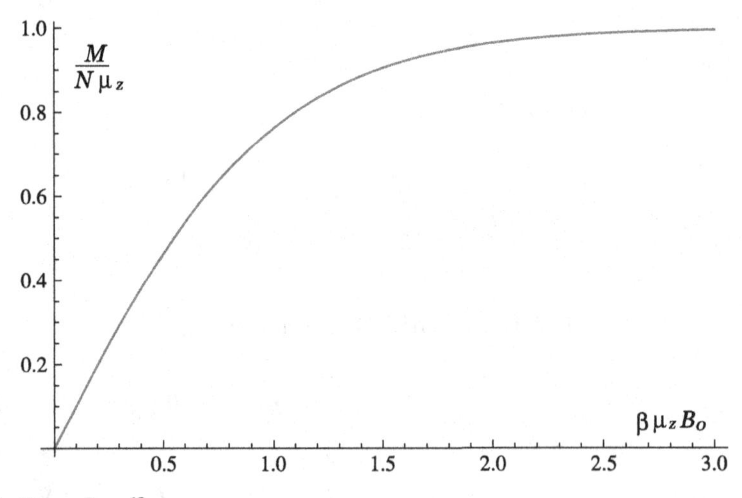

Fig. 11.2 Magnetization vs. $\beta \mu_{\frac{1}{2}} \mathcal{B}_0$.

The linear region where $\beta\mu_{\frac{1}{2}}\mathcal{B}_0 \ll 1$ is called the *Curie regime*. In that case

$$\tanh\left(\beta\mu_{\frac{1}{2}}\mathcal{B}_{0,z}\right) \approx \beta\mu_{\frac{1}{2}}\mathcal{B}_{0,z} \tag{11.57}$$

and

$$\langle \boldsymbol{M} \rangle \approx N\beta\mu_{\frac{1}{2}}^2\mathcal{B}_{0,z}. \tag{11.58}$$

The *average energy* (including the magnetization-independent energy) is

$$\langle \mathcal{H}_{op} \rangle = \frac{\sum_{n_+,n_-}^{N} \frac{N!}{n_+!n_-!}[NE_0 + (n_+ - n_-)\mu_{\frac{1}{2}}\mathcal{B}_{0,z}]e^{-\beta[NE_0 + (n_+ - n_-)\mu_{\frac{1}{2}}\mathcal{B}_{0,z}]}}{\mathcal{Z}_M} \tag{11.59}$$

so that *internal energy* is

$$\mathcal{U} = -\frac{\partial}{\partial\beta}\ln\mathcal{Z}_M + \mathcal{B}_{0,z}\langle \boldsymbol{M} \rangle, \tag{11.60}$$

which is simply

$$\mathcal{U} = NE_0. \tag{11.61}$$

Comparing Eq. 11.53 with Eq. 11.27 we see that the *uniform field Gibbs potential* is found from the uniform field partition function, Eq. G.23,

$$G_M = -\frac{1}{\beta}\ln\mathcal{Z}_M, \tag{11.62}$$

as discussed in Appendix G.

From Eq. 11.27 the *entropy* is

$$\mathcal{S} = -\left(\frac{\partial G_M}{\partial T}\right)_{p,\mathcal{B}_{0,z}} \tag{11.63}$$

$$= k_B\beta^2\left(\frac{\partial G_M}{\partial\beta}\right)_{p,\mathcal{B}_{0,z}} \tag{11.64}$$

$$= Nk_B\left\{\ln 2 + \ln\left[\cosh\left(\beta\mu_{\frac{1}{2}}\mathcal{B}_{0,z}\right)\right] - \beta\mu_{\frac{1}{2}}\mathcal{B}_{0,z}\tanh\left(\beta\mu_{\frac{1}{2}}\mathcal{B}_{0,z}\right)\right\}. \tag{11.65}$$

The entropy is represented in Figure 11.4 below. Note that as $\beta\mu_{\frac{1}{2}}\mathcal{B}_{0,z} \rightarrow 0$ the entropy attains its maximum value $\mathcal{S}_{max} = Nk_B\ln 2$, reflecting the original zero field two-fold degeneracy of the atom states.

The relevant *heat capacities* for paramagnets are those for which $\mathcal{B}$ or $\boldsymbol{M}$ are maintained constant. As can be derived from Eq. 11.27 the heat capacity at constant $\mathcal{B}$ is

$$\mathcal{C}_{\mathcal{B}} = \left(\frac{\partial H}{\partial T}\right)_{\mathcal{B}} \tag{11.66}$$

or in terms of entropy $\mathcal{S}$

$$\mathcal{C}_{\mathcal{B}} = T\left(\frac{\partial \mathcal{S}}{\partial T}\right)_{\mathcal{B}} . \tag{11.67}$$

Using Eqs. 11.65 and 11.67

$$\mathcal{C}_{\mathcal{B}} = Nk_B\left(\beta\mu_{\frac{1}{2}}\mathcal{B}_{0,z}\right)^2 \operatorname{sech}^2\left(\beta\mu_{\frac{1}{2}}\mathcal{B}_{0,z}\right), \tag{11.68}$$

a result which is shown in Figure 11.3. For $\beta\mu_{\frac{1}{2}}\mathcal{B}_0 \ll 1$ it shows quadratic behavior

$$\mathcal{C}_{\mathcal{B}} \approx Nk_B\left(\mu_{\frac{1}{2}}\beta\mathcal{B}_0\right)^2 . \tag{11.69}$$

The heat capacity at constant $\boldsymbol{M}$, as derived from Eq. 11.27, is

$$\mathcal{C}_{M} = \left(\frac{\partial \mathcal{U}}{\partial T}\right)_{M} \tag{11.70}$$

or

$$\mathcal{C}_{M} = T\left(\frac{\partial \mathcal{S}}{\partial T}\right)_{M} . \tag{11.71}$$

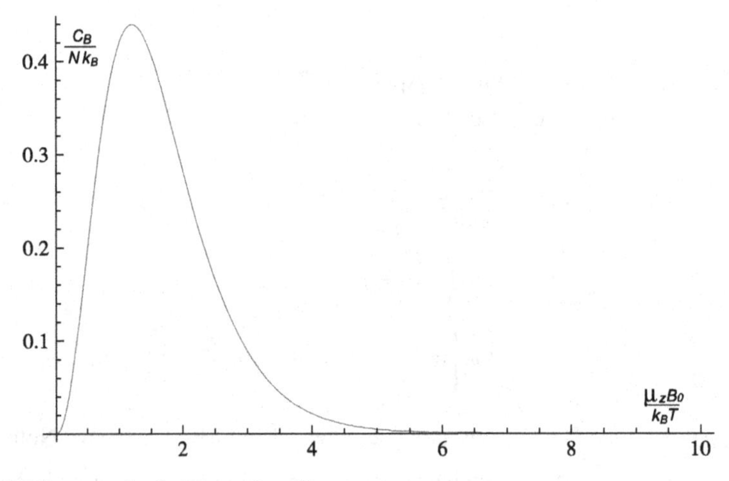

Fig. 11.3 Constant field heat capacity C_B/Nk_B vs. $\beta\mu_{\frac{1}{2}}\mathcal{B}_0$.

This time Eqs. 11.70 and 11.61 are used and obviously give[26]

$$\mathcal{C}_M = 0 \tag{11.72}$$

11.7 Magnetization fluctuations

Magnetization "fluctuations" $\Delta \boldsymbol{M}$ are defined by

$$\Delta \boldsymbol{M} \equiv \boldsymbol{M} - \langle \boldsymbol{M} \rangle \tag{11.73}$$

and "mean square magnetic fluctuations" (uncertainty) are[27,28]

$$\left\langle (\Delta \boldsymbol{M})^2 \right\rangle = \langle (\boldsymbol{M} - \langle \boldsymbol{M} \rangle)^2 \rangle \tag{11.74}$$

$$= \langle \boldsymbol{M}^2 \rangle - \langle \boldsymbol{M} \rangle^2, \tag{11.75}$$

where

$$\langle \boldsymbol{M}^2 \rangle = -\frac{\sum\limits_{n_-,n_+}^{N} \frac{N!}{n_+!n_-!} \left[\mu_{\frac{1}{2}} (n_+ - n_-) \right]^2 e^{-\beta \left[N E_0 + \mu_{\frac{1}{2}} (n_+ - n_-) \mathcal{B}_{0,z} \right]}}{\mathcal{Z}_M}. \tag{11.76}$$

Taking this result together with $\mathcal{Z}_M$ and $\langle \boldsymbol{M} \rangle$ (as calculated in Eq. 11.52), the mean square fluctuations are

$$\langle \boldsymbol{M}^2 \rangle - \langle \boldsymbol{M} \rangle^2 = \frac{1}{\beta^2} \frac{\partial^2}{\partial \mathcal{B}_0^2} \ln \mathcal{Z}_M \tag{11.77}$$

$$= N \mu_{\frac{1}{2}}^2 \operatorname{sech}^2 \left(\beta \mu_{\frac{1}{2}} \mathcal{B}_0 \right). \tag{11.78}$$

Expressed as dimensionless "root mean square fluctuations"

$$\frac{\sqrt{\langle \boldsymbol{M}^2 \rangle - \langle \boldsymbol{M} \rangle^2}}{\langle \boldsymbol{M} \rangle} = \frac{1}{\sqrt{N} \sinh \left(\beta \mu_{\frac{1}{2}} \mathcal{B}_{0.z} \right)}, \tag{11.79}$$

[26] Alternatively, the general relation $\mathcal{C}_M - \mathcal{C}_{\mathcal{B}} = T \left(\frac{\partial \mathcal{B}}{\partial T} \right)_M \left(\frac{\partial M}{\partial T} \right)_{\mathcal{B}}$, which can be simplified to $\mathcal{C}_M - \mathcal{C}_{\mathcal{B}} = -T \left[\left(\frac{\partial M}{\partial T} \right)_{\mathcal{B}} \right]^2 \left[\left(\frac{\partial M}{\partial \mathcal{B}} \right)_T \right]^{-1}$ with its more straightforward partial derivatives, confirms the zero result. Derivation of these results is assigned as a problem.

[27] Generally speaking, vanishingly small fluctuations assure meaningful thermodynamic descriptions.

[28] Fluctuations are formally associated with thermal state functions that have quantum operator representations. For example, temperature does not have well-defined fluctuations since there is no quantum temperature operator. See e.g. C. Kittel, "Temperature fluctuation: an oxymoron", *Physics Today* **41**(5), 93 (1988).

which decrease rapidly with increasing field $\mathcal{B}_0$, decreasing temperature T and with increasing N.

11.7.1 Example: Adiabatic (isentropic) demagnetization

A paramagnetic needle immersed in liquid He^4 initially at temperature $T_0 > T_\lambda$ is placed in a weak external field $\mathcal{B}_0$ directed along the needle's long axis.[29] The magnetic field is suddenly lowered to a value $\mathcal{B}_\ell$.

What is the change in temperature of the paramagnetic needle?

This sudden process corresponds to an adiabatic (isentropic) demagnetization – too fast for immediate heat exchange.

The solution to the problem is obvious from Eq. 11.65 where at any constant value of $\mathcal{S}$ the product $\beta\mu_{\frac{1}{2}}\mathcal{B}_0$ is also constant. Therefore for the adiabatic demagnetization process

$$\frac{\mathcal{B}_0}{T} = \text{constant}, \tag{11.80}$$

so that as $\mathcal{B}_0$ falls, T falls along with it. This is called the *magnetocaloric* effect.

Alternatively, we can begin with

$$dT = \left(\frac{\partial T}{\partial \mathcal{B}_0}\right)_{\mathcal{S}} d\mathcal{B}_0 + \left(\frac{\partial T}{\partial \mathcal{S}}\right)_{\mathcal{B}_0} d\mathcal{S}, \tag{11.81}$$

which for this isentropic process pares down to

$$dT = \left(\frac{\partial T}{\partial \mathcal{B}_0}\right)_{\mathcal{S}} d\mathcal{B}_0. \tag{11.82}$$

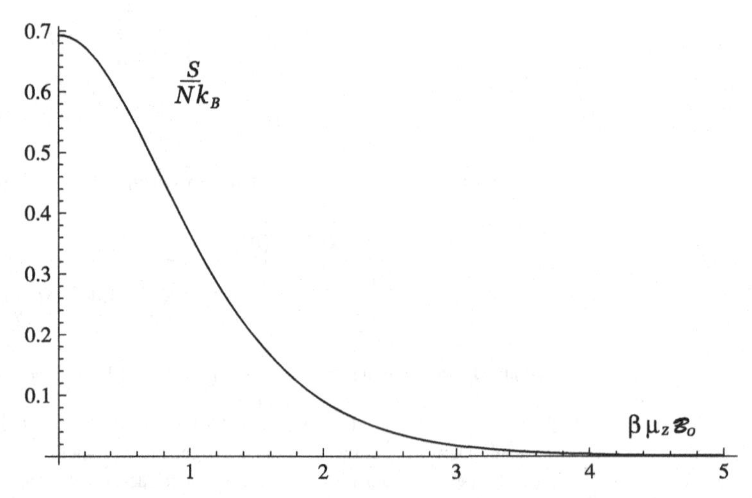

Fig. 11.4 Entropy vs. $\beta\mu_{\frac{1}{2}}\mathcal{B}_0$.

[29] In this configuration the demagnetization factor η is zero, which simplifies the situation.

Applying the cyclic chain rule (see Chapter 4)

$$\left(\frac{\partial T}{\partial \mathcal{B}_0}\right)_S = -\frac{\left(\frac{\partial \mathcal{S}}{\partial \mathcal{B}_0}\right)_T}{\left(\frac{\partial \mathcal{S}}{\partial T}\right)_{\mathcal{B}_0}} \tag{11.83}$$

$$= -\frac{T}{\mathcal{C}_{\mathcal{B}}}\left(\frac{\partial \mathcal{S}}{\partial \mathcal{B}_0}\right)_T. \tag{11.84}$$

Using the Gibbs potential expression as given in Eq. 11.27, and applying Euler's theorem, we have a Maxwell relation

$$\left(\frac{\partial \mathcal{S}}{\partial \mathcal{B}_0}\right)_T = \left(\frac{\partial \boldsymbol{M}}{\partial T}\right)_{\mathcal{B}_0} \tag{11.85}$$

so that Eq. 11.82 is now

$$\mathrm{d}T = -\frac{T}{\mathcal{C}_B}\left(\frac{\partial \boldsymbol{M}}{\partial T}\right)_{\mathcal{B}_0} \mathrm{d}\mathcal{B}_0. \tag{11.86}$$

Inserting $\mathcal{C}_{\mathcal{B}}$ from Eq. 11.69 and $\boldsymbol{M}$ from Eq. 11.58

$$\frac{\mathrm{d}T}{T} = \frac{\mathrm{d}\mathcal{B}_0}{\mathcal{B}_0} \tag{11.87}$$

which is integrated to finally give

$$T_f = \left(\frac{\mathcal{B}_\ell}{\mathcal{B}_0}\right) T_0, \tag{11.88}$$

i.e. the needle cools.

11.8 A model for ferromagnetism

In the previous sections, paramagnetism is modeled as N independent local magnetic moments. But in general these moments can interact to produce, approximately, an internal magnetic field acting in addition to the external field resulting in an average total effective field $\mathcal{B}^*$.

Short-range, nearest-neighbor magnetic moment coupling is frequently described by the Heisenberg exchange interaction,[30]

$$\mathcal{H}_{ex}(i,i') = -\frac{1}{2}\boldsymbol{m}_{op}(i)\,\mathcal{K}_{i,i'}\,\boldsymbol{m}_{op}(i'), \tag{11.89}$$

[30] W. Heisenberg, "Mehrkörperproblem und Resonanz in der Quantenmechanik", *Zeitschrift für Physik* **38**, 441 (1926).

where $\boldsymbol{m}_{op}(i)$ is the magnetic moment operator for the ith site and $\mathcal{K}_{i,i'}$ is an interaction which couples the moment at i with the moment at a nearest-neighbor site, i'. The macroscopic Heisenberg hamiltonian[31] is taken to be

$$\mathcal{H} = -\sum_{i=1}^{N} \mathcal{B}_0 \cdot \boldsymbol{M}_{op}(i) - \frac{1}{2}\sum_{i=1}^{N}\sum_{i'=1}^{z} \boldsymbol{M}_{op}(i) \cdot \mathcal{K}_{i,i'} \cdot \boldsymbol{M}_{op}(i'), \tag{11.90}$$

where z is the total number of nearest-neighbor moments. The double sum[32] includes only terms with $i \neq i'$.

Apart from in one or two dimensions, this many-body problem has, generally, no analytic solution. But an approximation – a *mean field approximation* (MFA)[33] – can be applied to replace the many-body model by an effective "one-body" model and plausibly account for the phenomenon of ferromagnetism.

11.9 A mean field approximation

In preparation for applying the MFA, rewrite Eq. 11.90 as

$$\mathcal{H} = -\sum_{i=1}^{N}\left\{\mathcal{B}_0 + \frac{1}{2}\sum_{i'=1}^{z}\mathcal{K}_{i,i'}\,\boldsymbol{M}_{op}(i')\right\} \cdot \boldsymbol{M}_{op}(i), \tag{11.91}$$

where, assuming an isotropic system, all z nearest neighbors can be treated as identical, i.e. $\mathcal{K}_{i,i'} \to \mathcal{K}$.

The essence of a mean field approximation is the identity

$$\begin{aligned}\boldsymbol{m}_{op}(i')\,\boldsymbol{m}_{op}(i) &= \left[\boldsymbol{m}_{op}(i') - \langle \boldsymbol{m}_{op}\rangle\right]\left[\boldsymbol{m}_{op}(i) - \langle \boldsymbol{m}_{op}\rangle\right] \\ &\quad + \boldsymbol{m}_{op}(i')\,\langle \boldsymbol{m}_{op}\rangle + \langle \boldsymbol{m}_{op}\rangle\,\boldsymbol{m}_{op}(i) - \langle \boldsymbol{m}_{op}\rangle\langle \boldsymbol{m}_{op}\rangle,\end{aligned} \tag{11.92}$$

where $\langle \boldsymbol{m}_{op}\rangle$ is the average magnetic moment, i.e. $\boldsymbol{M}/N$, the magnetization per site. The MFA neglects the first term, i.e. the product of fluctuations around the magnetization, whereas the last term contributes a constant value. The hamiltonian of Eq. 11.91 can then be written in its "mean field" form[34]

$$\mathcal{H} = \frac{1}{2}zN\mathcal{K}\langle \boldsymbol{m}_{op}\rangle^2 - \sum_{i=1}^{N}\left\{\mathcal{B}_0 + z\,\langle \boldsymbol{m}_{op}\rangle\,\mathcal{K}\right\} \cdot \boldsymbol{m}_{op}(i), \tag{11.93}$$

[31] Curiously, the Heisenberg "magnetic" interaction does not originate from magnetic arguments. Its source is strictly interatomic electronic interactions, in particular from the electron exchange interaction in the hydrogen molecule.

[32] The factor $\frac{1}{2}$ compensates for the ultimate double counting by the double sum.

[33] P Weiss, "L'hypothèse du champ moleculaire et la propriète ferrmognetique," *J. Phys. (Paris)* **6**, 661 (1907).

[34] The doubled mean field sum over i and i' (terms 2 and 3 in Eq. 11.92) cancels the factor $\frac{1}{2}$.

which includes the constant term from the MFA (see Eq. 11.92). The external field $\mathcal{B}_0$ is now supplemented by an "internal" field, $\mathcal{B}_{int}$

$$\mathcal{B}_{int} = z \langle \boldsymbol{m}_{op} \rangle \mathcal{K} \tag{11.94}$$

$$= \frac{z}{N} \boldsymbol{M} \mathcal{K} \tag{11.95}$$

so a total "effective" field $\mathcal{B}^*$ at the ith site is

$$\mathcal{B}^* (i) = \mathcal{B}_0 (i) + \mathcal{B}_{int} (i) \tag{11.96}$$

$$= \mathcal{B}_0 (i) + \frac{z}{N} \mathcal{K} \boldsymbol{M} (i) \tag{11.97}$$

giving rise to a mean field hamiltonian

$$\mathcal{H} = \frac{1}{2} \boldsymbol{M} \cdot \mathcal{B}_{int} - \sum_{i=1}^{N} \mathcal{B}^* (i) \cdot \boldsymbol{m}_{op} (i) . \tag{11.98}$$

This has the effect of replacing $\mathcal{B}_0$ in the paramagnet partition function of Eq. 11.51 by $\mathcal{B}^*$, in which case

$$\mathcal{Z}_{M^*} = e^{-\frac{1}{2}\beta \boldsymbol{M} \cdot \mathcal{B}_{int}} \left[2 \cosh \left(\beta \mu_{\frac{1}{2}} \mathcal{B}^* \right) \right]^N , \tag{11.99}$$

with a Gibbs potential[35]

$$\tilde{G}^* = -\frac{1}{\beta} \ln \mathcal{Z}_{M^*} \tag{11.100}$$

$$= \frac{1}{2} \boldsymbol{M} \cdot \mathcal{B}_{int} - \frac{N}{\beta} \ln \left(2 \cosh \beta \mu_{\frac{1}{2}} \mathcal{B}^* \right) . \tag{11.101}$$

In the absence of an external field, i.e. $\mathcal{B}_0 = 0$, we find from Eqs. 11.53, 11.97 and 11.99

$$\boldsymbol{M} = N \mu_{\frac{1}{2}} \tanh \left(\beta \mu_{\frac{1}{2}} \frac{z\mathcal{K}}{N} \boldsymbol{M} \right) , \tag{11.102}$$

which is a transcendental equation in $\boldsymbol{M}$ that describes the possibility of finite magnetization even in the absence of an external field.

[35] The constant term from the MFA has interesting thermodynamic consequences to be discussed in Appendix G (see Eq. G.30).

11.10 Spontaneous magnetization

Rewriting the $S = \frac{1}{2}$ result, Eq. 11.102, as a self-consistent expression in a dimensionless order parameter $\mathscr{M}$,

$$\mathscr{M} = \frac{M}{N\mu_{\frac{1}{2}}}, \tag{11.103}$$

Eq. 11.102 becomes

$$\mathscr{M} = \tanh\left(\frac{T_c}{T}\mathscr{M}\right), \tag{11.104}$$

where

$$T_c = \frac{\mu_{\frac{1}{2}}{}^2 z\mathcal{K}}{k_B}. \tag{11.105}$$

T_c is called the *Curie temperature* (see Table 11.1). Solving Eq. 11.104 graphically we see that for $T < T_c$, $\mathscr{M} > 0$ and magnetic moments spontaneously align (ferromagnetism). For $T > T_c$, $\mathscr{M} = 0$ and spontaneous alignment is destroyed, characterizing T_c as the transition temperature at which a phase transition from an ordered ($\mathscr{M} > 0$) to a disordered ($\mathscr{M} = 0$) state takes place.[36] (See Figure 11.5.)

Table 11.1 Sample values of Curie temperatures in ferromagnetic materials.

	Curie temperature, K
Fe	1043
Co	1388
Ni	627
Gd	293
Dy	85
$CrBr_3$	37
EuO	77
MnAs	318
MnBi	670
Fe_2B	1015
$GdCl_3$	2.2

[36] This is referred to as *symmetry breaking*.

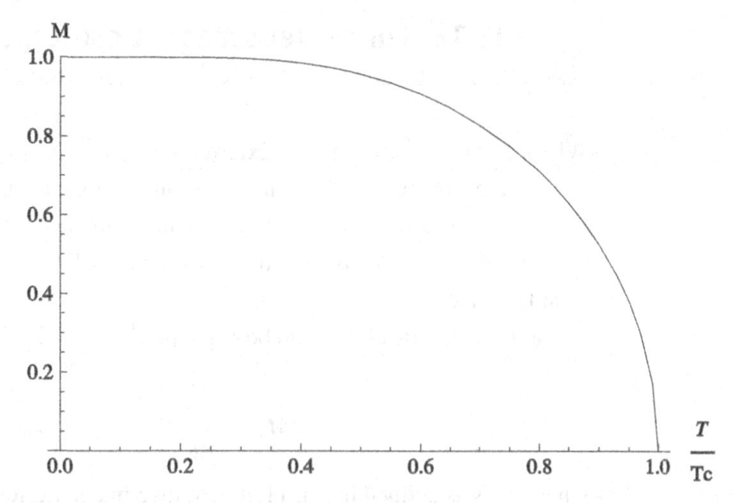

Fig. 11.5 A graphical solution of the self-consistent equation, Eq. 11.104. The sharp decline of the order parameter $\mathcal{M}$ as $T \to T_c^-$ (the Curie temperature) is followed by a slope discontinuity at $T = T_c$. This is the general characteristic of a magnetic phase transition. When $T > T_c$ the only solution to Eq. 11.104 is $\mathcal{M} = 0$.

11.11 Critical exponents

As T approaches T_c with $T < T_c$, the magnetic order parameter shows the power-law behavior

$$\mathcal{M} \approx \left(\frac{T_c}{T} - 1\right)^{\beta_c}, \tag{11.106}$$

where β_c is called a *critical exponent.* The value of β_c from the MFA is found by first inverting Eq. 11.104

$$\frac{T_c}{T}\mathcal{M} = \tanh^{-1}\mathcal{M} \tag{11.107}$$

and then expanding $\tanh^{-1}\mathcal{M}$ for small $\mathcal{M}$,

$$\frac{T_c}{T}\mathcal{M} = \mathcal{M} + \frac{1}{3}\mathcal{M}^3 + \ldots, \tag{11.108}$$

to give

$$\mathcal{M} \approx \sqrt{3}\left(\frac{T_c}{T} - 1\right)^{1/2}. \tag{11.109}$$

The $S = \frac{1}{2}$ mean field critical exponent is therefore

$$\beta_c = \frac{1}{2}. \tag{11.110}$$

11.12 Curie–Weiss magnetic susceptibility ($T > T_c$)

When $T > T_c$ and with no external field, i.e. $\mathcal{B}_0 = 0$, nearest-neighbor interactions are no longer sufficient to produce spontaneous magnetization. However, upon re-introduction of an external field $\mathcal{B}_0$ induced paramagnetic moments will still contribute internal fields, so that within an MFA a total internal field is again $\mathcal{B}^*$, as in Eq. 11.97.

For $T \gg T_c$ Eq. 11.102 can be expanded[37] and solved for $\langle \boldsymbol{M} \rangle$ to give

$$\langle \boldsymbol{M} \rangle = \frac{\mu_{\frac{1}{2}}^2}{k_B} (T - T_c)^{-1} \mathcal{B}_0, \tag{11.111}$$

where T_c is as defined in Eq. 11.105. With a magnetic susceptibility χ_M defined as[38]

$$\langle \boldsymbol{M} \rangle = \chi_M \mathcal{B}_0, \tag{11.112}$$

we have

$$\chi_M = \frac{\mu_{\frac{1}{2}}^2}{k_B} (T - T_c)^{-1}, \tag{11.113}$$

which is called the *Curie–Weiss law*. When $T \gg T_c$, this is a satisfactory description for magnetic susceptibility. But it fails near $T \approx T_c$, where the formula displays a singularity.[39]

The Curie–Weiss law is often expressed in terms of $\dfrac{1}{\chi_M}$, which is (advantageously) linear in $T - T_c$.

11.13 Closing comment

The study of magnetic matter remains a vast and varied topic that drives contemporary research, both fundamental and applied. The examples discussed in this chapter (paramagnetism and ferromagnetism) are but introductory samples of the role played by quantum mechanics in understanding macroscopic magnetism.

[37] $\tanh(x) \approx x - \frac{1}{3}x^3$.

[38] Unlike magnetization, magnetic susceptibility has no strict thermodynamic definition. In the case of non-linear materials an isothermal susceptibility $\chi_M = \left(\frac{\partial M}{\partial \mathcal{B}_0}\right)_T$ is a more practical definition.

[39] T_c experimentally determined from Curie–Weiss behavior is usually higher than T_c determined from the ferromagnetic phase transition.

12 Open systems

If to any homogeneous mass in a state of hydrostatic stress we suppose an infinitesimal quantity of any substance to be added, the mass remaining homogeneous and its entropy and volume remaining unchanged, the increase of the energy of the mass divided by the quantity of the substance added is the (chemical) potential for that substance in the mass considered.

J.W. Gibbs, "On the equilibrium of heterogeneous substances", *Trans. Conn. Acad.*, vol. III, 108–248 (1876)

12.1 Variable particle number

J. W. Gibbs ingeniously extended thermodynamics to include processes involving systems with variable particle number[1,2] – processes that include:

1. *Phase transitions*: Macroscopic matter transforms – abruptly – into different physically distinctive phases, usually accompanied by changes in symmetry and discontinuities in properties. Examples of phase changes are:

   (a) melting: $A_{solid} \rightleftarrows A_{liquid}$;
   (b) magnetization: paramagnet $\rightleftarrows$ ferromagnet;
   (c) evaporation: $A_{liquid} \rightleftarrows A_{vapour}$;
   (d) sublimation: $A_{solid} \rightleftarrows A_{vapour}$;
   (e) normal $\rightleftarrows$ superconductor;
   (f) normal He^4 $\rightleftarrows$ Bose–Einstein condensate.

2. Atoms, molecules or particles combine and recombine as different chemical units:

   (a) chemical reactions:

$$n_A A + n_B B \ldots \rightleftarrows n_x X + n_Y Y \ldots \tag{12.1}$$

[1] Willard Gibbs, "On the equilibrium of heterogeneous substances", *Trans. Conn. Acad.*, vol. III, 343–524 (1878).

[2] Willard Gibbs, "On the equilibrium of heterogeneous substances", *Amer. Jour. of Sci.* (3), vol. XVI, 441–458 (1878).

where $n_A, n_B, \ldots, n_X, n_Y, \ldots$ are integer numbers of the chemical participants, $A, B, \ldots, X, Y, \ldots$;

(b) fundamental particle interactions – electron–positron annihilation:

$$e^+ + e^- \to \gamma + \gamma. \tag{12.2}$$

12.2 Thermodynamics and particle number

The hamiltonians $\mathcal{H}$ and state functions $|\Psi\rangle$ in Schrödinger's quantum mechanics define microscopic dynamics for systems with a fixed number of particles (closed systems). These, so far, have been our quantum reference in understanding macroscopic systems. However, for many systems or processes this is physically or mathematically inadequate. By introducing into the hamiltonian a particle number operator $\mathcal{N}_{op}$ and its eigenvalue equation

$$\mathcal{N}_{op}|\bar{N}\rangle = N|\bar{N}\rangle \tag{12.3}$$

with eigenvalues $N = 0, 1, 2, \ldots$ and eigenfunctions $|\bar{N}\rangle$, together with a particle number fluctuation operator defined as

$$\Delta N_{op} = \mathcal{N}_{op} - \langle\mathcal{N}_{op}\rangle, \tag{12.4}$$

quantum mechanics (and thermodynamics) can logically and formally be extended to include variable particle number – the defining property of *open systems*. The number operator is not part of Schrödinger mechanics but belongs to quantum field theory, a quantum formulation far beyond any plan for this book. Yet a number operator and its eigenvalue property (see Eq. 12.3) are the basis of Gibbs' prescient "Grand Canonical" thermodynamics, a topic discussed and applied in this and several remaining chapters.

12.3 The open system

12.3.1 Formalities

The hamiltonian $\hat{\mathcal{H}}_{op}$, which constitutes the basis for open system thermodynamics, is

$$\hat{\mathcal{H}}_{op} = \mathbf{h}_{op} - \mu\mathcal{N}_{op}, \tag{12.5}$$

which now includes "particle" work $d\mathcal{W}_{\mathcal{N}} = -\mu\, d\langle\mathcal{N}_{op}\rangle$, where μ is the chemical potential for a single component species. In the case where there are M chemically distinct components,

$$\hat{\mathcal{H}}_{op} = \mathbf{h}_{op} - \sum_{i}^{M} \mu_i \left\langle \mathcal{N}_{op}^{i} \right\rangle, \tag{12.6}$$

where μ_i is a chemical potential for the ith chemical component.

While not significantly modifying the canonical formalism – in particular the commutation relation

$$\left[\hat{\rho}_{op}^{\tau}, \hat{\mathcal{H}}_{op}\right] = 0 \tag{12.7}$$

is still satisfied – introducing $\mu\mathcal{N}_{op}$ does have formal and thermodynamic consequences.

The single-species open system thermal Lagrangian which incorporates Eq. 12.5 now becomes

$$\begin{aligned}\mathcal{L} = -k_B &\sum_{N=0,1,2,\ldots} \sum_{s} \left\{\hat{\mathbf{P}}[\epsilon_s(N), N] \ln \hat{\mathbf{P}}[\epsilon_s(N), N]\right\} \\ -\frac{1}{T} &\sum_{N=0,1,2,\ldots} \left\{\sum_{s} \hat{\mathbf{P}}[\epsilon_s(N), N][\epsilon_s(N) - N\mu]\right\} \\ -\lambda_0 &\sum_{N=0,1,2\ldots} \left\{\sum_{s} \hat{\mathbf{P}}[\epsilon_s(N), N]\right\},\end{aligned} \tag{12.8}$$

where $\hat{\mathbf{P}}[\epsilon_s(N), N]$ is the normalized probability for:

- N particles;
- N-particle eigen-energies $\epsilon_s(N)$.

The "hat" worn by $\hat{\mathbf{P}}$ distinguishes it from closed system (canonical) probabilities.

Maximizing $\mathcal{L}$ with respect to variations in $\hat{\mathbf{P}}[\epsilon_s(N), N]$ gives

$$\hat{\mathbf{P}}[\epsilon_s(N), N] = \frac{e^{-\beta[\epsilon_s(N) - \mu N]}}{\sum_{N=0,1,2,\ldots} \left\{\sum_{s} e^{-\beta[\epsilon_s(N) - \mu N]}\right\}}, \tag{12.9}$$

where the normalizing denominator

$$\mathcal{Z}_{gr} = \sum_{N=0,1,2,\ldots} \left\{\sum_{s} e^{-\beta[\epsilon_s(N) - \mu N]}\right\} \tag{12.10}$$

is the *grand partition function*. As in the canonical case, the s-sum covers all N-particle eigenstates $|\epsilon_s(N)\rangle$ whereas the newly introduced N-sum covers all particle numbers $N = 0, 1, 2, \ldots$ (i.e. eigenvalues of $\mathcal{N}_{op}$).

Re-parsing Eq. 12.10 we can equivalently write

$$\mathcal{Z}_{gr} = \sum_{N=0,1,2,\ldots} e^{\beta\mu N}\left\{\sum_s e^{-\beta\,\epsilon_s(N)}\right\}, \tag{12.11}$$

which emphasizes the underlying form

$$\mathcal{Z}_{gr} = \sum_{N=0,1,2,\ldots} e^{\beta\mu N}\,\mathcal{Z}(N), \tag{12.12}$$

where

$$\mathcal{Z}\,(N) = \sum_s e^{-\beta\,\epsilon_s(N)} \tag{12.13}$$

is an N-particle partition function.

12.3.2 Grand thermodynamics

The quantum formulation having been adapted for variable particle number, resulting in probabilities $\hat{\mathbf{P}}[\epsilon_s(N), N]$, its consequences for thermodynamics can now be examined.

1. Average system particle number $\langle\mathcal{N}_{op}\rangle$,

$$\langle\mathcal{N}_{op}\rangle = \frac{\sum\limits_{N=0,1,2,\ldots} N\sum\limits_s e^{-\beta[\epsilon_s(N)-\mu N]}}{\sum\limits_{N=0,1,2,\ldots}\sum\limits_s e^{-\beta[\epsilon_s(N)-\mu N]}}, \tag{12.14}$$

is written in terms of Eq. 12.10, the grand partition function, as[3]

$$\langle\mathcal{N}_{op}\rangle = \frac{1}{\beta}\left(\frac{\partial}{\partial\mu}\ln\mathcal{Z}_{gr}\right)_{T,V}. \tag{12.15}$$

2. System internal energy $\mathcal{U}$,

$$\mathcal{U} = \frac{\sum\limits_{N=0,1,2,\ldots}\sum\limits_s \epsilon_s(N)e^{-\beta[\epsilon_s(N)-\mu N]}}{\sum\limits_{N=0,1,2,\ldots}\sum\limits_s e^{-\beta[\epsilon_s(N)-\mu N]}} \tag{12.16}$$

can be rearranged as

$$\mathcal{U} = -\left(\frac{\partial}{\partial\beta}\ln\mathcal{Z}_{gr}\right)_{\mu,V} + \mu\langle\mathcal{N}_{op}\rangle. \tag{12.17}$$

[3] $\langle\mathcal{N}_{op}\rangle$ is a continuous state variable.

3. Assuming a volume dependence $\epsilon_s(N, V)$, pressure p is

$$p = \frac{\sum\limits_{N=0,1,2,\ldots} \sum\limits_{s} \left[-\frac{\partial \epsilon_s(N, V)}{\partial V} \right] e^{-\beta[\epsilon_s(N) - \mu N]}}{\sum\limits_{N=0,1,2,\ldots} \sum\limits_{s} e^{-\beta[\epsilon_s(N) - \mu N]}} \tag{12.18}$$

which in terms of the grand partition function is

$$p = \frac{1}{\beta} \left(\frac{\partial}{\partial V} \ln \mathcal{Z}_{gr} \right)_{T,\mu}. \tag{12.19}$$

4. Entropy $\mathcal{S}$,

$$\mathcal{S} = -k_B \sum_{N=0,1,2,\ldots} \sum_{s} \left\{ \hat{\mathbf{P}}[\epsilon_s(N), N] \ln \hat{\mathbf{P}}[\epsilon_s(N), N] \right\}, \tag{12.20}$$

becomes, with Eqs. 12.9 and 12.10,

$$\mathcal{S} = -k_B \sum_{N=0,1,2,\ldots} \sum_{s} \left\{ \left[\frac{e^{-\beta(\epsilon_s - \mu N)}}{\mathcal{Z}_{gr}} \right] \left[-\ln \mathcal{Z}_{gr} - \beta(\epsilon_s - \mu N) \right] \right\} \tag{12.21}$$

$$= -k_B \, \beta \left(-\mathcal{U} - \frac{1}{\beta} \ln \mathcal{Z}_{gr} + \mu \langle \mathcal{N}_{op} \rangle \right). \tag{12.22}$$

Then with Eq. 12.17

$$\mathcal{S} = -k_B \beta^2 \left[\frac{\partial}{\partial \beta} \left(\frac{1}{\beta} \ln \mathcal{Z}_{gr} \right)_{\mu,V} \right]. \tag{12.23}$$

12.3.3 Grand potential

In addition to the grand partition function an open system *grand potential*, $\boldsymbol{\Omega}_{gr}$, is defined

$$\boldsymbol{\Omega}_{gr} = -\frac{1}{\beta} \ln \mathcal{Z}_{gr}. \tag{12.24}$$

Using $\boldsymbol{\Omega}_{gr}$ Eq. 12.22 can be rearranged as

$$\boldsymbol{\Omega}_{gr} = \mathcal{U} - T\mathcal{S} - \mu\langle\mathcal{N}_{op}\rangle. \tag{12.25}$$

With the Helmholtz potential $F = \mathcal{U} - T\mathcal{S}$ we also have

$$\boldsymbol{\Omega}_{gr} = F - \mu\langle\mathcal{N}_{op}\rangle \tag{12.26}$$

and in terms of the Gibbs potential $G = U - TS + pV$

$$\boldsymbol{\Omega}_{gr} = G - \mu\langle\mathcal{N}_{op}\rangle - p\,V. \tag{12.27}$$

Summarizing the grand potential's role in the thermodynamics of open systems, Eqs. 12.15, 12.17, 12.23 and 12.19 can be re-expressed as

$$\begin{aligned}
\langle\mathcal{N}_{op}\rangle &= -\left(\frac{\partial \boldsymbol{\Omega}_{gr}}{\partial \mu}\right)_{T,V}, \\
\mathcal{U} &= \left[\frac{\partial}{\partial \beta}(\beta\boldsymbol{\Omega}_{gr})\right]_{\mu,V} + \mu\langle\mathcal{N}_{op}\rangle, \\
p &= -\left(\frac{\partial \boldsymbol{\Omega}_{gr}}{\partial V}\right)_{T,\mu}, \\
\mathcal{S} &= k_B\,\beta^2\left(\frac{\partial \boldsymbol{\Omega}_{gr}}{\partial \beta}\right)_{\mu,V}.
\end{aligned} \tag{12.28}$$

Finally, differentiating $\boldsymbol{\Omega}_{gr} = \boldsymbol{\Omega}_{gr}(T,\ V,\ \mu)$,

$$\mathrm{d}\boldsymbol{\Omega}_{gr} = \left(\frac{\partial \boldsymbol{\Omega}_{gr}}{\partial T}\right)_{V,\mu} \mathrm{d}T + \left(\frac{\partial \boldsymbol{\Omega}_{gr}}{\partial V}\right)_{T,\mu} \mathrm{d}V + \left(\frac{\partial \boldsymbol{\Omega}_{gr}}{\partial \mu}\right)_{T,V} \mathrm{d}\mu \tag{12.29}$$

implies, from Eq. 12.28, the thermodynamic relation

$$\mathrm{d}\boldsymbol{\Omega}_{gr} = -\mathcal{S}\,\mathrm{d}T - p\ \mathrm{d}V - \langle\mathcal{N}_{op}\rangle\,\mathrm{d}\mu. \tag{12.30}$$

12.3.4 G, $\boldsymbol{\Omega}_{gr}$ and Euler's theorem

Combining Eqs. 12.27 and 12.30 the differential of the Gibbs potential is

$$dG = -\mathcal{S}\, dT + V\, dp + \mu\, d\langle \mathcal{N}_{op} \rangle \tag{12.31}$$

so that its natural variables are T, p and $\langle \mathcal{N}_{op} \rangle$. With $\langle \mathcal{N}_{op} \rangle$ the only extensive variable among them we write $G(T, \langle \mathcal{N}_{op} \rangle, p)$ in the Euler form,

$$G\left(T, \lambda\langle \mathcal{N}_{op} \rangle, p\right) = \lambda\, G\left(T, \langle \mathcal{N}_{op} \rangle, p\right), \tag{12.32}$$

indicating that G is homogeneous in $\langle \mathcal{N}_{op} \rangle$ of degree 1. Applying Euler's homogeneity theorem (see Chapter 4) gives

$$\langle \mathcal{N}_{op} \rangle \left(\frac{\partial G}{\partial \langle \mathcal{N}_{op} \rangle}\right)_{p,T} = G\left(T, \langle \mathcal{N}_{op} \rangle, p\right). \tag{12.33}$$

But from Eq. 12.31

$$\left(\frac{\partial G}{\partial \langle \mathcal{N}_{op} \rangle}\right)_{p,T} = \mu \tag{12.34}$$

which leads to

$$G\left(T, \langle \mathcal{N}_{op} \rangle, p\right) = \mu \langle \mathcal{N}_{op} \rangle. \tag{12.35}$$

For a multispecies system, say $A_1, A_2, A_3, \ldots$, this is generalized as

$$G = \sum_{i=1} \mu_{A_i} \langle \mathcal{N}_{op} \rangle_{A_i}. \tag{12.36}$$

According to Eq. 12.30 the natural variables of $\boldsymbol{\Omega}_{gr}$ are T, V and μ, of which only V is extensive. Writing $\boldsymbol{\Omega}_{gr} = \boldsymbol{\Omega}_{gr}(T, V, \mu)$ (see Eq. 12.30) in its Euler form

$$\boldsymbol{\Omega}_{gr}(T, \lambda V, \mu) = \lambda\, \boldsymbol{\Omega}_{gr}(T, V, \mu) \tag{12.37}$$

(denoting that $\boldsymbol{\Omega}_{gr}$ is homogeneous in V of degree 1). Applying Euler's homogeneous function theorem

$$V\left(\frac{\partial \boldsymbol{\Omega}_{gr}}{\partial V}\right)_{T,\mu} = \boldsymbol{\Omega}_{gr}, \tag{12.38}$$

which, upon substituting from Eq. 12.28, becomes

$$\boldsymbol{\Omega}_{gr} = -pV. \tag{12.39}$$

This – as we will see in the next section – is a useful result for deriving equations of state.

12.4 A "grand" example: the ideal gas

The grand partition function is applied to the ideal gas. Examples are then given that build on the ideal gas to model more complex physical systems.

Using results from Chapter 7 together with Eq. 12.12,

$$\mathcal{Z}_{gr} = \sum_{N=0,1,2,\ldots} e^{\beta\mu N} \mathcal{Z}(N) \tag{12.40}$$

$$= \sum_{N=0,1,2,\ldots} e^{\beta\mu N} \frac{1}{N!} (n_Q V)^N \tag{12.41}$$

$$= \sum_{N=0,1,2,\ldots} \zeta^N \frac{1}{N!} (n_Q V)^N , \tag{12.42}$$

where

$$\zeta = e^{\beta\mu} \tag{12.43}$$

is called the *fugacity* and where, as defined in Chapter 7,

$$n_Q = \left(\frac{m}{2\pi\hbar^2\beta} \right)^{3/2}. \tag{12.44}$$

Summing Eq. 12.42 gives the ideal gas grand partition function[4]

$$\mathcal{Z}_{gr} = e^{\zeta n_Q V} \tag{12.45}$$

and the ideal gas grand potential

$$\boldsymbol{\Omega}_{gr} = -\frac{\zeta n_Q V}{\beta}. \tag{12.46}$$

[4] The ideal gas, even in this "grand" picture, is still semi-classical requiring the ad-hoc corrections of Chapter 7.

12.4.1 Ideal gas thermodynamic properties

1. From Eqs. 12.28 and 12.46 the average particle number is

$$\langle \mathcal{N}_{op} \rangle = \zeta n_Q V, \tag{12.47}$$

which is solved to obtain the ideal gas chemical potential[5]

$$\mu = \frac{1}{\beta} \ln \frac{\langle \mathcal{N}_{op} \rangle}{n_Q V}. \tag{12.48}$$

2. Similarly, from Eq. 12.28 the internal energy $\mathcal{U}$ is

$$\mathcal{U} = \frac{3\langle \mathcal{N}_{op} \rangle}{2\beta}. \tag{12.49}$$

3. Again using Eq. 12.28 the entropy is

$$\mathcal{S} = \langle \mathcal{N}_{op} \rangle \left[\frac{5}{2} k_B - \ln \frac{\langle \mathcal{N}_{op} \rangle}{n_Q V} \right] \tag{12.50}$$

which is the Sackur–Tetrode equation of Chapter 7.

4. Using Eqs. 12.35 and 12.48 the Gibbs potential for the ideal gas is

$$G\left(T, \langle \mathcal{N}_{op} \rangle, p\right) = \frac{\langle \mathcal{N}_{op} \rangle}{\beta} \ln \frac{\langle \mathcal{N}_{op} \rangle}{n_Q V}, \tag{12.51}$$

which is identical with

$$G\left(T, \langle \mathcal{N}_{op} \rangle, p\right) = \mu \langle \mathcal{N}_{op} \rangle. \tag{12.52}$$

5. Finally, using Eq. 12.39 and the ideal gas value for $\boldsymbol{\Omega}_{gr}$, i.e. Eq. 12.46,

$$\frac{\zeta n_Q V}{\beta} = pV, \tag{12.53}$$

[5] Since n_Q (the quantum concentration) depends on $\hbar$, even the so-called "classical" ideal gas has a quantum mechanical marker, suggesting "classical thermodynamics" is an oxymoron.

which, with $\zeta = e^{\beta\mu}$ and μ from Eq. 12.48, gives

$$pV = \langle \mathcal{N}_{op} \rangle k_B T, \tag{12.54}$$

the ideal gas law.

12.5 Van der Waals' equation

An ideal gas (classical or quantum) is defined by neglecting interactions between particles. More-realistic models (theoretical or computational simulations) are generally based on pairwise forces with a long-range attractive and short-range repulsive component. In 1873 van der Waals pursued these ideas in his Ph.D. thesis, replacing the ideal gas law with a far-reaching equation of state that bears his name.[6] For this he was awarded the 1910 Nobel Prize in Physics.

In contrast to an ideal gas, the van der Waals equation postulates an inter-molecular interaction energy per particle $U(\mathbf{r}' - \mathbf{r})$ that displays "hard-core" repulsion as well as a long-range attraction, as pictured in Figure 12.1. The hard core ensures that a gas of atoms or molecules can occupy only a finite volume.

Assume an interaction energy

$$\mathcal{V}_{int} = -\frac{1}{2} \int d\mathbf{r} \int d\mathbf{r}' [\rho(\mathbf{r}) U(\mathbf{r}' - \mathbf{r}) \rho(\mathbf{r}')], \tag{12.55}$$

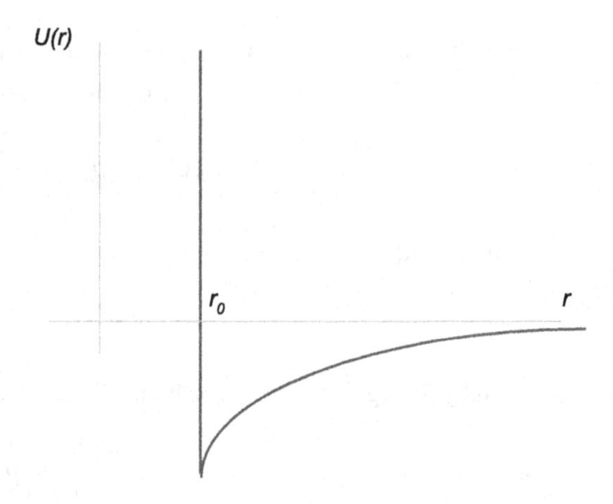

Fig. 12.1 Intermolecular potential.

[6] The microscopic reality of atoms was still being challenged by influential skeptics.

with the density in units of particle number per unit volume. This many-particle intermolecular interaction is manageable in the MFA,[7,8] in which case

$$\mathscr{V}_{int} = -\frac{1}{2}\int d\mathbf{r}\int d\mathbf{r}'\{2\rho(\mathbf{r}') - \langle\tilde{\rho}\rangle\}\langle\tilde{\rho}\rangle U(\mathbf{r}-\mathbf{r}') \tag{12.57}$$

with $\langle\tilde{\rho}\rangle$ a uniform particle density,

$$\langle\tilde{\rho}\rangle = \frac{N}{V}. \tag{12.58}$$

The first term in Eq. 12.57 evaluates to

$$-\langle\tilde{\rho}\rangle\int d\mathbf{r}\int d\mathbf{r}'\rho(\mathbf{r}')U(\mathbf{r}-\mathbf{r}') = -V\langle\tilde{\rho}\rangle\int d\mathbf{r}U(\mathbf{r})\rho(\mathbf{r}) \tag{12.59}$$

which, with $\rho(\mathbf{r}) = \langle\mathcal{N}_{op}(\mathbf{r})\rangle/V$ and assuming gas uniformity, becomes (noting Figure 12.1)

$$= -4\pi\frac{N}{V}\langle\mathcal{N}_{op}\rangle\int_{r_0}^{\infty} U(r)r^2\,dr. \tag{12.60}$$

Representing the integral as

$$a = 4\pi\int_{r_0}^{\infty} U(r)r^2\,dr, \tag{12.61}$$

and noting that the second term in Eq. 12.57 evaluates to an ignorable constant, the mean field interaction potential is[9]

$$\mathscr{V}_{int} = -Na\frac{\langle\mathcal{N}_{op}\rangle}{V}, \tag{12.62}$$

and the grand partition function is

$$\mathcal{Z}_{gr} = \sum_{N=0,1,2\ldots} e^{\beta\left[\mu + a\frac{\langle\mathcal{N}_{op}\rangle}{V}\right]N}\mathcal{Z}(N) \tag{12.63}$$

$$= \exp\left\{n_Q V e^{\beta\left[\mu + a\frac{\langle\mathcal{N}_{op}\rangle}{V}\right]}\right\}. \tag{12.64}$$

[7] The essence of a mean field approximation is the identity

$$\rho(\mathbf{x}')\rho(\mathbf{x}) = [\rho(\mathbf{x}') - \langle\rho\rangle][\rho(\mathbf{x}) - \langle\rho\rangle] + \rho(\mathbf{x}')\langle\rho\rangle + \langle\rho\rangle\rho(\mathbf{x}) - \langle\rho\rangle\langle\rho\rangle \tag{12.56}$$

where $\langle\rho\rangle$ is the average density. The MFA neglects the first term, i.e. fluctuations around the average density.

[8] Peter Palffy-Muhoray, "The single particle potential in mean field theory", *Am. J. Phys.* **70**, 433–437 (2002).

[9] This term is the inter-particle contribution to the internal energy.

From this we find

$$\langle \mathcal{N}_{op} \rangle = \frac{1}{\beta} \left(\frac{\partial}{\partial \mu} \ln \mathcal{Z}_{gr} \right)_{T,V} \tag{12.65}$$

$$= \frac{1}{\beta} \left(\frac{\partial}{\partial \mu} \right) n_Q V e^{\beta \left[\mu + a \frac{\langle \mathcal{N}_{op} \rangle}{V} \right]} \tag{12.66}$$

$$= n_Q V e^{\beta \left[\mu + a \frac{\langle \mathcal{N}_{op} \rangle}{V} \right]}, \tag{12.67}$$

with the van der Waals chemical potential

$$\mu = -a \frac{\langle \mathcal{N}_{op} \rangle}{V} + \frac{1}{\beta} \ln \frac{\langle \mathcal{N}_{op} \rangle}{n_Q V}. \tag{12.68}$$

Finally, the pressure is

$$p = \frac{1}{\beta} \left(\frac{\partial}{\partial V} \ln Z_{gr} \right)_{T,\mu} \tag{12.69}$$

$$= \frac{\langle \mathcal{N}_{op} \rangle}{\beta V} - a \left(\frac{\langle \mathcal{N}_{op} \rangle}{V} \right)^2, \tag{12.70}$$

with an equation of state, thus far,

$$p + a \left(\frac{\langle \mathcal{N}_{op} \rangle}{V} \right)^2 = \frac{\langle \mathcal{N}_{op} \rangle}{\beta V}. \tag{12.71}$$

The final step in the van der Waals argument assigns a minimum gas volume $V_{min} = b\langle \mathcal{N}_{op} \rangle$ as the fully packed space occupied by the gas molecules, where $b = \frac{\pi}{6} r_0^3$ is the hard-core restricted volume per molecule. Thus, the van der Waals equation of state becomes

$$\left[p + a \left(\frac{\langle \mathcal{N}_{op} \rangle}{V} \right)^2 \right] [V - \langle \mathcal{N}_{op} \rangle b] = \frac{\langle \mathcal{N}_{op} \rangle}{\beta}. \tag{12.72}$$

$[V - \langle \mathcal{N}_{op} \rangle b]$ is usually labeled V^*, the "effective volume".

12.6 A star is born

Star formation takes place inside cold ($T \sim 10$ K), dense interstellar regions of molecular hydrogen (H_2), carbon monoxide (CO) and dust, called *giant molecular clouds* (GMCs). The process of star formation is one of gravitational collapse, triggered by mutual gravitational attraction within the GMC, to form a region called a *protostar*. GMCs have typical masses of 6×10^6 solar masses, densities of 100 particles per cm^3

and diameters of 9.5×10^{14} km. The process of collapse resembles a phase transition in that its onset occurs at critical GMC values of temperature (T_c), mass (m_c) and pressure (p_c). Once begun, the collapse continues until the GMC is so hot and compact that nuclear fusion begins and a star is born ($\approx$10–15 $\times$ 10^6 years).

12.6.1 Thermodynamic model

A thermodynamic model for a GMC is an ideal gas with mutual gravitational attraction between the gas particles. Were it not for gravitational effects the GMC equation of state would simply be $pV = Nk_BT$. But an attractive gravitational potential (proportional to $1/|\mathbf{r}|$) contributes the additional interaction energy

$$\mathscr{V}_{grav} = -\frac{1}{2}G\int d\mathbf{r}\int d\mathbf{r}'\left[\rho(\mathbf{r})\frac{1}{|\mathbf{r}-\mathbf{r}'|}\rho(\mathbf{r}')\right] \tag{12.73}$$

with $\rho(\mathbf{r})$ the mass density of the cloud and G the universal gravitational constant. The many-particle gravitational interaction is manageable for this application in a mean field approximation (MFA), in which each particle sees the average potential of all the other particles as an effective interaction, i.e.

$$\mathscr{V}_{eff} = -\frac{1}{2}\int d\mathbf{r}\langle\tilde{\rho}\rangle\Phi(\mathbf{r}). \tag{12.74}$$

Applying an MFA (see footnote Eq. 12.56) the potential $\Phi(\mathbf{r})$ is

$$\Phi(\mathbf{r}) = G\int d\mathbf{r}'\frac{\{2\rho(\mathbf{r}') - \langle\tilde{\rho}\rangle\}}{|\mathbf{r}-\mathbf{r}'|}, \tag{12.75}$$

with $\langle\tilde{\rho}\rangle$ a uniform particle mass density

$$\langle\tilde{\rho}\rangle = \frac{Nm_H}{V} \tag{12.76}$$

$$= \frac{3Nm_H}{4\pi R^3}, \tag{12.77}$$

and where

$$\rho(\mathbf{r}) = m_H\langle\mathcal{N}_{op}(\mathbf{r})\rangle/V \tag{12.78}$$

is the particle density function. Here N is the number of molecules in the GMC, with m_H the mass of a hydrogen molecule (assumed to be the dominant species in the GMC) and V the cloud's volume.

Poisson's equation under the MFA (see Eq. 12.75) is

$$\nabla^2\Phi(\mathbf{r}) = 4\pi G\{2\rho(\mathbf{r}) - \langle\tilde{\rho}\rangle\}, \tag{12.79}$$

which suggests Gauss' gravitational flux law for integrating Eq. 12.75. In this "bare bones" model all GMC radial mass dependence is ignored, the cloud being assumed spherical with uniform density.[10,11] Under this uniformity assumption $\rho(\mathbf{r}) \to \langle\tilde{\rho}\rangle$ and using Gauss' flux law

$$\Phi(r) = \frac{2\pi G}{3}\langle\tilde{\rho}\rangle(r^2 - 3R^2), \qquad 0 < r \preceq R, \tag{12.80}$$

so that

$$\mathcal{V}_{eff} = \frac{1}{2}\langle\tilde{\rho}\rangle \int_0^R \mathrm{d}r\, 4\pi r^2 \Phi(r) \tag{12.81}$$

$$= -\frac{3GM^2}{5R} \tag{12.82}$$

$$= -N\, m_H \frac{3GM}{5R}. \tag{12.83}$$

As in Eq. 12.40, a grand partition function may now be written

$$\mathcal{Z}_{gr} = \sum_{N=0,1,2,\ldots} e^{\beta\left[\mu + \frac{3m_H GM}{5R}\right]N} \mathcal{Z}(N), \tag{12.84}$$

which from Eq. 12.63 is[12]

$$\mathcal{Z}_{gr} = \exp\left\{ n_Q V\, e^{\beta\left[\mu + \frac{3m_H GM}{5R}\right]} \right\}. \tag{12.85}$$

Using this we find:

1. the mean number of hydrogen molecules in the GMC:

$$\langle\mathcal{N}_{op}\rangle = \frac{1}{\beta}\left(\frac{\partial}{\partial\mu}\ln \mathcal{Z}_{gr}\right)_{T,V} \tag{12.86}$$

$$= \frac{1}{\beta}\left(\frac{\partial}{\partial\mu}\right) n_Q V\, e^{\beta\left[\mu + m_H\left(\frac{3GM}{5R}\right)\right]} \tag{12.87}$$

$$= n_Q V\, e^{\beta\left[\mu + m_H\left(\frac{3GM}{5R}\right)\right]}; \tag{12.88}$$

2. the chemical potential:

$$\mu = -m_H\left(\frac{3GM}{5R}\right) + \frac{1}{\beta}\ln\frac{\langle\mathcal{N}_{op}\rangle}{n_Q V}; \tag{12.89}$$

[10] This simplification has obvious shortcomings. Nevertheless, main features of gravitational collapse are preserved.

[11] W.B. Bonnor, "Boyle's law and gravitational instability", *Mon. Notices Roy. Astron. Soc.* **116**, 351 (1956).

[12] The factor 3/5 is due to assumed spherical symmetry of the GMC.

3. the pressure in the GMC:

$$p = \frac{1}{\beta}\left(\frac{\partial}{\partial V}\ln Z_{gr}\right)_{T,\mu}, \tag{12.90}$$

which, with $R = \left(\frac{3V}{4\pi}\right)^{1/3}$ and Eq. 12.89, gives the GMC equation of state

$$p = \frac{\langle \mathcal{N}_{op}\rangle k_B T}{V} - \frac{\langle \mathcal{N}_{op}\rangle m_H}{5}\left(\frac{4\pi}{3}\right)^{1/3} GMV^{-4/3}. \tag{12.91}$$

Restoring $M = \langle \mathcal{N}_{op}\rangle\, m_H$ the equation of state becomes

$$p = \frac{\langle \mathcal{N}_{op}\rangle k_B T}{V} - \frac{\langle \mathcal{N}_{op}\rangle^2 m_H^2}{5}\left(\frac{4\pi}{3}\right)^{1/3} GV^{-4/3}. \tag{12.92}$$

12.6.2 Collapse criterion

A criterion for collapse is that the cloud's adiabatic compressibility

$$\kappa_S = -\frac{1}{V}\left(\frac{\partial V}{\partial p}\right)_S \tag{12.93}$$

$$= \frac{1}{V\gamma}\left(\frac{\partial V}{\partial p}\right)_T, \tag{12.94}$$

becomes infinite.[13] At that critical point the cloud becomes gravitationally unstable, spontaneously shrinking until some new pressure source (nuclear fusion) overcomes gravity.[14] As a first step towards visualizing Eq. 12.92, set $\gamma = \frac{1}{5}\left(\frac{4\pi}{3}\right)^{1/3}$. Then form the dimensionless variables

$$\hat{p} = (G^3 M^2 \beta^4 m_H^4)p \tag{12.98}$$

[13] As previously defined, $\gamma = \mathcal{C}_P/\mathcal{C}_V$.

[14] This is equivalent to the Jeans criterion, [J. H. Jeans, "The stability of a spherical nebula", *Phil. Trans. Royal Soc. (London). Series A* **199**, 1 (1902)] that the velocity of sound c_s in the GMC becomes *imaginary*! Since the velocity of sound in a gas is

$$c_s^2 = \frac{\mathrm{K}_S}{\rho}, \tag{12.95}$$

where K_S is the gas' adiabatic bulk modulus,

$$\mathrm{K}_S = -V\left(\frac{\partial p}{\partial V}\right)_S \tag{12.96}$$

$$= \gamma V\left(\frac{\partial p}{\partial V}\right)_T, \tag{12.97}$$

and ρ is its density, the Jeans criterion is that the adiabatic bulk modulus (the reciprocal of compressibility) becomes zero.

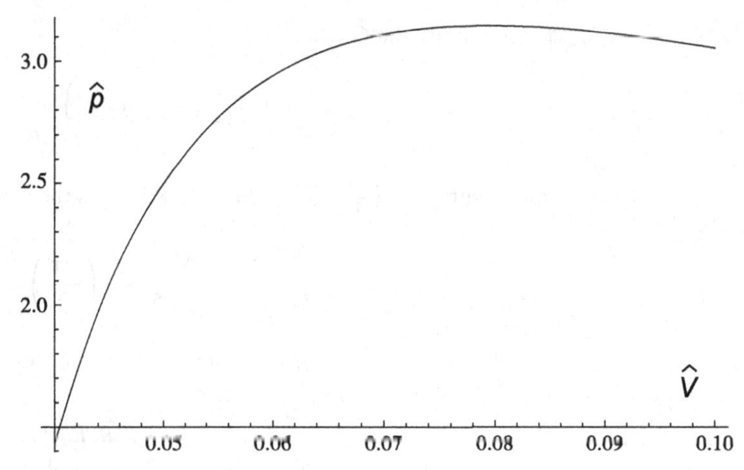

Fig. 12.2 Reduced equation of state for GMC. $\hat{V}_c = 0.0794$. $\hat{p}_c = 3.147$.

and

$$\hat{V} = (GM\beta m_H)^{-3} V \tag{12.99}$$

to give a reduced GMC equation of state

$$\hat{p} = \frac{1}{\hat{V}} - \frac{\gamma}{\hat{V}^{4/3}}. \tag{12.100}$$

This is plotted in Figure 12.2. The slope of the curve is zero at $\hat{V}_c = 0.0794$ and $\hat{p}_c = 3.147$, the reduced coordinates of instability.

Alternatively, from Eqs. 12.92 and 12.94 the adiabatic compressibility is

$$k_S = \frac{1}{\gamma V}\left[\langle \mathcal{N}_{op}\rangle k_B T\, V^{-2} - \frac{4}{15}\left(\frac{4\pi}{3}\right)^{1/3} \langle \mathcal{N}_{op}\rangle^2 m_H^2 G V^{-7/3}\right]^{-1} \tag{12.101}$$

so that with $\rho = \langle \mathcal{N}_{op}\rangle / V$ the critical radius for collapse, r_c, is

$$r_c = \left(\frac{45}{16\pi}\right)^{1/2} \sqrt{\frac{k_B T}{m_H^2 G \rho}} \tag{12.102}$$

and the critical pressure is

$$p_c = \frac{\rho k_B T}{4}. \tag{12.103}$$

Star formation is triggered by GMC compression to $p \approx p_c$ by collisions with other GMCs or compression by a nearby supernova event. Alternatively, galactic collisions can trigger bursts of star formation as the gas clouds in each galaxy are compressed.

12.7 Closing comment

Beginning with a thermal Lagrangian, Eq. 12.8, the grand canonical picture unfolds, introducing the indispensable chemical potential and bringing a workable methodology to open systems – physical situations in which particle number is variable and particle interactions can be included. The next chapter discusses the chemical potential's role in diffusion within inhomogeneous systems and extends μ's application to chemical reactants and distinct physical phases.

Problems and exercises

12.1 Find the internal energy of a van der Waals gas.

12.2 Find $C_p - C_V$ for the van der Waals gas.

13 The amazing chemical potential

Gibbs' "Equilibrium of Heterogeneous Substances" has practically unlimited scope. It spells out the fundamental thermodynamic theory of gases, mixtures, surfaces, solids, phase changes, chemical reactions, electrochemical cells, sedimentation and osmosis. It has been called, without exaggeration, the Principia of Thermodynamics.

William H. Cropper, *Great Physicists*, Oxford University Press US (2004)

13.1 Introduction

The *chemical potential* μ, a creation of J. W. Gibbs,[1] is the essential state variable for studying the thermodynamics of open systems, in particular chemical reactions, phase transitions, non-uniform systems, surfaces and other cases which can be characterized by varying particle number.[2,3,4,5] Although sometimes regarded as vague,[6] when Schrödinger's fixed particle number theory is extended by including $\mathcal{H}_{open} = -\mu\mathcal{N}_{op}$, μ provides both utility and clarity. Thermal Lagrangians now provide unambiguous rules governing a variety of diffusive processes.

In order to improve understanding of the thermodynamics of reactions, phase transformations and heterogeneous diffusion, this chapter includes supplementary examples and applications of the chemical potential.

[1] J. W. Gibbs, "A method of geometrical representation of the thermodynamic properties of substances by means of surfaces", *Transactions of the Connecticut Academy* (1873).

[2] G. Cook and R. H. Dickerson, "Understanding the chemical potential", *Am. J. Phys.* **63**, 73742 (1995).

[3] T. A. Kaplan, "The chemical potential", *J. of Stat. Physics.* **122**, 1237–1260 (2006).

[4] R. Baierlein, "The elusive chemical potential", *Am. J. Phys.* **69**, 423 (2001).

[5] G. Job and F. Herrmann, "Chemical potential – a quantity in search of recognition", *Eur. J. Phys.* **27**, 353–371 (2006).

[6] In the preface to Introduction to Solid State Physics (Wiley, New York, 1971), C. Kittel writes: "A vague discomfort at the thought of the chemical potential is still characteristic of a physics education. This intellectual gap is probably due to the obscurity of the writings of J. Willard Gibbs who discovered and understood the matter 100 years ago."

13.2 Diffusive equilibrium

Consider a single system (equivalently referred to as a *phase*) at temperature T and pressure p composed of, say, two distinct species A and B. Allow the species to initially be diffusively isolated and chemically non-reactive. This species composite can be described in terms of an open system macroscopic hamiltonian

$$\hat{\mathcal{H}}_{composite} = \mathbf{h}_{op} - \left[\mu^A \mathcal{N}_{op}^A + \mu^B \mathcal{N}_{op}^B\right] \tag{13.1}$$

with a corresponding thermal Lagrangian

$$\mathcal{L} = -k_B \hat{\mathbf{P}} \ln \hat{\mathbf{P}} - \lambda_0 \hat{\mathbf{P}} - \frac{1}{T}\langle \mathbf{h}_{op} \rangle + \frac{1}{T}\left[\mu^A \langle \mathcal{N}_{op}^A \rangle + \mu^B \langle \mathcal{N}_{op}^B \rangle\right], \tag{13.2}$$

where for total species isolation $\mu^A \neq \mu^B$ and where abbreviations

$$\hat{\mathbf{P}} \ln \hat{\mathbf{P}} = \sum_{N=0,1,2,\ldots} \left\{\sum_s \hat{\mathbf{P}}\left[\epsilon_s(N), N\right] \ln \hat{\mathbf{P}}\left[\epsilon_s(N), N\right]\right\} \tag{13.3}$$

$$\hat{\mathbf{P}} = \sum_{N=0,1,2,\ldots} \left\{\sum_s \hat{\mathbf{P}}\left[\epsilon_s(N), N\right]\right\} \tag{13.4}$$

imply sums over all species, as does $\langle h_{op} \rangle$.

If the two species A and B are in complete diffusive contact, at equilibrium the individual $\langle \mathcal{N}_{op} \rangle$ and $\langle \mathcal{N}_{op}^B \rangle$ are no longer distinguishable being replaced by the cumulative $\langle \mathcal{N}_{op} \rangle = \langle \mathcal{N}_{op}^A \rangle + \langle \mathcal{N}_{op}^B \rangle$ and, therefore, by the corresponding thermal Lagrangian

$$\mathcal{L} = -k_B \hat{\mathbf{P}} \ln \hat{\mathbf{P}} - \lambda_0 \hat{\mathbf{P}} - \frac{1}{T}\langle \mathbf{h}_{op} \rangle + \frac{\mu}{T}\left[\langle \mathcal{N}_{op}^A \rangle + \langle \mathcal{N}_{op}^B \rangle\right], \tag{13.5}$$

i.e. if species A and B freely diffuse, A and B must have the same chemical potential, $\mu^A = \mu^B = \mu$.

13.2.1 Examples

In the following examples of diffusive equilibrium the ideal gas provides an easily understood model for illustrating the versatile role of the chemical potential.

Ideal gas in the Earth's gravitational field

At a distance z above – but close to – the Earth's surface, ideal gas molecules experience a quasi-classical gravitational potential energy per particle $PE = mgz$. Here g

is the Earth's gravitational constant and m the molecular mass. As introduced earlier (see Table 5.1 and Chapter 12), a thermal Lagrangian, and hence $\mathcal{Z}_{gr}$, can be written to reflect this additional gravitational potential (work) contribution:

$$\mathcal{Z}_{gr} = \sum_{N=0,1,\ldots} e^{\beta(\mu - mgz)N} \mathcal{Z}(N), \tag{13.6}$$

where $\mathcal{Z}(N)$ is the N-particle (canonical) partition function for the ideal gas (see Chapter 7)

$$\mathcal{Z}(N) = \frac{1}{N!}\left(n_Q V\right)^N. \tag{13.7}$$

Upon summing N Eq. 13.6 becomes

$$\mathcal{Z}_{gr} = \exp\left[e^{\beta(\mu - mgz)} n_Q V\right], \tag{13.8}$$

and then, as also shown in Chapter 12 (see Eq. 12.15), the average particle number at a distance $z > 0$ above the Earth's surface is

$$\langle \mathcal{N}_{op}(z) \rangle = \frac{1}{\beta}\left(\frac{\partial}{\partial \mu} \ln \mathcal{Z}_{gr}\right)_{T,V} \tag{13.9}$$

$$= \frac{1}{\beta}\left(\frac{\partial}{\partial \mu}\right) e^{\beta(\mu - mgz)} n_Q V \tag{13.10}$$

$$= e^{\beta\mu} e^{-\beta mgz} n_Q V. \tag{13.11}$$

The chemical potential of gas molecules at any height z is, therefore,

$$\mu(z) = \frac{1}{\beta} \ln \frac{\langle \mathcal{N}_{op}(z) \rangle}{n_Q V} + mgz. \tag{13.12}$$

Gas particles (at height z) can freely diffuse to any adjacent level, say $z \pm \Delta z$, so that in diffusive equilibrium μ must be uniform throughout the atmosphere, i.e. independent of z. At ground level, i.e. $z = 0$, the chemical potential for ideal gas molecules is

$$\mu(0) = \frac{1}{\beta} \ln \frac{\langle \mathcal{N}_{op}(0) \rangle}{n_Q V}, \tag{13.13}$$

so in diffusive equilibrium Eqs. 13.12 and 13.13 are equated to give

$$\langle \mathcal{N}_{op}(z) \rangle = \langle \mathcal{N}_{op}(0) \rangle e^{-\beta mgz}, \tag{13.14}$$

the well-known "barometric" equation.

Charged ideal gas between capacitor plates

Ideal gas molecules, each with charge q, lie between a pair of large, parallel capacitor plates placed a distance d apart and charged to a potential $\mathcal{V} = E_x d$, where E_x is a uniform electric field between the plates. An electric potential energy per particle, $PE = -qxE_x$ adds, quasi-classically, to the N-particle ideal gas eigen-energies so that, as in Eq. 13.11, the number of charged molecules a distance x from the positive plate is

$$\langle \mathcal{N}_{op}(x) \rangle = e^{\beta\mu} e^{\beta q x E_x} n_Q V. \tag{13.15}$$

Solving for the chemical potential,

$$\mu(x) = \frac{1}{\beta} \ln \frac{\langle \mathcal{N}_{op}(x) \rangle}{n_Q V} - q\, x\, E_x. \tag{13.16}$$

When $x = 0$ (the positive plate),

$$\mu(0) = \frac{1}{\beta} \ln \frac{\langle \mathcal{N}_{op}(0) \rangle}{n_Q V}. \tag{13.17}$$

For diffusive equilibrium, in the region between the plates μ is independent of x and

$$\langle \mathcal{N}_{op}(x) \rangle = \langle \mathcal{N}_{op}(0) \rangle e^{\beta q x E_x}. \tag{13.18}$$

At $x = d$, i.e. the negative plate,

$$\langle \mathcal{N}_{op}(d) \rangle = \langle \mathcal{N}_{op}(0) \rangle e^{\beta q \mathcal{V}}, \tag{13.19}$$

where $\mathcal{V} = E_x d$ is the electric potential difference between the plates.

Ideal gas in a rotating cylinder

Ideal gas particles, mass m, in a cylinder rotating with angular frequency ω, acquire a rotational potential energy per particle (in the rotating frame), which, quasi-classically, is $PE_{rot} = -\frac{1}{2} m\omega^2 r^2$. Here r is a particle's radial distance from the cylinder axis. The grand partition function is

$$\mathcal{Z}_{gr} = \sum_{N=0,1,\ldots} e^{\beta\left(\mu + \frac{1}{2} m\omega^2 r^2\right) N} \mathcal{Z}(N), \tag{13.20}$$

where $\mathcal{Z}(N)$ is, as in Eq. 13.7, the N-particle (canonical) ideal gas partition function. After summing N, at a distance r from the axis of rotation the average particle number is

$$\langle \mathcal{N}_{op}(r) \rangle = \frac{1}{\beta} \left(\frac{\partial}{\partial \mu} \ln \mathcal{Z}_{gr} \right)_{T,V} \tag{13.21}$$

$$= \frac{1}{\beta} \left(\frac{\partial}{\partial \mu} \right) e^{\beta\left(\mu - \frac{1}{2} m\omega^2 r^2\right)} n_Q V \tag{13.22}$$

$$= e^{\beta\mu} e^{\frac{\beta}{2} m\omega^2 r^2} n_Q V. \tag{13.23}$$

Hence, the chemical potential at r is

$$\mu(r) = \frac{1}{\beta} \ln \frac{\langle \mathcal{N}_{op}(r) \rangle}{n_Q V} - \frac{1}{2} m \omega^2 r^2, \tag{13.24}$$

whereas the chemical potential at $r = 0$ is

$$\mu(0) = \frac{1}{\beta} \ln \frac{\langle \mathcal{N}_{op}(0) \rangle}{n_Q V}. \tag{13.25}$$

In diffusive equilibrium μ must be uniform throughout the cylinder, i.e. independent of r, giving the radial distribution of particles in a centrifuge

$$\langle \mathcal{N}_{op}(r) \rangle = \langle \mathcal{N}_{op}(0) \rangle \exp\left(\frac{\beta}{2} m \omega^2 r^2 \right). \tag{13.26}$$

13.3 Thermodynamics of chemical equilibrium

Consider the concrete example of a system at temperature T containing three species A, B and AB. Initially allow all three species to be non-interacting so that a thermal Lagrangian is

$$\mathcal{L} = -k_B \hat{\mathbf{P}} \ln \hat{\mathbf{P}} - \lambda_0 \hat{\mathbf{P}} - \frac{1}{T} \langle \mathcal{H}_{op} \rangle + \frac{1}{T} \left[\mu^A \langle \mathcal{N}_{op}^A \rangle + \mu^B \langle \mathcal{N}_{op}^B \rangle + \mu^{AB} \langle \mathcal{N}_{op}^{AB} \rangle \right], \tag{13.27}$$

where μ^A, μ^B and μ^{AB} are the chemical potentials for each of the isolated species.

Now let the species react according to the chemical equation

$$A + B \leftrightarrows AB. \tag{13.28}$$

In such a chemical reaction all reactants and products coexist with small fluctuations about mean equilibrium concentrations of the participants.[7] This balanced chemical equation defines "atom" diffusion such that at equilibrium only the average atom sums

$$\langle \mathcal{N}_{op}^A \rangle + \langle \mathcal{N}_{op}^{AB} \rangle = \nu_A \tag{13.29}$$

and

$$\langle \mathcal{N}_{op}^B \rangle + \langle \mathcal{N}_{op}^{AB} \rangle = \nu_B \tag{13.30}$$

[7] There are a few rare chemical reactions where concentrations of reactants and products oscillate in time so that "equilibrium" is never attained.

are still identifiable. These constraints coordinate with a diffusive equilibrium condition

$$\mu^A + \mu^B = \mu^{AB} \tag{13.31}$$

to give the thermal Lagrangian

$$\mathcal{L} = -k_B \hat{\mathbf{P}} \ln \hat{\mathbf{P}} - \lambda_0 \hat{\mathbf{P}} - \frac{1}{T}\langle h_{op}\rangle + \frac{1}{T}\left[\mu^A\left(\langle \mathcal{N}_{op}^A\rangle + \langle \mathcal{N}_{op}^{AB}\rangle\right) + \mu^B\left(\langle \mathcal{N}_{op}^B\rangle + \langle \mathcal{N}_{op}^{AB}\rangle\right)\right]. \tag{13.32}$$

As a second example consider a hypothetical reaction among the diatoms A_2, B_2 and the molecule AB,

$$A_2 + B_2 \leftrightarrows 2\,AB. \tag{13.33}$$

Assuming uniform reaction temperature T and pressure p, consider the chemical species as initially non-reactive. In that case the thermal Lagrangian is

$$\mathcal{L} = -k_B \hat{\mathbf{P}} \ln \hat{\mathbf{P}} - \lambda_0 \hat{\mathbf{P}} - \frac{1}{T}\langle h_{op}\rangle + \frac{1}{T}\left[\mu^{A_2}\langle \mathcal{N}_{op}^{A_2}\rangle + \mu^{B_2}\langle \mathcal{N}_{op}^{B_2}\rangle + \mu^{AB}\langle \mathcal{N}_{op}^{AB}\rangle\right]. \tag{13.34}$$

According to Eq. 13.33 the atom-reaction constraints for this case are

$$\langle \mathcal{N}_{op}^{A_2}\rangle + \frac{\langle \mathcal{N}_{op}^{AB}\rangle}{2} = \nu_A \tag{13.35}$$

and

$$\langle \mathcal{N}_{op}^{B_2}\rangle + \frac{\langle \mathcal{N}_{op}^{AB}\rangle}{2} = \nu_B, \tag{13.36}$$

where ν_A and ν_B are the still identifiable equilibrium atom numbers. These reaction constraints coordinate with chemically reactive equilibrium

$$\mu^{A_2} + \mu^{B_2} = 2\mu^{AB}, \tag{13.37}$$

resulting in the thermal Lagrangian

$$\mathcal{L} = -k_B \hat{\mathbf{P}} \ln \hat{\mathbf{P}} - \lambda_0 \hat{\mathbf{P}} - \frac{1}{T}\langle h_{op}\rangle + \frac{1}{T}\left[\mu^{A_2}\left(\langle \mathcal{N}_{op}^{A_2}\rangle + \frac{1}{2}\langle \mathcal{N}_{op}^{AB}\rangle\right) + \mu^{B_2}\left(\langle \mathcal{N}_{op}^{B_2}\rangle + \frac{1}{2}\langle \mathcal{N}_{op}^{AB}\rangle\right)\right] \tag{13.38}$$

derived from Eq. 13.34.

These results can be generalized to any chemical reaction, e.g.

$$a_1\, A_1 + a_2\, A_2 + a_3\, A_3 \leftrightarrows z_1\, Z_1 + z_2\, Z_2 + z_3\, Z_3, \tag{13.39}$$

in which case the atom reaction constraints are

$$\langle \mathcal{N}_{op}^{A_1} \rangle + \frac{a_1}{z_3} \langle \mathcal{N}_{op}^{Z_3} \rangle = \nu_{a_1}, \tag{13.40}$$

$$\langle \mathcal{N}_{op}^{A_2} \rangle + \frac{a_2}{z_3} \langle \mathcal{N}_{op}^{Z_3} \rangle = \nu_{a_2}, \tag{13.41}$$

$$\langle \mathcal{N}_{op}^{A_3} \rangle + \frac{a_3}{z_3} \langle \mathcal{N}_{op}^{Z_3} \rangle = \nu_{a_3}, \tag{13.42}$$

$$\langle \mathcal{N}_{op}^{Z_1} \rangle - \frac{z_1}{z_3} \langle \mathcal{N}_{op}^{Z_3} \rangle = \nu_{z_1}, \tag{13.43}$$

$$\langle \mathcal{N}_{op}^{Z_2} \rangle - \frac{z_2}{z_3} \langle \mathcal{N}_{op}^{Z_3} \rangle = \nu_{z_2}, \tag{13.44}$$

where ν_{a_1}, ν_{a_2}, ν_{a_3}, ν_{z_1} and ν_{z_2} are still identifiable atom numbers in accord with the chemically diffusive equilibrium condition

$$a_1\, \mu^{A_1} + a_2\, \mu^{A_2} + a_3\, \mu^{A_2} - z_1\, \mu^{Z_1} - z_2\, \mu^{Z_2} - z_3\, \mu^{Z_3} = 0 \tag{13.45}$$

to generate the thermal Lagrangian.

$$\mathcal{L} = -k_B \hat{\mathbf{P}} \ln \hat{\mathbf{P}} - \lambda_0 \hat{\mathbf{P}} - \frac{1}{T}\langle h_{op} \rangle + \frac{1}{T}\Bigg[\mu^{A_1}\left(\langle \mathcal{N}_{op}^{A_1} \rangle + \frac{a_1}{z_3} \langle \mathcal{N}_{op}^{Z_3} \rangle \right)$$

$$+ \mu^{A_2}\left(\langle \mathcal{N}_{op}^{A_2} \rangle + \frac{a_2}{z_3} \langle \mathcal{N}_{op}^{Z_3} \rangle \right) + \ldots + \mu^{Z_2}\left(\langle \mathcal{N}_{op}^{Z_2} \rangle - \frac{z_2}{z_3} \langle \mathcal{N}_{op}^{Z_3} \rangle \right) \Bigg]. \tag{13.46}$$

13.4 A law of mass action

Chemical reactions such as (hypothetically) defined in Eqs. 13.28, 13.33, 13.39 do not usually go to completion. A sealed reaction "vessel" contains an equilibrium mixture of reactants and products. For example, the gaseous reaction

$$3H_2 + N_2 \leftrightarrows 2\,NH_3 \tag{13.47}$$

in which at thermal equilibrium

$$3\mu^{H_2} + \mu^{N_2} - 2\mu^{NH_3} = 0, \tag{13.48}$$

yields only a small amount of NH_3 (ammonia), with reactants H_2 and N_2 still having a substantial presence. The equilibrium concentrations of reactants and products in the closed reaction "vessel" is, however, well described by a "law of mass action" which is a consequence of Eq. 13.48. This "law" is arrived at by first exponentiating Eq. 13.48:

$$\frac{\exp\left(3\mu^{H_2}\right) \times \exp\left(\mu^{N_2}\right)}{\exp\left(2\mu^{NH_3}\right)} = 1 \tag{13.49}$$

or

$$\frac{\left[\exp\left(\mu^{H_2}\right)\right]^3 \times \left[\exp\left(\mu^{N_2}\right)\right]}{\left[\exp\left(\mu^{NH_3}\right)\right]^2} = 1. \tag{13.50}$$

Continuing in the spirit of the example, assume all participating gases g are ideal and at sufficiently low temperature that no internal modes of the reactive molecules (rotational, vibrational or electronic) are excited. In which case

$$\beta\mu^g = \ln\left[\frac{\mathscr{C}^{(g)}}{n_Q^{(g)}}\right] \tag{13.51}$$

$$= \ln\left\{\left[\frac{m^{(g)}}{2\pi\hbar^2\beta}\right]^{-3/2}\mathscr{C}^{(g)}\right\}, \tag{13.52}$$

where $m^{(g)}$ is the molecular mass of the gas g and

$$\mathscr{C}^{(g)} = \frac{\langle \mathcal{N}_{op}^{(g)} \rangle}{V} \tag{13.53}$$

is the gas concentration. From this follows a typical "mass action" result

$$\frac{\left[\mathscr{C}^{NH_3}\right]^2}{\left[\mathscr{C}^{H_2}\right]^3\left[\mathscr{C}^{N_2}\right]} = \frac{\left\{n_Q^{H_2}\right\}^3\left\{n_Q^{N_2}\right\}}{\left\{n_Q^{NH_3}\right\}^2} \tag{13.54}$$

$$= \mathscr{K}(T), \tag{13.55}$$

where $\mathscr{K}$ depends only on T. Specific mass action constants $\mathscr{K}$ are associated with different chemical reactions or molecular models. If, for example, in the previous model temperature is raised sufficiently to excite molecular rotational and vibration (internal modes) the N-particle canonical partition function becomes

$$\check{Z}(N) = \frac{1}{N!}\left[Z_{int}\right]^N\left[n_Q V\right]^N \tag{13.56}$$

where Z_{int} is the partition function for the internal (rotational and vibrational) modes. In that case we redefine n_Q for the molecular gas g

$$n_Q^{(g)} \rightarrow Z_{int}^{(g)} n_Q^{(g)} = n_{int}^{(g)} \tag{13.57}$$

so that the mass action expression becomes

$$\frac{\left[\mathscr{C}^{NH_3}\right]^2}{\left[\mathscr{C}^{H_2}\right]^3\left[\mathscr{C}^{N_2}\right]} = \mathscr{K}_{int}(T), \tag{13.58}$$

where

$$\mathscr{K}_{int}(T) = \frac{\left\{n_{int}^{H_2}\right\}^3\left\{n_{int}^{N_2}\right\}}{\left\{n_{int}^{NH_3}\right\}^2}. \tag{13.59}$$

13.5 Thermodynamics of phase equilibrium

A *phase* is a chemically and physically uniform macroscopic state of matter. Vapor, liquid and solid are among the familiar examples of phases. Other examples include different coexisting crystal structures with the same chemical composition or the more exotic He_4–Bose–Einstein condensate phases. Phase equilibrium describes a situation in which two or more homogeneous regions (phases) coexist in thermodynamic equilibrium within a rigid container but are separated by physical boundaries. An example are the three phases of H_2O – ice, liquid water and water vapor – in which any two or all three can coexist. Distinct phases may also exist within a given state, such as in solid iron alloys and in the several phases that can coexist for liquid states. Those phases that are possible depend on temperature and pressure. If two coexisting phases, designated α and β, composed of a single species, freely exchange particles at temperature T within a fixed volume $(V^\alpha + V^\beta)$, then a thermal Lagrangian is[8]

$$\begin{aligned}\mathcal{L}_{E,N,V} = &-k_B\left[\hat{\mathbf{P}}_\alpha \ln\hat{\mathbf{P}}_\alpha + \hat{\mathbf{P}}_\beta \ln\hat{\mathbf{P}}_\beta\right] - \lambda_{0\alpha}\hat{\mathbf{P}}_\alpha - \lambda_{0\beta}\hat{\mathbf{P}}_\beta\\ &-\frac{1}{T}\left[\langle\mathbf{h}_{op}^\alpha\rangle + \langle\mathbf{h}_{op}^\beta\rangle\right] - \frac{1}{T}\langle\boldsymbol{p}_{op}\rangle\left(V^\alpha + V^\beta\right) + \frac{\mu}{T}\left[\langle\mathcal{N}_{op}^\alpha\rangle + \langle\mathcal{N}_{op}^\beta\rangle\right],\end{aligned} \tag{13.60}$$

with

$$\begin{aligned}T_\alpha &= T_\beta = T,\\ \mu^\alpha &= \mu^\beta = \mu,\\ \langle\boldsymbol{p}_{op}^\alpha\rangle &= \langle\boldsymbol{p}_{op}^\beta\rangle = \langle\boldsymbol{p}_{op}\rangle.\end{aligned} \tag{13.61}$$

[8] Mechanical energy, $p\,V$, is included in the Lagrangian.

13.6 Gibbs–Duhem relation

Long before he published his monumental treatise on statistical mechanics Gibbs had already extended the fundamental equation of thermodynamics to include particle transfer. For a single phase and a single species the fundamental equation becomes

$$T\,\mathrm{d}\mathcal{S} = \mathrm{d}\mathcal{U} + p\,\mathrm{d}V - \mu\,\mathrm{d}\langle\mathcal{N}_{op}\rangle. \tag{13.62}$$

When $\mathcal{U}$ is expressed in terms of its natural variables

$$\mathcal{U} = \mathcal{U}\big(\mathcal{S}, V, \langle\mathcal{N}_{op}\rangle\big), \tag{13.63}$$

all of which are extensive, then in the Euler form

$$\mathcal{U}\left(\lambda\mathcal{S}, \lambda V, \lambda\langle\mathcal{N}_{op}\rangle\right) = \lambda\,\mathcal{U}\left(\mathcal{S}, V, \mathcal{N}\right). \tag{13.64}$$

Euler's homogeneous function theorem gives

$$\mathcal{S}\left(\frac{\partial\mathcal{U}}{\partial\mathcal{S}}\right)_{V,\langle\mathcal{N}_{op}\rangle} + V\left(\frac{\partial\mathcal{U}}{\partial V}\right)_{\mathcal{S},\langle\mathcal{N}_{op}\rangle} + \langle\mathcal{N}_{op}\rangle\left(\frac{\partial\mathcal{U}}{\partial\langle\mathcal{N}_{op}\rangle}\right)_{V,\mathcal{S}} = \mathcal{U}\left(\mathcal{S}, V, \langle\mathcal{N}_{op}\rangle\right). \tag{13.65}$$

Comparing this with Eq. 13.62, the fundamental equation is effectively integrated to get

$$\mathcal{U} - T\mathcal{S} + Vp - \mu\langle\mathcal{N}_{op}\rangle = 0. \tag{13.66}$$

Taking the total differential

$$\mathrm{d}\mathcal{U} - T\,\mathrm{d}\mathcal{S} - \mathcal{S}\,\mathrm{d}T + V\,\mathrm{d}p + p\,\mathrm{d}V - \mu\,\mathrm{d}\langle\mathcal{N}_{op}\rangle - \langle\mathcal{N}_{op}\rangle\,\mathrm{d}\mu = 0 \tag{13.67}$$

and combining it with the fundamental equation Eq. 13.62, we have for a single phase with one species

$$-\mathcal{S}\,\mathrm{d}T + V\,\mathrm{d}p - \langle\mathcal{N}_{op}\rangle\,\mathrm{d}\mu = 0. \tag{13.68}$$

This is called the *Gibbs–Duhem equation*. In the case of a single phase with M species Gibbs–Duhem becomes

$$-\mathcal{S}\,\mathrm{d}T + V\,\mathrm{d}p - \sum_i^M \langle\mathcal{N}_{op}\rangle_i\,\mathrm{d}\mu_i = 0, \tag{13.69}$$

showing that intensive variables $T,\ p,\ \mu_i$ are not independent.

13.7 Multiphase equilibrium

Gibbs–Duhem opens the door to basic understanding of coexistence between different phases, such as between liquid water and water vapor, or between ice and liquid water, etc.

Begin by rewriting Eq. 13.68 in the form

$$d\mu = -s\, dT + v\, dp \tag{13.70}$$

where intensive quantities

$$s = \frac{\mathcal{S}}{\langle \mathcal{N}_{op} \rangle} \qquad v = \frac{V}{\langle \mathcal{N}_{op} \rangle} \tag{13.71}$$

have been defined. Taking advantage of p and T as natural variables, for two phases, α and β, at the same T and p and composed of only a single species (pure phases), the chemical potentials are

$$\mu^{(\alpha)} = \mu^{(\alpha)}(p, T), \tag{13.72}$$

$$\mu^{(\beta)} = \mu^{(\beta)}(p, T). \tag{13.73}$$

If these phases are to coexist, then

$$\mu^{(\alpha)}(p, T) = \mu^{(\beta)}(p, T), \tag{13.74}$$

which is a "system" of one equation with two unknowns, T and p. Although there is no unique solution, T can be found in terms of p so that the α and β phases can coexist along some curve in the $\{p, T\}$-plane.

Now suppose there are three pure phases, α, β and γ. Then for coexistence between the three phases

$$\mu^{(\alpha)}(p, T) = \mu^{(\beta)}(p, T), \tag{13.75}$$

$$\mu^{(\alpha)}(p, T) = \mu^{(\gamma)}(p, T), \tag{13.76}$$

which is a "system" of two equations and two unknowns with a unique solution – a single point in the $\{p, T\}$-plane. The three phases can coexist at what is called a *triple point*.[9] Figure 13.1 is an example of a phase diagram for a single component system, e.g. water[10] in which temperature and pressure form the coordinate axes. As shown in the diagram, only certain phases are possible at a particular temperature and pressure, with each phase separated from the others by a curve in the $\{p, T\}$

[9] Triple point of water: $T_{tp} = 273.16\,\text{K}$, $p_{tp} = 0.006\,04\,\text{atm}$ ($1\,\text{atm} = 101.3\,\text{k Pa}$).

[10] The experimental solid–liquid boundary for water has an anomalous negative slope so the diagram is only an approximate representation for water. This is due to the peculiarity of ice being less dense than liquid water.

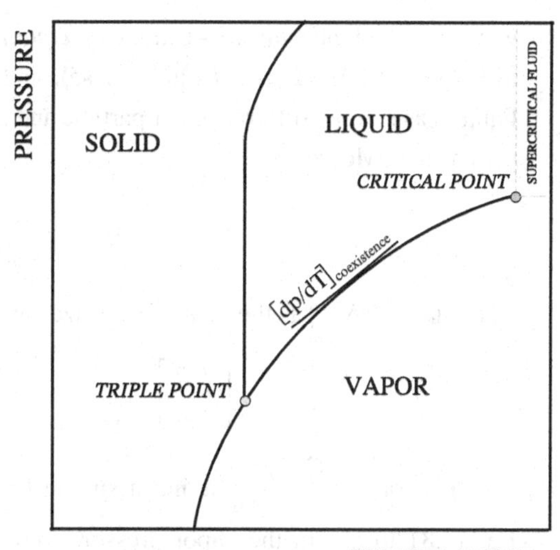

Fig. 13.1 Single-species phase diagram. The point where the solid–liquid and liquid–vapor phase lines meet is called the *triple point*. At that point all three phases – solid, liquid and vapor – can coexist.

plane called a *phase boundary*, which indicates the values of T and p along which two phases can coexist.

A unique coexistence property is the point at which the phase boundary between liquid and vapor abruptly disappears. This is called a *critical point*. Beyond this point liquid phase and vapor phase lose their usual meaning, becoming instead a *supercritical fluid* having properties of both vapor and liquid.[11]

13.8 The Clausius–Clapeyron equation

For the case of two coexisting pure phases α and β ($\mu^{(\alpha)} = \mu^{(\beta)}$), the Gibbs–Duhem relation gives

$$-s^{(\alpha)}\,\mathrm{d}T + v^{(\alpha)}\,\mathrm{d}p = -s^{(\beta)}\,\mathrm{d}T + v^{(\beta)}\,\mathrm{d}p, \tag{13.77}$$

so that the slope of a coexistence curve is (see Figure 13.1)

$$\left[\frac{\mathrm{d}p}{\mathrm{d}T}\right]_{coexistence} = \frac{s^{(\alpha)} - s^{(\beta)}}{v^{(\alpha)} - v^{(\beta)}} \tag{13.78}$$

$$= \frac{\Delta s}{\Delta v}, \tag{13.79}$$

[11] Critical point of water: $T_c = 647\,\mathrm{K}$, $p_c = 217.7\,\mathrm{atm}$. Critical point of CO_2: $T_c = 304\,\mathrm{K}$, $p_c = 72.8\,\mathrm{atm}$.

which is called the *Clausius–Clapeyron equation*. As a concrete example that relates to Figure 13.1 let $\alpha \equiv \ell$ (liquid phase), and $\beta \equiv g$ (gas phase). Then Δs is the liquid–gas entropy difference per particle and Δv is the liquid–gas volume difference per particle. Moreover

$$\Delta s = \frac{\mathcal{Q}_{vap}}{\langle \mathcal{N}_{op} \rangle T}, \tag{13.80}$$

where $\mathcal{Q}_{vap}/\langle \mathcal{N}_{op} \rangle$ is the heat of vaporization per particle,[12] so that

$$\left[\frac{dp}{dT}\right]_{coexistence} = \frac{\mathcal{Q}_{vap}}{T \langle \mathcal{N}_{op} \rangle \Delta v}. \tag{13.81}$$

In general $\dfrac{\mathcal{Q}_{vap}}{T \langle \mathcal{N}_{op} \rangle \Delta v}$ is not a simple function of p and T so that integrating Eq. 13.81 to obtain the vapor pressure may not be straightforward. On the other hand, assuming an ideal gas phase and since, in general, $v^{(gas)} \gg v^{(\ell)}$, the differential equation simplifies to

$$\left[\frac{dp}{dT}\right]_{coexistence} = \frac{p\, \mathcal{Q}_{vap}}{k_B T^2}. \tag{13.82}$$

Since generally $v^{(solid)} < v^{(\ell)}$ for most systems, as suggested in Figure 13.1, the solid–liquid coexistence curve is nearly vertical but positively sloped. In the unusual case of water for which $v^{(solid)} > v^{(\ell)}$ the curve is also nearly vertical, but now slightly negatively sloped.

13.9 Surface adsorption: Langmuir's model

In the process of film growth or surface doping, atoms or molecules in gas phase or in dilute solution bind to the film surface (Figure 13.2). A theory of solid surface coverage by these molecules was formulated by Irving Langmuir (1916), the acknowledged pioneer of surface chemistry.

Assumptions of the model are as follows.

1. The solid surface is in contact with ideal monatomic gas atoms at temperature T and pressure p_0.
2. There are a fixed number of sites N on the surface available for bonding.
3. Each surface site can be only singly occupied (monolayer coverage).
4. Adsorption at a given site is independent of occupation of neighboring sites (no interactions).

[12] The heat required to evaporate one particle.

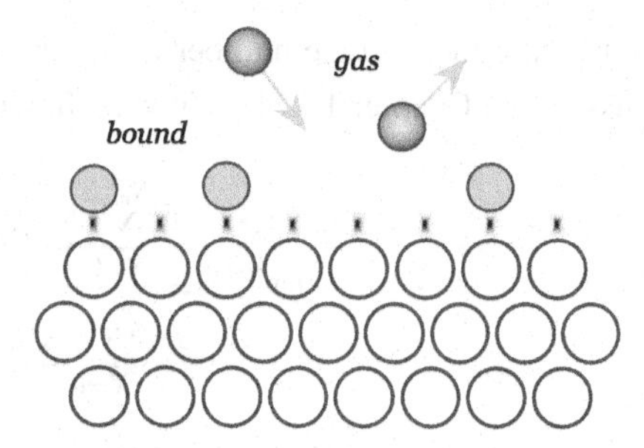

Fig. 13.2 Gas molecules interacting with solid surface.

5. A gas molecule bound to the surface has the non-degenerate eigen-energy

$$E_{bound} = -|\epsilon| \tag{13.83}$$

with respect to the energy of a free gas atom.

6. The energy difference between adsorbed gas atoms and free gas atoms is independent of surface coverage.
7. Atoms bound to the surface (A_{bd}) are in thermal and diffusive (phase) equilibrium with gas atoms (A_{gas}) as described by the reaction

$$A_{gas} \leftrightarrows A_{bd}. \tag{13.84}$$

With Langmuir's model the fractional surface coverage

$$\theta = \frac{\langle \mathcal{N}_{op}^{bd} \rangle}{N_{max}} \tag{13.85}$$

can be determined, where $\langle \mathcal{N}_{op}^{bd} \rangle$ is the average number of gas atoms bound to N_{max} accessible solid surface sites.

Diffusive (phase) equilibrium in the surface phenomenon of Eq. 13.84 is described by the thermal Lagrangian

$$\mathcal{L}_{E,N} = -k_B \left[\hat{\mathbf{P}}_{bd} \ln \hat{\mathbf{P}}_{bd} + \hat{\mathbf{P}}_{gas} \ln \hat{\mathbf{P}}_{gas} \right] - \lambda_{0_{bd}} \hat{\mathbf{P}}_{bd} - \lambda_{0_{gas}} \hat{\mathbf{P}}_{gas} \\ - \frac{1}{T} \left[\langle \mathbf{h}_{op}^{bd} \rangle + \langle \mathbf{h}_{op}^{gas} \rangle \right] - \frac{\mu}{T} \left[\langle \mathcal{N}_{op}^{bd} \rangle + \langle \mathcal{N}_{op}^{gas} \rangle \right], \tag{13.86}$$

which correlates with

$$\mu_{gas} = \mu_{bd} = \mu, \tag{13.87}$$

where μ_{bd} is the chemical potential of atoms bound to surface sites and μ_{gas} is the chemical potential for ideal gas atoms.

Finding the average number of atoms bound to surface sites $\langle \mathcal{N}_{op}^{bd} \rangle$ follows discussions in Chapter 12 where it was shown that

$$\langle \mathcal{N}_{op}^{bd} \rangle = \frac{\sum\limits_{N=0}^{N_{max}} \sum\limits_{s} N e^{-\beta\{[E_s(N)] - \mu_{bd} N\}}}{\sum\limits_{N=0}^{N_{max}} \sum\limits_{s} e^{-\beta\{[E_s(N)] - \mu_{bd} N\}}}, \tag{13.88}$$

with the number of bound atoms limited by N_{max} and where $E_s(N) = -N|\epsilon|$. The denominator in Eq. 13.88 is the grand partition function $\mathcal{Z}_{gr}$,

$$\mathcal{Z}_{gr} = \sum_{N=0}^{N_{max}} e^{\beta \mu_{bd} N} \mathcal{Z}(N), \tag{13.89}$$

where $\mathcal{Z}(N)$ is the (canonical) partition function for N-bound atoms

$$\mathcal{Z}(N) = g(N, N_{max})\, e^{N\beta|\epsilon|}\,. \tag{13.90}$$

Since the bound atom's only eigen-energy (see Eq. 13.83) has the configurational degeneracy

$$g(N, N_{max}) = \frac{N_{max}!}{(N_{max} - N)!\,N!} \tag{13.91}$$

the grand partition function is

$$\mathcal{Z}_{gr} = \sum_{N=0}^{N_{\max}} \frac{N_{\max}!}{(N_{\max} - N)!\,N!} e^{N\beta(\mu_{bd} + |\epsilon|)} \tag{13.92}$$

$$= \left[1 + e^{\beta(\mu_{bd} + |\epsilon|)}\right]^{N_{\max}} \tag{13.93}$$

and the fractional surface coverage is (see Eq. 13.88)

$$\theta = \frac{\langle \mathcal{N}_{op}^{bd} \rangle}{N_{\max}} \tag{13.94}$$

$$= \frac{1}{N_{\max}} \times \frac{1}{\beta} \frac{\partial}{\partial \mu_{bd}} \ln \mathcal{Z}_{gr} \tag{13.95}$$

$$= \frac{e^{\beta\mu_{bd}} e^{\beta|\epsilon|}}{1 + e^{\beta\mu_{bd}} e^{\beta|\epsilon|}}. \tag{13.96}$$

The ideal gas in contact with the surface has the chemical potential (see Eq. 12.48)

$$\mu_{gas} = \frac{1}{\beta} \ln \frac{\langle \mathcal{N}_{op}^{gas} \rangle}{n_Q V}. \tag{13.97}$$

Equating the chemical potentials (see Eq. 13.87) (diffusive equilibrium) and using the ideal gas equation of state $p_0 V = \frac{\langle \mathcal{N}_{op}^{gas} \rangle}{\beta}$ gives the Langmuir fractional surface coverage

$$\theta = \frac{p_0}{\Pi_0 + p_0}, \tag{13.98}$$

where Π_0 is the temperature-dependent factor

$$\Pi_0 = \frac{n_Q \, e^{-\beta|\epsilon|}}{\beta}. \tag{13.99}$$

This is known as the *Langmuir isotherm.*

13.10 Dissociative adsorption

A diatomic gas molecule, say A_2, may not simply bind to a surface site, but may dissociate, with each component atom binding to a single site. This is referred to as *dissociative adsorption* and is described by the "chemical reaction"

$$A_2 \leftrightharpoons 2A_{bound}. \tag{13.100}$$

Therefore, the thermal equilibrium chemical potentials satisfy

$$\mu_{A_2} - 2\mu_A = 0, \tag{13.101}$$

where μ_{A_2} is the chemical potential of the ideal diatomic gas and μ_A is the chemical potential of surface bound atoms. The ideal diatomic gas in contact with the surface has the chemical potential

$$\mu_{A_2} = \frac{1}{\beta} \ln \frac{\langle \mathcal{N}_{op}^{A_2} \rangle}{n_{int}^{A_2} V}, \tag{13.102}$$

which includes the effect of internal (rotational and vibrational) modes (see Eq. 13.56). Once again taking $E_{bound} = -|\epsilon|$ the fractional surface occupation is, according to Eq. 13.96,

$$\theta = \frac{e^{\beta\mu_A} e^{\beta|\epsilon|}}{1 + e^{\beta\mu_A} e^{\beta|\epsilon|}}. \tag{13.103}$$

Now applying Eq. 13.101 the surface coverage is

$$\theta = \frac{e^{(1/2)\beta\mu_{A_2}} e^{\beta|\epsilon|}}{1 + e^{(1/2)\beta\mu_{A_2}} e^{\beta|\epsilon|}}. \tag{13.104}$$

Using the ideal gas result, Eq. 13.102, we finally have

$$\theta = \frac{\sqrt{p}}{\Pi_D + \sqrt{p}}, \tag{13.105}$$

where p is the gas pressure and the Langmuir isotherm is

$$\Pi_D = \sqrt{\frac{n_{\text{int}}^{A_2}}{\beta}}\, e^{-\beta|\varepsilon|}. \tag{13.106}$$

13.11 Crystalline bistability

In an ordered crystal it is common for atoms to be bistable, i.e. atoms at normal crystalline sites can migrate to "abnormal" sites (usually accompanied by a lattice distortion), where they have a different binding energy. If a crystal at a temperature T has "normal" sites and "displaced" sites, the latter lying on an interstitial sublattice (see Figure 13.3), after a long time a certain fraction of the atoms will occupy "displaced" sites. Since the total number of atoms – normal plus defect – is conserved, the atom constraint is

$$\langle \mathcal{N}_{op}^{norm} \rangle + \langle \mathcal{N}_{op}^{dis} \rangle = N_{normal}\,(\text{the total number of normal sites})\,. \tag{13.107}$$

The migration process can be interpreted as the "chemical reaction"

$$A_{norm} \leftrightarrows A_{dis}\,. \tag{13.108}$$

Following details of Langmuir's model and assuming:

1. an equal number of "normal" and "displaced" sites, N_s;
2. "displaced" atoms have energy $\epsilon_0 > 0$ with respect to "normal" sites;
3. only single occupancy of "normal" and "displaced" sites is permitted;

the grand partition function for single occupancy of a "normal" site is

$$\mathcal{Z}_{gr}^{norm} = \sum_{N=0}^{N_s} e^{N\beta\mu_n} \left\{ \frac{N_s!}{(N_s - N)!N!} \right\} \tag{13.109}$$

$$= \left(1 + e^{\beta\mu_n}\right)^{N_s}\,, \tag{13.110}$$

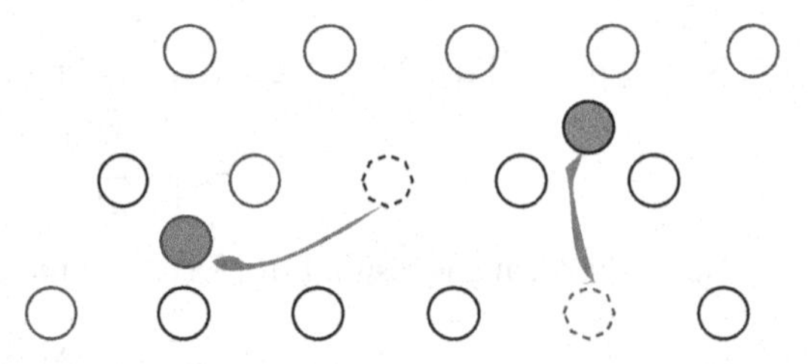

Fig. 13.3 Lattice atoms migrating from "normal" to "displaced" sites.

while the grand partition function for single occupancy of a "displaced" site is

$$\mathcal{Z}_{gr}^{inter} = \sum_{N=0}^{N_s} e^{N\beta\mu_{dis}} \left\{ \frac{N_s!}{(N_s - N)!N!} e^{N\beta\epsilon_0} \right\} \tag{13.111}$$

$$= \left(1 + e^{\beta\epsilon_0} e^{\beta\mu_{dis}}\right)^{N_s}. \tag{13.112}$$

From grand partition functions, Eqs. 13.110 and 13.112,

$$\langle \mathcal{N}_{op}^{norm} \rangle = \frac{1}{\beta} \frac{\partial}{\partial \mu_{norm}} \ln \mathcal{Z}_{gr}^{norm} \tag{13.113}$$

$$= N_s \frac{\lambda_{norm}}{1 + \lambda_{norm}} \tag{13.114}$$

and

$$\langle \mathcal{N}_{op}^{dis} \rangle = \frac{1}{\beta} \frac{\partial}{\partial \mu_{dis}} \ln \mathcal{Z}_{gr}^{dis} \tag{13.115}$$

$$= N_s \frac{\lambda_{dis} e^{-\beta\epsilon_0}}{1 + \lambda_{dis} e^{-\beta\epsilon_0}}, \tag{13.116}$$

where

$$\lambda_{norm} = e^{\beta\mu_{norm}} \tag{13.117}$$

and

$$\lambda_{dis} = e^{\beta\mu_{dis}}. \tag{13.118}$$

At thermal equilibrium

$$\mu_{norm} = \mu_{dis} \tag{13.119}$$

or

$$\lambda_{norm} = \lambda_{dis} = \lambda. \tag{13.120}$$

Applying Eq. 13.107 ($N_{normal} = N_s$) together with Eqs. 13.114, 13.116, 13.120,

$$\frac{\lambda}{1+\lambda} + \frac{\lambda e^{-\beta\epsilon_0}}{1 + \lambda e^{-\beta\epsilon_0}} = 1, \tag{13.121}$$

we find

$$\lambda = e^{\beta\epsilon_0/2}. \tag{13.122}$$

Therefore, according to Eq. 13.116 the average fractional occupation of "displacement" sites is

$$\frac{\langle \mathcal{N}_{op}^{dis} \rangle}{N_s} = \frac{1}{1 + \exp(\beta\epsilon_0/2)} \tag{13.123}$$

as shown in Figure 13.4.

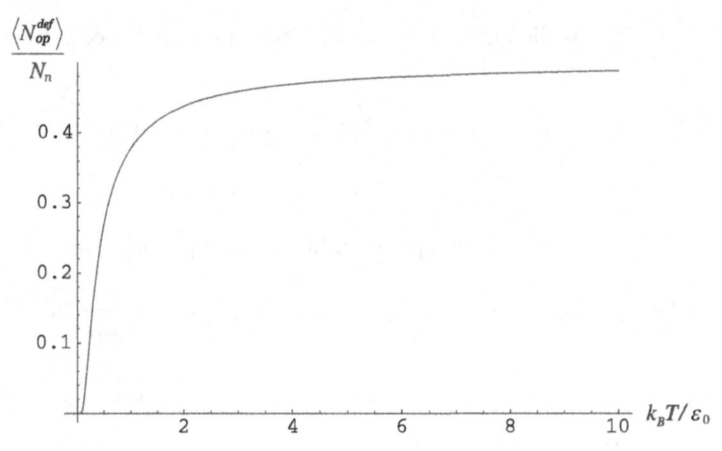

Fig. 13.4 Fractional occupation number of displaced sites.

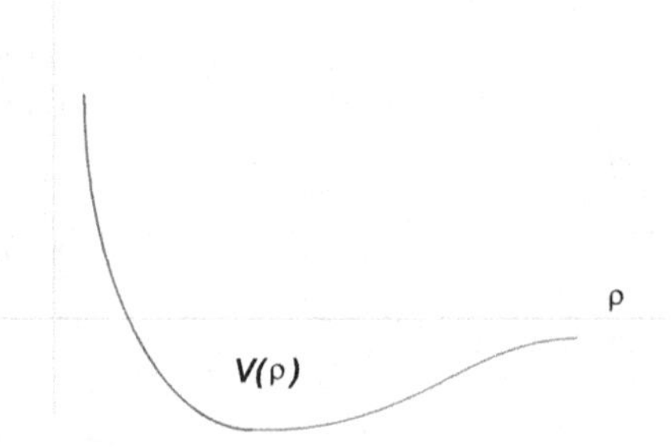

Fig. 13.5 Two-dimensional potential.

Problems and exercises

13.1 An absorbed surface layer of atoms uniformly covers an area A. The atoms, although free to move, interact with a two-dimensional potential $V(\rho)$, where ρ is the interatomic separation. (See Figure 13.5.)

Find the surface pressure p_S of this two-dimensional gas as a function of atom density $n = \langle \mathcal{N}_{op} \rangle / A$ up to order n^2.

13.2 Atoms are bound to solid surface sites with binding energy ϵ_0 relative to vapor (free, ideal gas) atoms with which they are in equilibrium. Bound atoms are modeled as two-dimensional Einstein oscillators with frequency ω_0 so that the eigen-energies of the αth bound surface atom are

$$\mathcal{E} = -\epsilon_0 + \hbar\omega_0 \left(n_{\alpha,x} + n_{\alpha,y} \right), \tag{13.124}$$

where $n_{\alpha,x},\ n_{\alpha,y} = 0, 1, 2, 3, \ldots, \infty$.

a. Find the grand potential $\boldsymbol{\Omega}$ that describes bound surface atoms.
b. Find the chemical potential μ for the bound surface atoms.
c. Find an expression for the pressure of the ideal gas vapor that is in equilibrium with surface atoms.

13.3 A particular mesh surface is employed as a catalyst for removal of ozone O_3 from exhaust gases by converting it to oxygen molecules O_2 according to the reaction

$$2O_3 \rightleftarrows 3O_2. \tag{13.125}$$

If the binding energy of an oxygen molecule to a surface site is $E_B = -\epsilon$ and the ozone pressure is p_{O_3}, what is the covering fraction

$$\theta = \frac{\langle n_{O_2} \rangle}{N}, \tag{13.126}$$

where $\langle n_{O_2} \rangle$ is the average number of oxygen molecules bound to the surface per unit area and N is the number of possible O_2 binding sites per unit mesh area?

14 Thermodynamics of radiation

A new scientific truth does not triumph by convincing its opponents and making them see the light, but because its opponents eventually die, and a new generation grows up that is familiar with it.

Max Planck, *Scientific Autobiography and Other Papers*,
Copyright: Philosophical Library

14.1 Introduction

In 1900 Max Planck discovered that the temperature-dependent law of radiating bodies could not be derived solely from Maxwellian electrodynamics according to which the energy of an electromagnetic field is[1]

$$E = \frac{1}{2}\int d\mathbf{x}\left[\mathcal{E}(\mathbf{x},\mathbf{t})^2 + \mathcal{B}(\mathbf{x},\mathbf{t})^2\right]. \tag{14.1}$$

Planck arrived, instead, at results consistent with the relevant electromagnetic experiments by treating radiation of a given frequency ν as though it consisted of "packets" of energy – "photons" – each with energy $h\nu$, with a corresponding electromagnetic field energy

$$E_{Planck} = nh\nu, \tag{14.2}$$

where $n = 0, 1, 2, \ldots$ is the number of "photons" in the packet and h is Planck's universal constant.

Planck's hypothesis[2] was the initial link in the chain of 20th-century discoveries that is quantum physics.

[1] In rationalized cgs units.
[2] Likely inspired by Boltzmann's concept of "microstates".

14.2 Electromagnetic eigen-energies

Proceeding directly to the thermodynamics of radiation, Eq. 14.2 is replaced by the electromagnetic quantum eigen-energies for photons of a single mode **k** having polarization λ

$$E_{\mathbf{k},\lambda} = h\nu_{\mathbf{k},\lambda}\left[n\left(\mathbf{k},\lambda\right) + \frac{1}{2}\right], \tag{14.3}$$

where $n\left(\mathbf{k},\lambda\right) = 0, 1, 2, \ldots$ is the number of photons. **k** and λ are quantum numbers for a single photon of frequency $\nu_{\mathbf{k},\lambda}$.[3]

Photons are fundamental particle-like "excitations" of the electromagnetic field. They have zero mass and are their own antiparticle. They can be created and absorbed (destroyed) without number conservation and therefore have zero chemical potential. They appear in an almost limitless variety of atomic, molecular or nuclear processes as well as continuous radiation (synchrotron radiation) associated with kinematic acceleration of particles. The photon carries "spin" angular momentum $\pm\hbar$ corresponding to right and left circular polarization.

The Maxwellian total electromagnetic energy (see Eq. 14.1) is similarly replaced by the quantum average

$$\langle\mathcal{H}_{EM}\rangle = \sum_{\lambda=1,2}\sum_{\mathbf{k}} h\nu_{\mathbf{k},\lambda}\left[\langle n_{op}\left(\mathbf{k},\lambda\right)\rangle + \frac{1}{2}\right] \tag{14.4}$$

where $\mathcal{H}_{EM}$ is the electromagnetic hamiltonian,

$$\mathcal{H}_{EM} = \sum_{\lambda=1,2}\sum_{\mathbf{k}} h\nu_{\mathbf{k},\lambda}\left[n_{op}(\mathbf{k},\lambda) + \frac{1}{2}\right]. \tag{14.5}$$

The "photon" mode number operator $n_{op}(\mathbf{k},\lambda)$ is identified by a three-component wave vector **k**, with $\mathbf{k} \equiv k_x, k_y, k_z$, and two mutually perpendicular polarization directions $\lambda \equiv \lambda_1, \lambda_2$. The eigenvalues of $n_{op}\left(\mathbf{k},\lambda\right)$ are $n(\mathbf{k},\lambda) = 0, 1, 2, \ldots$ and $\langle n_{op}(\mathbf{k},\lambda)\rangle$ is the average number of photons in that mode. The wave vector **k** is the direction of propagation of the photon and the polarization λ is the vector direction of the concurrent electric field $\mathcal{E}$ (see Figure H.1). In free space the two-component polarization vector is perpendicular (transverse) to **k**. This follows from the Maxwell equation[4] $\nabla\cdot\mathcal{E} = 0$. For electromagnetic radiation in free space, the energy of a

[3] Arriving at Planck's result from Maxwell's equations is the realm of quantum field theories, the details of which are well beyond the scope of this book. Nevertheless, for completeness an outline of the method is discussed in Appendix H.

[4] Since there is no analog to this Maxwell equation in crystalline elastic equations of motion, there remain three *phonon* polarizations.

photon associated with a particular mode $(\mathbf{k}, \lambda)$ is

$$h\nu_{\mathbf{k},\lambda} = \frac{h}{2\pi} c\,|\mathbf{k}|\,, \tag{14.6}$$

which depends only on the magnitude $|\mathbf{k}|$ and is independent of the polarization λ. Here c is the speed of light in vacuum. As indicated in Appendix H, in free space $\langle n_{op}(\mathbf{k}, \lambda)\rangle$, the average number of photons in the mode $(\mathbf{k}, \lambda)$ also depends only on $|\mathbf{k}|$ and is independent of the polarization λ.

14.3 Thermodynamics of electromagnetism

Based on macroscopic electromagnetic eigen-energies given by Eq. 14.3 (see Appendix H), a thermal electromagnetic Lagrangian $\mathcal{L}_{EM}$ is constructed:

$$\begin{aligned}\mathcal{L}_{EM} = &-k_B \sum_{n_{\mathbf{k},\lambda}} \mathbf{P}(n_{k,\lambda}) \ln \mathbf{P}(n_{k,\lambda}) - \lambda_0 \sum_{n_{\mathbf{k},\lambda}} \mathbf{P}(n_{k,\lambda}) \\ &- \frac{1}{T} \sum_{n_{\mathbf{k},\lambda}} \mathbf{P}(n_{k,\lambda}) \sum_{\substack{\mathbf{k} \\ \lambda=1,2}} \left\{ h\nu_{\mathbf{k},\lambda} \left(n_{\mathbf{k},\lambda} + \frac{1}{2} \right) \right\},\end{aligned} \tag{14.7}$$

where $\mathbf{P}\left(n_{k,\lambda}\right)$ is the probability that $n_{\mathbf{k},\lambda}$ photons are in the mode $\mathbf{k}$ with polarization λ. The sum

$$\sum_{n_{\mathbf{k},\lambda}} \tag{14.8}$$

ranges over $n_{\mathbf{k},\lambda} = 0, 1, 2, \ldots$ for each mode $\mathbf{k}$ and both polarizations $(\lambda = 1, 2)$. The Lagrange multiplier λ_0 assures normalized probabilities. Following familiar procedures, we find from Eq. 14.7

$$\mathbf{P}\left(n_{\mathbf{k},\lambda}\right) = \frac{\exp\left\{ -\beta \sum\limits_{\substack{\mathbf{k} \\ \lambda=1,2}} h\nu_{\mathbf{k},\lambda} \left(n_{\mathbf{k},\lambda} + \frac{1}{2} \right) \right\}}{\sum\limits_{n_{\mathbf{k},\lambda}} \exp\left\{ -\beta \sum\limits_{\substack{\mathbf{k} \\ \lambda=1,2}} h\nu_{\mathbf{k},\lambda} \left(n_{\mathbf{k},\lambda} + \frac{1}{2} \right) \right\}}, \tag{14.9}$$

where the normalizing denominator is the thermal equilibrium electromagnetic partition function,

$$\mathcal{Z}_{EM} = \sum_{n_{\mathbf{k},\lambda}} \exp\left\{ -\beta \sum_{\lambda=1,2} \sum_{\mathbf{k}} h\nu_{\mathbf{k},\lambda} \left(n_{\mathbf{k},\lambda} + \frac{1}{2} \right) \right\}. \tag{14.10}$$

14.3.1 Thermodynamics for a single photon mode

To simplify evaluation of Eq. 14.10 we first examine the partition function for a single mode $\mathbf{k}_\alpha$

$$\mathcal{Z}_{EM}^{\alpha} = \sum_{n_{\mathbf{k}_\alpha,\lambda}=0,1,2,3,\ldots} \exp\left\{-\beta \sum_{\lambda=1,2} h\nu_{\mathbf{k}_\alpha,\lambda}\left(n_{\mathbf{k}_\alpha,\lambda}+\frac{1}{2}\right)\right\}, \tag{14.11}$$

where

$$\sum_{n_{\mathbf{k}_\alpha,\lambda}=0,1,2,3,\ldots} \tag{14.12}$$

is the sum over all integer photon numbers for a single mode $(\mathbf{k}_\alpha, \lambda)$. Explicitly summing over both polarizations in the exponent of Eq. 14.11,

$$\mathcal{Z}_{EM}^{\alpha} = e^{-\beta h\nu_{\mathbf{k}_\alpha}} \sum_{n_{\mathbf{k}_\alpha,1}} \exp\left[-\beta h\nu_{\mathbf{k}_\alpha}\left(n_{\mathbf{k}_\alpha,1}\right)\right] \sum_{n_{\mathbf{k}_\alpha,2}} \exp\left[-\beta h\nu_{\mathbf{k}_\alpha}\left(n_{\mathbf{k}_\alpha,2}\right)\right] \tag{14.13}$$

$$= e^{-\beta h\nu_{\mathbf{k}_\alpha}} \left\{\sum_{n_{\mathbf{k}_\alpha,1}} \exp\left[-\beta h\nu_{\mathbf{k}_\alpha}\left(n_{\mathbf{k}_\alpha,1}\right)\right]\right\}^2, \tag{14.14}$$

where the factor $e^{-\beta h\nu_{\mathbf{k}_\alpha}}$ results from the vacuum radiation term in the eigenenergies. The sum over photon numbers $n_{\mathbf{k}_\alpha,1} = 0, 1, 2, \ldots$ is just a geometric series, giving for the single mode $\mathbf{k}_\alpha$ and both polarizations

$$\mathcal{Z}_{EM}^{\alpha} = \left[e^{-\frac{\beta h\nu_{\mathbf{k}_\alpha}}{2}} \frac{1}{1-e^{-\beta h\nu_{\mathbf{k}_\alpha}}}\right]^2. \tag{14.15}$$

14.3.2 Average photon number

From Eq. 14.9 the average number of photons in the single mode $\mathbf{k}_\alpha$ with a single polarization $\lambda = 1$, say $\langle n_{\mathbf{k}_\alpha,1}\rangle$, is

$$\langle n_{\mathbf{k}_\alpha,1}\rangle = \frac{\sum_{n_{\mathbf{k}_\alpha,1}} n_{\mathbf{k}_\alpha,1} \exp\left\{-\beta h\nu_{\mathbf{k}_\alpha,1} n_{\mathbf{k}_\alpha,1}\right\}}{\sum_{n_{\mathbf{k}_\alpha,1}} \exp\left\{-\beta h\nu_{\mathbf{k}_\alpha,1} n_{\mathbf{k}_\alpha,1}\right\}}. \tag{14.16}$$

This is identical with

$$\langle n_{k_\alpha,1}\rangle = \left[\frac{1}{2h\nu_{k_\alpha,1}} \frac{\partial}{\partial\beta} \ln\left(-\mathcal{Z}_{EM}^{\alpha}\right)\right] - \frac{1}{2} \tag{14.17}$$

$$= \frac{1}{e^{\beta h\nu_{k_\alpha,1}} - 1}, \tag{14.18}$$

which is called a *Planck distribution function.*

14.3.3 Helmholtz potential and internal energy

The Helmholtz potential F_α for a single mode $\mathbf{k}_\alpha$ is

$$F_\alpha = -\frac{1}{\beta}\log \mathcal{Z}^{\alpha}_{EM} \tag{14.19}$$

$$= h\nu_{\mathbf{k}_\alpha} + \frac{2}{\beta}\log\left(1 - e^{-\beta h\nu_{\mathbf{k}_\alpha}}\right). \tag{14.20}$$

Similarly, for a single mode $\mathbf{k}_\alpha$ (counting both polarizations) the radiation field internal energy $\mathcal{U}_\alpha$ is

$$\mathcal{U}_\alpha = -\frac{\partial}{\partial\beta}\ln Z^{\alpha}_{EM} \tag{14.21}$$

$$= h\nu_{\mathbf{k}_\alpha}\left(\frac{e^{\beta h\nu_{\mathbf{k}_\alpha}}+1}{e^{\beta h\nu_{\mathbf{k}_\alpha}}-1}\right) \tag{14.22}$$

$$= 2h\nu_{\mathbf{k}_\alpha}\left(\frac{1}{2} + \frac{1}{e^{\beta h\nu_{\mathbf{k}_\alpha}}-1}\right) \tag{14.23}$$

$$= 2h\nu_{\mathbf{k}_\alpha}\left(\frac{1}{2} + \langle n_{\mathbf{k}_\alpha}\rangle\right). \tag{14.24}$$

14.4 Radiation field thermodynamics

The thermal electromagnetic field assembles all photon modes $\mathbf{k}$ with both transverse polarizations $\lambda = 1, 2$, requiring the full evaluation of Eq. 14.10. Rewriting $\mathcal{Z}_{EM}$ as

$$\mathcal{Z}_{EM} = \left[\sum_{n_{\mathbf{k}_1}=0,1,2,\ldots} e^{-\beta h\nu_{\mathbf{k}_1}\left(n_{\mathbf{k}_1}+1/2\right)}\right]^2 \left[\sum_{n_{\mathbf{k}_2}=0,1,2,\ldots} e^{-\beta h\nu_{\mathbf{k}_2}\left(n_{\mathbf{k}_2}+1/2\right)}\right]^2 \cdots \tag{14.25}$$

$$= \prod_{\mathbf{k}}\left[e^{-\beta h\nu_{\mathbf{k}}}\left(\frac{1}{1-e^{-\beta h\nu_{\mathbf{k}}}}\right)^2\right], \tag{14.26}$$

in analogy with Eqs. 14.19 and 14.24 the Helmholtz potential F_{EM} is

$$F_{EM} = -\frac{1}{\beta}\sum_{\mathbf{k}}\ln\left[e^{-\beta h\nu_{\mathbf{k}}}\left(\frac{1}{1-e^{-\beta h\nu_{\mathbf{k}}}}\right)^2\right] \tag{14.27}$$

$$= \sum_{\mathbf{k}}\left[h\nu_{\mathbf{k}} + \frac{2}{\beta}\ln\left(1-e^{-\beta h\nu_{\mathbf{k}}}\right)\right] \tag{14.28}$$

and the internal energy $\mathcal{U}_{EM}$ is

$$\mathcal{U}_{EM} = 2\sum_{\mathbf{k}} h\nu_{\mathbf{k}}\left(\frac{1}{2} + \langle n_{\mathbf{k}}\rangle\right). \tag{14.29}$$

14.5 Stefan–Boltzmann, Planck, Rayleigh–Jeans laws

In the limit of macroscopic volume V the sum over $\mathbf{k}$ in Eq. 14.29 is replaced by (see Appendix H)

$$\sum_{\mathbf{k}}^{\infty} \Rightarrow \frac{V}{(2\pi)^3} \int_{-\infty}^{\infty} \mathrm{d}k_x \int_{-\infty}^{\infty} \mathrm{d}k_y \int_{-\infty}^{\infty} \mathrm{d}k_z, \tag{14.30}$$

so that ignoring the divergent, constant vacuum energy

$$\mathcal{U}_{EM} = 2\frac{V}{(2\pi)^3} \int_{-\infty}^{\infty} \mathrm{d}k_x \int_{-\infty}^{\infty} \mathrm{d}k_y \int_{-\infty}^{\infty} \mathrm{d}k_z \left(\frac{h\nu_{\mathbf{k}}}{e^{\beta h\nu_{\mathbf{k}}} - 1} \right). \tag{14.31}$$

But in vacuum we have Eq. 14.6 so that the integrals can be carried out in spherical coordinates, in which case

$$\int_{-\infty}^{\infty} \mathrm{d}k_x \int_{-\infty}^{\infty} \mathrm{d}k_y \int_{-\infty}^{\infty} \mathrm{d}k_z = \int_0^{2\pi} \mathrm{d}\varphi \int_0^{\pi} \mathrm{d}\theta \sin\theta \int_0^{\infty} \mathrm{d}\,|\mathbf{k}|\, |\mathbf{k}|^2 \tag{14.32}$$

and the internal energy $\mathcal{U}_{EM}$ is (with $h = 2\pi\hbar$)

$$\mathcal{U}_{EM} = \frac{\hbar c V}{\pi^2} \int_0^{\infty} \mathrm{d}\,|\mathbf{k}|\, |\mathbf{k}|^3 \left(\frac{1}{e^{\beta\hbar c|\mathbf{k}|} - 1} \right). \tag{14.33}$$

The integral can be brought into a more standard form with the substitution

$$|\mathbf{k}| = \frac{x}{\beta\hbar c} \tag{14.34}$$

so that

$$\mathcal{U}_{EM} = \frac{\hbar c V}{\pi^2 (\beta\hbar c)^4} \int_0^{\infty} \mathrm{d}x\, x^3 \left(\frac{1}{e^x - 1} \right). \tag{14.35}$$

The integral in Eq. 14.35 is one of several similar integrals that appear in thermal radiation theory. They are somewhat subtle to evaluate but they can be found in comprehensive tables or computed with Mathematica with the result

$$\int_0^{\infty} \mathrm{d}x\, x^3 \left(\frac{1}{e^x - 1} \right) = \frac{\pi^4}{15} \tag{14.36}$$

so that the radiation energy density is

$$\frac{\mathcal{U}_{EM}}{V} = \frac{\pi^2}{15\beta^4 (\hbar c)^3}. \tag{14.37}$$

This is called the *Stefan–Boltzmann radiation law.*

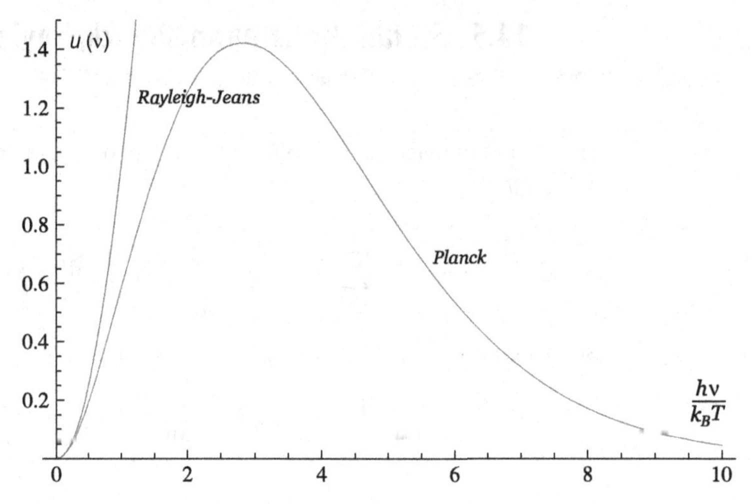

Fig. 14.1 Planck radiation law $u(\nu)$ and Rayleigh–Jeans approximation.

If in Eq. 14.35 we substitute $x = \beta h\nu$, the radiation energy density may be written

$$\frac{\mathcal{U}_{EM}}{V} = \int_0^\infty d\nu\, u(\nu), \tag{14.38}$$

where

$$u(\nu) = \frac{8\pi h}{c^3} \frac{\nu^3}{e^{\beta h\nu} - 1}, \tag{14.39}$$

which is the frequency distribution of thermal radiation at any temperature T. It is called the *spectral density* or *Planck's radiation law*. In the high temperature limit, $\beta h\nu \ll 1$, the Planck law becomes the classical *Rayleigh–Jeans law* (see Figure 14.1):

$$u_{classical}(\nu) = \frac{8\pi k_B T}{c^3}\nu^2. \tag{14.40}$$

The cancellation of Planck's constant removes any quantum reference.

Example

What is the average number density of thermal photons – all modes, both polarizations – at temperature T?

Beginning with Eq. 14.17 the average photon number is

$$\langle n\rangle = \sum_{\substack{\mathbf{k} \\ \lambda=1,2}} \langle n_{\mathbf{k},\lambda}\rangle \tag{14.41}$$

$$= \frac{2V}{(2\pi)^3}\int d\mathbf{k}\frac{1}{e^{\beta h\nu_{\mathbf{k}}} - 1}. \tag{14.42}$$

In spherical coordinates this becomes (with $\nu = \frac{ck}{2\pi}$)

$$\langle n \rangle = \frac{2V}{(2\pi)^3} \int_0^{2\pi} \mathrm{d}\phi \int_0^{\pi} \mathrm{d}\theta \sin\theta \int_0^{\infty} \mathrm{d}k \frac{k^2}{e^{\beta\hbar ck} - 1} \tag{14.43}$$

$$= \frac{V}{\pi^2} \int_0^{\infty} \mathrm{d}k \frac{k^2}{e^{\beta\hbar ck} - 1}. \tag{14.44}$$

After the change of variable $\beta\hbar ck = x$ the number density is

$$\frac{\langle n \rangle}{V} = \frac{1}{\pi^2 (\beta\hbar c)^3} \int_0^{\infty} \mathrm{d}x \frac{x^2}{e^x - 1}. \tag{14.45}$$

This integral resembles Eq. 14.36, but unlike that integral this one has no result in terms of familiar constants. But it can be estimated by the following series of steps.

$$\int_0^{\infty} \mathrm{d}x \frac{x^2}{e^x - 1} = \int_0^{\infty} \mathrm{d}x \frac{x^2 e^{-x}}{1 - e^{-x}} \tag{14.46}$$

$$= \int_0^{\infty} \mathrm{d}x x^2 e^{-x} \sum_{s=0}^{\infty} e^{-sx} \tag{14.47}$$

$$= \sum_{s=0}^{\infty} \int_0^{\infty} \mathrm{d}x x^2 e^{-x(s+1)} \tag{14.48}$$

$$= \sum_{s=0}^{\infty} \frac{2}{(s+1)^3} \tag{14.49}$$

$$= 2 \sum_{s'=1}^{\infty} \frac{1}{s'^3}. \tag{14.50}$$

The last sum is called the *Riemann ζ function.*[5] In this case we have $\zeta(3)$ which can be evaluated by summing, term by term, to any desired accuracy, giving $\zeta(3) = 1.202\,06\ldots$

Therefore the photon number density is

$$\frac{\langle n \rangle}{V} = \frac{2\zeta(3)}{\pi^2 (\beta\hbar c)^3}. \tag{14.51}$$

[5] $\zeta(z) = \sum_{n=1}^{\infty} \frac{1}{n^z}$. For even integer values of z (but not for odd values) the ζ function can be found in closed form. In general, ζ is complex.

14.6 Wien's law

In stellar astronomy two revealing parameters are a star's surface temperature and its luminosity. The surface temperature is found from its color, which corresponds to the frequency peak in the spectral density curve of Eq. 14.39. Determining luminosity is more complex, involving observed brightness and the distance to the star.

These two pieces of information determine the star's location on the empirical but important Hertzsprung–Russell diagram (which you can read about in any introductory astronomy text). From this it is possible to accurately determine the chemistry of the star, its age and its stage of evolution.

The spectral peak is determined from Eq. 14.39 by differentiation, i.e.

$$W\left(\beta h\nu\right) = \frac{\mathrm{d}}{\mathrm{d}\nu}\left(\frac{\nu^3}{e^{\beta h\nu}-1}\right) = 0, \tag{14.52}$$

which gives

$$W\left(\beta h\nu\right) = 3e^{\beta h\nu} - 3 - \beta h\nu\, e^{\beta h\nu} = 0. \tag{14.53}$$

Solving graphically (see Figure 14.2)

$$h\nu_{max} = 2.82\, k_B\, T. \tag{14.54}$$

The frequency ν_{max} at which the peak of Planck's radiation curve is located is proportional to the absolute temperature of the radiating source. This is called *Wien's law*.

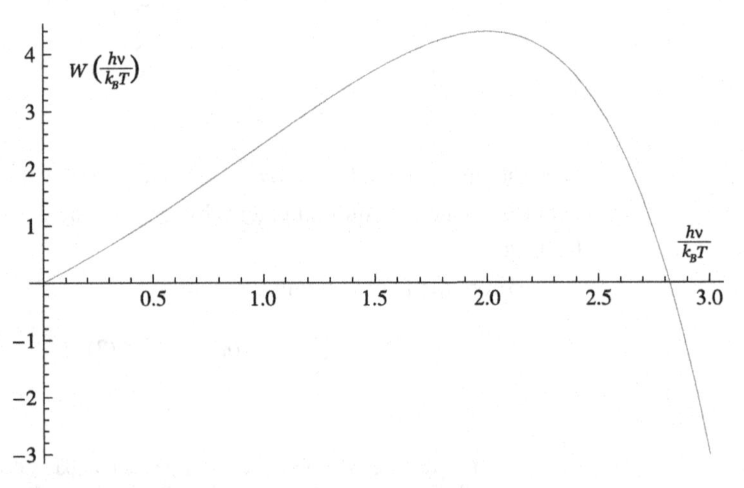

Fig. 14.2 Graphical determination of Wien's law constant.

14.7 Entropy of thermal radiation

The entropy of thermal radiation is

$$\mathcal{S}_{EM} = -k_B \sum_{n_{k,\lambda}=0,1,2,\ldots} \mathbf{P}(n_{\mathbf{k},\lambda}) \ln \mathbf{P}(n_{\mathbf{k},\lambda}), \tag{14.55}$$

with $\mathbf{P}(n_{\mathbf{k},\lambda})$ as found in Eq. 14.9. This is equivalent to

$$\mathcal{S}_{EM} = -k_B \beta^2 \frac{\partial}{\partial \beta} \ln \mathcal{Z}_{EM} \tag{14.56}$$

$$= k_B \beta^2 \left(\frac{\partial F_{EM}}{\partial \beta} \right)_V, \tag{14.57}$$

where $\mathcal{Z}_{EM}$ is the partition function (see Eq. 14.10) and F_{EM} is the electromagnetic Helmholtz potential (see e.g. Eq. 14.19). Completing the full radiation field calculation of Eq. 14.19 requires summing over all modes. Substituting $x = \beta h \nu$ into that equation and ignoring the vacuum field contribution (it does not depend on β), the Helmholtz potential is the integral

$$F_{EM} = \frac{8\pi V}{\beta^4 (hc)^3} \int_0^\infty \mathrm{d}x\, x^2 \ln\left(1 - e^{-x}\right). \tag{14.58}$$

Integrating by parts gives

$$F_{EM} = \frac{8\pi V}{3\beta^4 (hc)^3} \int_0^\infty \mathrm{d}x \frac{x^3}{(e^x - 1)} \tag{14.59}$$

$$= -\frac{V\pi^2}{45\hbar^3 c^3 \beta^4}, \tag{14.60}$$

where the integral Eq. 14.36 has been used. Finally, applying Eq. 14.56, the entropy is

$$\mathcal{S}_{EM} = \frac{4\pi^2 k_B V}{45 (\beta \hbar c)^3} \tag{14.61}$$

which is proportional to VT^3, consistent with a conjecture that with increasing "radius" the temperature of the "universe" falls proportional to $1/T$.

14.8 Stefan–Boltzmann radiation law

An object at temperature T radiates electromagnetic energy from its surface (see Figure 14.3). The energy radiated per unit area per unit time (energy current density)

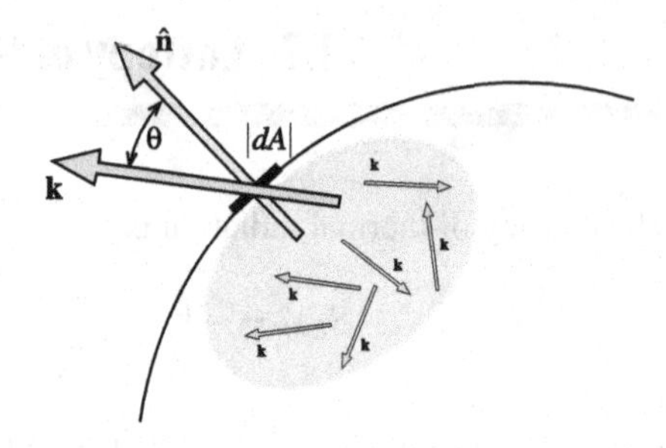

Fig. 14.3 Photon **k** leaving the surface of a radiating body.

is the *Poynting vector* **S**, which in the classical theory is

$$\mathbf{S}_{classical} = \frac{1}{8\pi} \int_V d\mathbf{x}\, \mathcal{E} \times \mathcal{B}. \tag{14.62}$$

The Poynting vector for photons **k**, λ is

$$\mathbf{S}_{\mathbf{k},\lambda} = V^{-1} h\nu_{\mathbf{k}} \langle n_{\mathbf{k},\lambda}\rangle\, c\, \hat{\mathbf{k}}, \tag{14.63}$$

where $\hat{\mathbf{k}}$ is the unit vector in the direction of propagation of the mode **k**, λ and

$$\langle n_{\mathbf{k},\lambda}\rangle = \frac{1}{e^{\beta h\nu_{\mathbf{k}}} - 1}. \tag{14.64}$$

(See Eq. 14.17.)

The differential radiation flux density $d\Phi_{\mathbf{k},\lambda}$ of the mode **k** with polarization λ from the element of area $d\mathbf{A} = \hat{\mathbf{n}}\, dA$ (see Figure 14.3) is

$$d\Phi_{\mathbf{k},\lambda} = \mathbf{S}_{\mathbf{k},\lambda} \cdot d\mathbf{A} \tag{14.65}$$

with the total radiation flux differential $d\Phi$

$$d\Phi = V^{-1} \sum_{\substack{\mathbf{k}\\ \lambda=1,2}} h\nu_{\mathbf{k}} \langle n_{\mathbf{k},\lambda}\rangle\, c\, \hat{\mathbf{k}} \cdot \hat{\mathbf{n}}\; dA, \tag{14.66}$$

where $\hat{\mathbf{n}}$ is the unit vector normal to the surface element dA (see Figure 14.3). The total radiation flux density per unit area is therefore

$$\frac{d\Phi}{dA} = V^{-1} \sum_{\substack{\mathbf{k}\\ \lambda=1,2}} h\nu_{\mathbf{k}} \langle n_{\mathbf{k},\lambda}\rangle\, c\, \hat{\mathbf{k}} \cdot \hat{\mathbf{n}}. \tag{14.67}$$

With $\hat{\mathbf{k}} \cdot \hat{\mathbf{n}} = \cos\theta$ the total outward $(0 < \theta < \pi/2)$ radiation flux density per unit area is

$$\frac{d\Phi}{dA} = \frac{2}{(2\pi)^3} \int_0^{2\pi} d\phi \int_0^{\pi/2} d\theta \sin\theta \cos\theta \int_0^{\infty} dk\, \frac{ch\nu_{\mathbf{k}} k^2}{e^{\beta h\nu_{\mathbf{k}}} - 1}, \tag{14.68}$$

which is integrated to give

$$\frac{d\Phi}{dA} = \frac{\pi^2 (k_B T)^4}{60\hbar^3 c^2} \tag{14.69}$$

or

$$\frac{d\Phi}{dA} = \sigma_B T^4, \tag{14.70}$$

where

$$\sigma_B = \frac{\pi^2 {k_B}^4}{60\hbar^3 c^2} \tag{14.71}$$

$$= 5.67 \times 10^{-8}\,\mathrm{W\,m^{-2}\,K^{-4}} \tag{14.72}$$

is called the *Stefan–Boltzmann constant.*

Example: Surface temperature of exoplanet Gliese 581d

Gliese 581, an M-class red-dwarf star, has an approximate radius of 2×10^5 km and an effective surface temperature about 3480 K. A recently discovered exoplanet, designated Gliese 581d, orbits 3.27×10^7 km from the star. Assuming the star and the exoplanet behave as perfect absorbers and radiators, estimate the surface temperature of Gliese 581d.

$$\underbrace{4\pi R_*^2 \sigma_B T_*^4}_{\text{energy rate radiated from star}} \times \overbrace{\left\{\frac{\pi r_p^2}{4\pi R_{*\to p}^2}\right\}}^{\text{fraction intercepted by planetary disk}} \tag{14.73}$$

$$= \underbrace{4\pi r_p^2 \sigma_B T_p^4}_{\text{energy rate reradiated from planet}} \tag{14.74}$$

where r_p is the exoplanet radius, T_p is the exoplanet temperature, R_* is the star radius and $R_{*\to p}$ is the exoplanet's orbital radius.

After simplification we have

$$T_p = \left(\frac{R_*^2}{4R_{*\to p}^2}\right)^{1/4} T_*. \tag{14.75}$$

In particular, for Gliese 581d

$$T_g = 192\,\mathrm{K}, \tag{14.76}$$

not too far off from the "Goldilocks" zone of supporting life.

14.9 Radiation momentum density

The momentum p of a photon (a massless particle) is related, by special relativity, to its energy E_γ by

$$p = \frac{E_\gamma}{c}. \tag{14.77}$$

Therefore the radiated momentum flux density Π per unit area from an object at temperature T is

$$\frac{\mathrm{d}\Pi}{\mathrm{d}A} = V^{-1} \sum_{\substack{\mathbf{k} \\ \lambda=1,2}} \hbar\nu_{\mathbf{k}} \langle n_{\mathbf{k},\lambda} \rangle \hat{\mathbf{k}} \cdot \hat{\mathbf{n}}, \tag{14.78}$$

which is evaluated, as in Eqs. 14.67–14.69, to give

$$\frac{\mathrm{d}\Pi}{\mathrm{d}A} = \frac{\pi^2 (k_B T)^4}{60\hbar^3 c^3}. \tag{14.79}$$

Example: Star dust blow-out

When stars are formed by gravitational collapse they are initially embedded in "cocoons" of tiny silicate dust particles, a dust so dense that at this stage the star can be detected only from secondary infrared radiation emitted by the heated dust. When the newly formed star reaches a sufficiently high temperature, the radiation from the star begins blowing out the dust cocoon, allowing the star to be seen in visible light.

Assuming that the radiation absorption cross-section of a dust particle is approximately half of the physical cross-section and the dust particles have radii $a < 10^{-7}$ m, at what stellar temperature will all cocoon dust particles be blown away from the star?

The total momentum density per unit time radiated by the star is

$$\Pi = \frac{\pi^2 (k_B T_*)^4}{60\hbar^3 c^3} \times 4\pi R_*^2, \tag{14.80}$$

where T_* is the star's temperature and R_* is its radius. Therefore the momentum density per unit time intercepted by a dust particle of radius a at a distance R_{dust} from the star is

$$\Pi_{dust} = \frac{\pi^2 (k_B T_*)^4}{60\hbar^3 c^3} \times 4\pi R_*^2 \times \left(\frac{\pi a^2}{4\pi R_{dust}^2} \right). \tag{14.81}$$

Taking into account the estimated 50% absorption cross-section, the momentum density per unit time absorbed by a dust particle is

$$\Pi^A_{dust} = \frac{\pi^2 (k_B T_*)^4}{60\hbar^3 c^3} \times 4\pi R_*^2 \times \left(\frac{\pi a^2}{4\pi R_{dust}^2}\right) \times (1/2). \tag{14.82}$$

On the other hand, the force of gravity acting on the particle is, by Newton's law of gravitation,

$$F_{grav} = \frac{GM_*}{R_{dust}^2} \times \left(\frac{4\pi a^3}{3}\rho_{dust}\right), \tag{14.83}$$

where G is the gravitational constant and $\rho_{dust} = 3 \times 10^3\ \mathrm{kg\,m^{-3}}$ is the approximate density of silicate dust.

For the radiation-momentum absorbing dust particle to be in gravitational equilibrium

$$\frac{\pi^2 (k_B T_*)^4}{60\hbar^3 c^3} \times 4\pi R_*^2 \times \left(\frac{\pi a^2}{4\pi R_{dust}^2}\right) \times (1/2) = \frac{GM_*}{R_{dust}^2} \times \left(\frac{4\pi a^3}{3}\rho_{dust}\right), \tag{14.84}$$

where M_* is the mass of the star, so that

$$T_*^4 = \frac{8a}{3} \times \frac{cGM_*\rho_{dust}}{\sigma_B R_*^2}, \tag{14.85}$$

where σ_B is the Stefan–Boltzmann constant (see Eq. 14.71).

Assuming, for numerical specificity, that the newly formed star is sun-like with 1 solar mass and 1 solar radius, i.e.

$$M_* = 1.99 \times 10^{30}\ \mathrm{kg} \tag{14.86}$$

$$R_* = 6.95 \times 10^{8}\ \mathrm{m} \tag{14.87}$$

$$G = 6.67 \times 10^{-11}\ \mathrm{N\,m^2\,kg^{-2}}, \tag{14.88}$$

the dust blow-out temperature is $T \approx 6000\ \mathrm{K}$, independent of the distance between the star and the dust particles.

Problems and exercises

14.1 Show that the work done by photons during an isentropic expansion is

$$\mathcal{W} = \left(\frac{\pi^2}{15\hbar^3 c^3}\right) k_B^4 \left(T_i - T_f\right) V_i T_i^3, \tag{14.89}$$

where the subscripts i and f refer to initial and final states.

14.2 The total energy radiated by the Sun and received per unit time per unit area at the Earth's orbital distance is K_S, the solar constant, which has the value $K_S = 0.136\,\mathrm{J\,s^{-1}\,cm^{-2}}$.

The mean Earth–Sun distance is $R_{ES} = 1.5 \times 10^{11}$ m and the radius of the Sun is approximately $R_\odot = 7 \times 10^8$ m.

Find the effective blackbody temperature of the Sun.

14.3 Show that the average square fluctuation of blackbody internal energy

$$\langle(\Delta\mathcal{U})^2\rangle = \langle(\mathcal{U} - \langle\mathcal{U}\rangle)^2\rangle \tag{14.90}$$

is equal to

$$\langle(\Delta\mathcal{U})^2\rangle = \frac{\partial^2}{\partial\beta^2}\ln\mathcal{Z}. \tag{14.91}$$

14.4 Gliese 581 is an M-class red dwarf star which lies in the constellation Libra. It has an approximate radius of 2×10^5 km and an effective surface temperature about 3480 K.

Estimate the peak frequency at which the star radiates energy.

14.5 Consider a volume V containing thermal equilibrium electromagnetic radiation at a temperature T. Given that the partition function for electromagnetic radiation is (neglecting zero-point energy)

$$\mathcal{Z} = \prod_{\mathbf{k}}\left\{\left(1 - e^{-\beta\hbar\omega_{\mathbf{k}}}\right)^{-2}\right\}, \tag{14.92}$$

with the product over all modes $\mathbf{k}$ and counting both polarizations, evaluate Eq. 14.91 and show that the average square fluctuation in electromagnetic energy is

$$\langle(\Delta\mathcal{U})^2\rangle = \frac{4}{15\hbar^3c^3}\left(k_BT\right)^5. \tag{14.93}$$

14.6 Find the average square fluctuation in blackbody radiation particle number. Hint: You may need the integrals

$$\int_0^\infty dx\,x^3\left(\frac{1}{e^x - 1}\right) = \frac{\pi^4}{15}, \qquad \int_0^\infty dx\,\frac{x^4e^x}{(e^x - 1)^2} = \frac{4\pi^4}{15}. \tag{14.94}$$

14.7 The ideal Carnot engine operates reversibly between an energy source, in which the working substance (usually an ideal gas) absorbs heat isothermally at high temperature, and an energy sink, to which it exhausts heat isothermally at a low temperature. The remaining curves which close the Carnot cycle are adiabats. The efficiency η of such an engine is defined as

$$\eta = \frac{\mathcal{W}_r}{\mathcal{Q}_H}, \tag{14.95}$$

where $\mathcal{W}_r$ is the work done in the cycle and $\mathcal{Q}_H$ is the heat absorbed from the high-temperature energy source.

If light is used as the working substance, find the Carnot engine efficiency.

14.8 When nucleons collide at very high energy, say by shooting a proton beam into a metallic target, they annihilate and produce an amount of energy W mainly in the form of short-lived subatomic pions (quark–antiquark pairs), which can be positively charged π^+, negatively charged π^- or neutral π^0. The π^+ and π^- are particle–antiparticle pairs while the π^0 is its own antiparticle. The pions have ultra-relativistic energies so their rest masses are negligible, i.e. they behave like "photons" with energies

$$E = \hbar c\,|\mathbf{k}| \tag{14.96}$$

with periodic boundary conditions

$$|\mathbf{k}|^2 = {k_x}^2 + {k_y}^2 + {k_z}^2 \tag{14.97}$$

where

$$k_j = \frac{2\pi}{L}\nu_j; \quad \nu_j = 0, \pm 1, \pm 2, \ldots \tag{14.98}$$

Whereas photons have two polarizations (λ_1 *and* λ_2), pions are regarded as having three polarizations (π^+, π^0, π^-). Assuming:

- the pions have *zero* chemical potential,
- thermal equilibrium is rapidly reached (times shorter than the pion lifetimes),
- all the collision energy is produced within a volume V_π,

a. find an expression, in terms of W and V_π, for the temperature of the pion "spark";

b. find an expression, in terms of W and V_π, for the mean number of pions produced in the collision.

15 Ideal Fermi gas

Why is it that particles with half-integer spin are Fermi particles whereas particles with integer spin are Bose particles? An explanation has been worked out by Pauli from complicated arguments from quantum field theory and relativity. He has shown that the two must necessarily go together . . . but we have not been able to reproduce his arguments on an elementary level. This probably means we do not have a complete understanding of the fundamental principle involved. . .

R. P. Feynman, R.B. Leighton and M. Sands, *Feynman Lectures on Physics*, Volume 3, Chapter 4, Section 1, Addison-Wesley, Reading, MA (1963)

15.1 Introduction

Particles with half-integer angular momentum obey the *Pauli exclusion principle* (PEP) – a restriction that a non-degenerate single-particle quantum state can have occupation number of only 0 or 1. This restriction was announced by W. Pauli in 1924 for which, in 1945, he received the Nobel Prize in Physics. Soon after Pauli, the exclusion principle was generalized by P. Dirac and E. Fermi who – independently – integrated it into quantum mechanics. As a consequence half-integer spin particles are called *Fermi–Dirac particles* or *fermions*. PEP applies to electrons, protons, neutrons, neutrinos, quarks – and their antiparticles – as well as composite fermions such as He^3 atoms. Thermodynamic properties of metals and semiconductors are largely determined by electron (fermion) behavior. Metals, for example, exhibit the uniquely fermionic low-temperature heat capacity $C_V \propto T$, while some Fermi–Dirac systems undergo a transition to a distinctive superconducting state with signature properties of magnetic flux exclusion (no interior $\mathcal{B}$ field) and zero electrical resistivity.

15.2 Ideal gas eigen-energies

The eigen-energies and eigenfunctions of any ideal (non-interacting) gas are constructed from the eigenvalue problem

$$\mathbf{h}_{op}|\epsilon_s\rangle = \epsilon_s|\epsilon_s\rangle \tag{15.1}$$

with ϵ_s the single-particle eigen-energies and $|\bar{\epsilon}_s\rangle$ the corresponding single-particle eigenstates. $\mathbf{h}_{op}$ is the single-particle hamiltonian operator for the ith particle[1]:

$$\mathbf{h}_{op}(i) = -\frac{\hbar^2}{2m}\nabla_i^2 + \mathscr{V}(i). \tag{15.2}$$

Generally $\mathscr{V}(i)$ represents some charge-neutralizing positive background potential, e.g. positive ions, so obtaining a solution to the single-particle problem, i.e. Eqs. 15.1 and 15.2, may be computationally intensive. It is often convenient to assume that the background ions are replaced by a uniform positive charge, the *free electron approximation*. Presuming a solution to this reduced problem is obtained, we can proceed to investigate Fermi–Dirac system thermodynamics.

15.3 Grand partition function

A many-particle (open system) thermodynamic hamiltonian is

$$\hat{\mathcal{H}}_{op} = \mathbf{h}_{op} - \mu\mathcal{N}_{op}, \tag{15.3}$$

where the many-particle hamiltonian $\hat{\mathcal{H}}_{op}$ is a sum of single-particle hamiltonians

$$\hat{\mathcal{H}}_{op} = \sum_{i=1}\left\{\mathbf{h}_{op}(i) - \mu\mathcal{N}_{op}(i)\right\}. \tag{15.4}$$

The macroscopic eigen-energies of $\hat{\mathcal{H}}_{op}$ are

$$\hat{E}(n_1, n_2, \dots) = \sum_s(\epsilon_s - \mu)\,n_s, \tag{15.5}$$

where $n_s = 0, 1, 2, \dots$ are the number of particles occupying the single-particle eigenstate $|\epsilon_s\rangle$ and μ is the system chemical potential. The sum over s is a sum over all states.[2]

However, temporarily ignoring spin, each (non-degenerate) single-particle state is restricted by the PEP to have occupation number 0 (no particles) or 1 (one particle), i.e. $n_s = 0$ *or* 1. Constructing a thermal Lagrangian[3]

[1] This applies to particles of either spin.

[2] Each single-particle state can be pictured as an "open system" into and out of which particles "diffuse" so that in thermodynamic (diffusive) equilibrium all particles have the same chemical potential.

[3] See Appendix B.

$$\mathcal{L}_{FD} = -k_B \sum_{n_1,n_2,\ldots=0,1} \hat{\mathbf{P}}_{FD}(n_1,n_2,\ldots) \ln \hat{\mathbf{P}}_{FD}(n_1,n_2,\ldots)$$
$$-\frac{1}{T} \sum_{n_1,n_2,\ldots=0,1} \hat{\mathbf{P}}_{FD}(n_1,n_2,\ldots) \sum_s (\epsilon_s - \mu) n_s$$
$$-\lambda_0 \sum_{n_1,n_2,\ldots=0,1} \hat{\mathbf{P}}_{FD}(n_1,n_2,\ldots), \tag{15.6}$$

where λ_0 insures normalization.

$\hat{\mathbf{P}}_{FD}(n_1,n_2,\ldots)$ is the probability there are n_1 particles in the single-particle state $|\epsilon_1\rangle$, n_2 particles in the single-particle state $|\epsilon_2\rangle$, etc. and where the occupation numbers are restricted by the PEP, so that the sums mean

$$\sum_{n_1,n_2,\ldots=0,1} \Rightarrow \sum_{n_1=0,1} \sum_{n_2=0,1} \cdots \tag{15.7}$$

Maximizing $\mathcal{L}_{FD}$ with respect to $\hat{\mathbf{P}}_{FD}(n_1,n_2,\ldots)$ we find the probabilities

$$\hat{\mathbf{P}}_{FD}(n_1,n_2,\ldots) = \frac{e^{-\beta\sum_s(\epsilon_s-\mu)n_s}}{\sum_{n_1,n_2,\ldots=0,1} e^{-\beta\sum_s(\epsilon_s-\mu)n_s}}. \tag{15.8}$$

The denominator in Eq. 15.8 is identified as the grand partition function,

$$\mathcal{Z}_{gr}^{FD} = \sum_{n_1,n_2,\ldots=0,1} e^{-\beta\sum_s(\epsilon_s-\mu)n_s}. \tag{15.9}$$

It is evaluated by rearranging the sum over states s in the exponential to give

$$\mathcal{Z}_{gr}^{FD} = \prod_s \sum_{n_s=0,1} e^{-\beta(\epsilon_s-\mu)n_s} \tag{15.10}$$

and summing $n_s = 0,\ 1$

$$\mathcal{Z}_{gr}^{FD} = \prod_s \left[1 + e^{-\beta(\epsilon_s-\mu)}\right]. \tag{15.11}$$

15.4 Electron spin

Taking electron spin into account, a pair of single-particle states, $|\epsilon_{s,\uparrow}\rangle$, $|\epsilon_{s,\downarrow}\rangle$ emerges with corresponding eigen-energies $\epsilon_{s,\uparrow}$ and $\epsilon_{s,\downarrow}$. Assuming them initially to be non-degenerate

$$\mathcal{Z}_{gr}^{FD} = \prod_s \left[1 + e^{-\beta(\epsilon_{s,\uparrow}-\mu)}\right]\left[1 + e^{-\beta(\epsilon_{s,\downarrow}-\mu)}\right]. \tag{15.12}$$

However, in zero magnetic field a two-fold degeneracy arises with $\epsilon_{s,\uparrow} = \epsilon_{s,\downarrow} = \epsilon_s$ producing the partition function

$$\mathcal{Z}_{gr}^{FD} = \prod_s \left[1 + e^{-\beta(\epsilon_s - \mu)}\right]^2. \tag{15.13}$$

15.5 Fermi–Dirac thermodynamics

Thermodynamic properties of Fermi–Dirac systems are determined from $\mathbf{\Omega}_{gr}^{FD}$, the grand potential,

$$\mathbf{\Omega}_{gr}^{FD} = -\frac{1}{\beta} \ln \mathcal{Z}_{gr}^{FD}, \tag{15.14}$$

which, with Eq. 15.13, is

$$\mathbf{\Omega}_{gr}^{FD} = -\frac{2}{\beta} \sum_s \ln\left[1 + e^{-\beta(\epsilon_s - \mu)}\right], \tag{15.15}$$

where the sum $\sum_s$ is over all states $|\epsilon_s\rangle$.

1. The average particle number $\langle \mathcal{N}_{op} \rangle$ follows from Eq. 15.8,

$$\langle \mathcal{N}_{op} \rangle = \frac{2 \prod_s \left[\sum_{n_s=0,1} n_s e^{-\beta(\epsilon_s - \mu) n_s} \right]}{\prod_s \left[\sum_{n_s=0,1} e^{-\beta(\epsilon_s - \mu) n_s} \right]}, \tag{15.16}$$

the factor 2 accounting for spin degeneracy. An equivalent result is

$$\langle \mathcal{N}_{op} \rangle = -\left(\frac{\partial \mathbf{\Omega}_{gr}^{FD}}{\partial \mu} \right)_{T,V} \tag{15.17}$$

$$= 2 \sum_s \left[\frac{1}{e^{\beta(\epsilon_s - \mu)} + 1} \right] \tag{15.18}$$

$$= 2 \sum_s \langle n_s^{FD} \rangle, \tag{15.19}$$

where $\langle n_s^{FD} \rangle$ is a single-particle average occupation number for the state $|\bar{\epsilon}_s\rangle$,

$$\langle n_s^{FD} \rangle = \left[\frac{1}{e^{\beta(\epsilon_s - \mu)} + 1} \right], \tag{15.20}$$

which is called the *Fermi–Dirac function*.

Table 15.1 Average electron density $\langle \mathcal{N}_{op} \rangle / V$ for several elemental metals. Densities are in units of 10^{28} m^{-3}.

Element	$\frac{\langle \mathcal{N}_{op} \rangle}{V}$
Cu	8.47
Ag	5.86
Au	5.90
Al	18.1
Be	24.7
Sn	14.8
Mg	8.61
Fe	17.0
Pb	13.2
Hg	8.65

Using Eq. 15.18, the chemical potential can be calculated as a function of $\langle \mathcal{N}_{op} \rangle / V$ and temperature T. (See the examples below.) In free electron models the average fermion (electron) density $\langle \mathcal{N}_{op} \rangle / V$ (see Table 15.1) is a parameter that distinguishes among different materials.[4]

2. The Fermi–Dirac internal energy $\mathcal{U}_{FD}$ is found from the general result of Eq. 12.28

$$\mathcal{U}_{FD} = \left[\frac{\partial}{\partial \beta} \left(\beta \mathbf{\Omega}_{gr}^{FD} \right) \right]_{V,\mu} + \mu \langle N_{op} \rangle, \tag{15.21}$$

which, using Eq. 15.15, yields the intuitively apparent

$$\mathcal{U}_{FD} = 2 \sum_{s} \epsilon_s \langle n_s^{FD} \rangle. \tag{15.22}$$

3. The equation of state of the Fermi–Dirac gas is

$$pV = -\mathbf{\Omega}_{gr}^{FD} \tag{15.23}$$

$$= -\frac{2}{\beta} \sum_{s} \ln \left(1 - \langle n_s^{FD} \rangle \right). \tag{15.24}$$

4. The entropy $\mathcal{S}$, as discussed in Chapter 6, is

$$\mathcal{S} = -k_B \sum_{n_1, n_2, \ldots = 0,1} \hat{\mathbf{P}}_{FD} (n_1, n_2, \ldots) \ln \hat{\mathbf{P}}_{FD} (n_1, n_2, \ldots) \tag{15.25}$$

[4] At $T = 0$ the electron chemical potential is referred to as the Fermi energy, ϵ_F: i.e. $\mu (T = 0) \equiv \epsilon_F$.

where $\hat{\mathbf{P}}_{FD}(n_1, n_2, \ldots)$ is as in Eq. 15.8. Equivalently,

$$\mathcal{S} = k_B\beta^2 \left[\frac{\partial \mathbf{\Omega}_{gr}^{FD}}{\partial \beta}\right]_{\mu,V} \tag{15.26}$$

$$= -2k_B \sum_s \left[\left(1 - \langle n_s^{FD}\rangle\right) \ln\left(1 - \langle n_s^{FD}\rangle\right) + \langle n_s^{FD}\rangle \ln \langle n_s^{FD}\rangle\right]. \tag{15.27}$$

15.6 Independent fermion model

The independent fermion approximation for metals usually includes interactions with a static background of positive ions (band structure approximation) but neglects both electron–electron repulsion (correlations) and interactions with vibrating ions (electron–phonon interactions).[5]

Independent fermion thermodynamics begins with $\mathbf{\Omega}_{gr}^{FD}$ (see Eq. 15.15) where, using Dirac's delta-function[6]

$$\int_{-\infty}^{\infty} d\omega f(\omega)\, \delta(\omega - \epsilon_s) = f(\epsilon_s), \tag{15.28}$$

$\mathbf{\Omega}_{gr}^{FD}$ may be strategically rewritten as

$$\mathbf{\Omega}_{gr}^{FD} = -\frac{2}{\beta}\int_0^{\infty} d\omega \sum_s \delta(\omega - \epsilon_s) \ln\left[1 + e^{-\beta(\omega-\mu)}\right]. \tag{15.29}$$

Identifying the density of single-particle states $\mathscr{D}(\omega)$ (see Appendix E)[7]

$$\mathscr{D}(\omega) = \sum_s \delta(\omega - \epsilon_s), \tag{15.30}$$

$\mathbf{\Omega}_{gr}^{FD}$ becomes

$$\mathbf{\Omega}_{gr}^{FD} = -\frac{2}{\beta}\int_0^{\infty} d\omega \mathscr{D}(\omega) \ln\left[1 + e^{-\beta(\omega-\mu)}\right]. \tag{15.31}$$

[5] These interactions can often be summarized for individual systems by a few parameters, such as an effective mass m^* and an electron–phonon coupling constant λ. The "Fermi gas" model has been successful in describing many metals and semiconductors.

[6] $\delta(\omega - \epsilon_s)$ is the Dirac delta function.

[7] Density of single-particle states defined here does not include a factor 2 for spin. (See Eq. E.21.)

Introducing a density of states also enables some experimental results to be used in electronic structure calculations.[8]

Using the δ function in an identical way, Eqs. 15.18 and 15.22 become

$$\langle \mathcal{N}_{op} \rangle = 2 \int_0^\infty d\omega \frac{\mathcal{D}(\omega)}{e^{\beta(\omega-\mu)}+1} \tag{15.32}$$

and

$$\mathcal{U}_{FD} = 2 \int_0^\infty d\omega\, \omega \frac{\mathcal{D}(\omega)}{e^{\beta(\omega-\mu)}+1}. \tag{15.33}$$

Evaluating, or even approximating, integrals Eqs. 15.32, 15.33 and related types that appear in degenerate fermion models ($\beta\mu \gg 1$) requires specialized techniques. The simplest of these is Sommerfeld's asymptotic approximation, which is discussed in Appendix I.[9,10]

15.7 The chemical potential ($T \neq 0$)

Thermodynamic properties of a degenerate Fermi gas ($\beta\mu \gg 1$) can be expressed as expansions[11] in T. In particular, the chemical potential is determined from the density $\frac{\langle \mathcal{N}_{op} \rangle}{V}$ by expanding the right-hand side of Eq. 15.32 about $\mu = \epsilon_F$ to linear order, where ϵ_F (Fermi energy) is the symbol assigned to the chemical potential at $T = 0$, to give

$$\langle \mathcal{N}_{op} \rangle = 2 \int_0^\infty d\omega \frac{\mathcal{D}(\omega)}{e^{\beta(\omega-\epsilon_F)}+1} + \frac{\beta(\mu-\epsilon_F)}{2} \int_0^\infty d\omega \mathcal{D}(\omega)\, \mathrm{sech}^2\left[\frac{\beta}{2}(\omega-\epsilon_F)\right]. \tag{15.34}$$

[8] Approximate densities of states can be obtained from optical or tunneling experiments.

[9] Sommerfeld's asymptotic approximation is not an expansion in the usual Taylor series sense. It takes advantage of the property that for $T \approx 0$ the Fermi–Dirac function $\langle n_s^{FD} \rangle$ approaches the unit step function at $\omega \approx E_f$ (see Figure 15.1).

[10] With increasing temperature, $\beta\mu \ll 1$, the steepness of the step declines over a width $\sim k_B T$, eventually taking the form of a smooth exponential $\frac{1}{e^{\beta(\omega-\mu)}+1} \to e^{-\beta(\omega-\mu)}$.

[11] These expansions are not the usual convergent Taylor series type but belong to a category called *asymptotic expansions*. These, generally, must be handled with care.

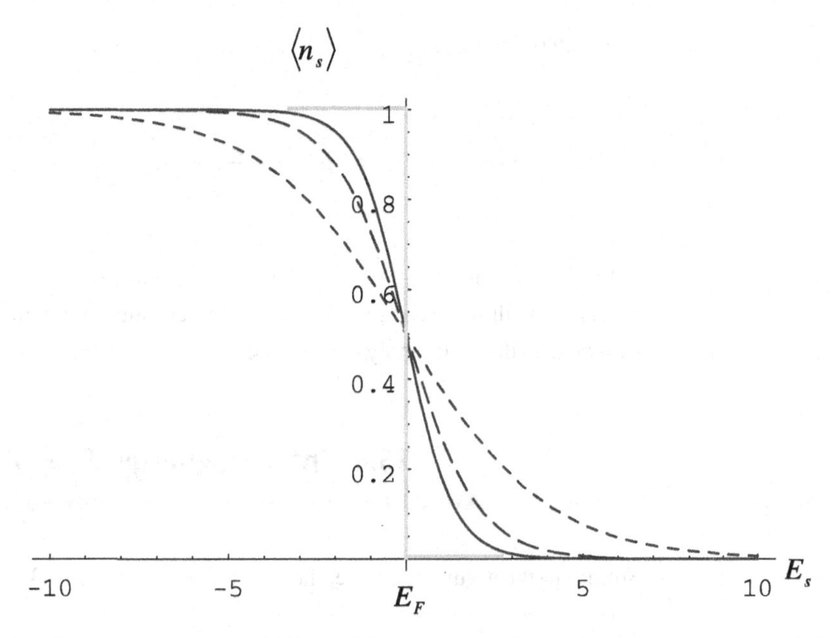

Fig. 15.1 The Fermi–Dirac function in Eq. 15.20. The solid line represents $\langle n_S \rangle$ at low temperature. The short dashed curve is representative of higher temperatures. With increasing temperature $\langle n_S \rangle \to \exp(-\beta E_S)$ – the semi-classical limit. At $T = 0$, $\langle n_S \rangle$ is a step function with unit jump at $E_S = E_F$.

The first integral on the right-hand side can be approximated by Sommerfeld's method (see Appendix I) to give

$$\int_0^\infty d\omega \frac{\mathscr{D}(\omega)}{e^{\beta(\omega-\epsilon_F)}+1} \approx \int_0^{\epsilon_F} d\omega \mathscr{D}(\omega) + \frac{\pi^2}{6\beta^2}\mathscr{D}'(\epsilon_F), \tag{15.35}$$

while the second integral is approximated by

$$\frac{\beta}{4}\int_0^\infty d\omega \mathscr{D}(\omega)\,\text{sech}^2\left[\frac{\beta}{2}(\omega-\epsilon_F)\right] \approx \mathscr{D}(\epsilon_F) + \frac{\pi^2}{6\beta^2}\mathscr{D}''(\epsilon_F), \tag{15.36}$$

where $\mathscr{D}(\omega)$ has been similarly expanded about $\mu = \epsilon_F$. Combining Eqs. 15.35 and 15.36 with Eq. 15.34, while noting that

$$2\int_0^{\epsilon_F} d\omega \mathscr{D}(\omega) = \langle \mathcal{N}_{op} \rangle, \tag{15.37}$$

gives

$$\mu = \epsilon_F - \frac{\dfrac{\pi^2}{6\beta^2}\mathscr{D}'(\epsilon_F)}{\mathscr{D}(\epsilon_F) + \dfrac{\pi^2}{6\beta^2}\mathscr{D}''(\epsilon_F)}, \tag{15.38}$$

which to order T^2 is

$$\mu = \epsilon_F - \frac{\pi^2}{6\beta^2} \frac{\mathscr{D}'(\epsilon_F)}{\mathscr{D}(\epsilon_F)}. \tag{15.39}$$

In the degenerate regime (where this expansion applies) chemical potential decreases with increasing temperature. At very high temperature (where this expansion clearly does not apply) $\mu \to -\infty$.

15.8 Internal energy ($T \neq 0$)

Similar to the argument above, the internal energy (Eq. 15.33) is first expanded about $\mu = \epsilon_F$,

$$\int_0^\infty d\omega \frac{\omega \mathscr{D}(\omega)}{e^{\beta(\omega-\mu)}+1} \approx \int_0^\infty d\omega \frac{\omega \mathscr{D}(\omega)}{e^{\beta(\omega-\epsilon_F)}+1} + \frac{\beta(\mu-\epsilon_F)}{4} \int_0^\infty d\omega\, \omega \mathscr{D}(\omega) \times \operatorname{sech}^2\left[\frac{\beta}{2}(\omega-\epsilon_F)\right]. \tag{15.40}$$

Applying Sommerfeld's approximation to the first term on the right-hand side,

$$\int_0^\infty d\omega \frac{\omega \mathscr{D}(\omega)}{e^{\beta(\omega-\epsilon_F)}+1} \approx \int_0^{\epsilon_F} d\omega\, \omega \mathscr{D}(\omega) + \frac{\pi^2}{6\beta^2} \left\{\frac{d}{d\omega'}\left[\omega' \mathscr{D}(\omega')\right]\right\}_{\omega'=\epsilon_F}, \tag{15.41}$$

while the second term becomes

$$\int_0^\infty d\omega\, \omega \mathscr{D}(\omega) \operatorname{sech}^2\left[\frac{\beta}{2}(\omega-\epsilon_F)\right] \approx \frac{4}{\beta}\epsilon_F \mathscr{D}(\epsilon_F). \tag{15.42}$$

Combining Eqs. 15.40, 15.41, 15.42 with 15.39

$$\mathcal{U} = 2\int_0^{\epsilon_F} d\omega\, \omega \mathscr{D}(\omega) + \frac{\pi^2}{3\beta^2} \mathscr{D}(\epsilon_F) \tag{15.43}$$

from which the constant volume heat capacity

$$\mathcal{C}_V = \left(\frac{\partial \mathcal{U}}{\partial T}\right)_V$$

is

$$\mathcal{C}_V = \frac{2\pi^2}{3} k_B^2 \mathscr{D}(\epsilon_F)\, T, \tag{15.44}$$

quite different from both the constant-valued classical result and the T^3 phonon contribution.[12] Macroscopic heat capacity is proportional to $\mathscr{D}(\epsilon_F)$, the density of states of the Fermi system evaluated at the Fermi energy, a microscopic quantum property. This particular form follows because the PEP allows only an effective number of particles near ϵ_F, i.e. $N_{eff} \sim N k_B T$, to participate in low-temperature thermal processes.

15.9 Pauli paramagnetic susceptibility

In the presence of an external magnetic field $\mathcal{B}_0$, the spin degeneracy of conduction electrons is lifted (Zeeman splitting) and the single-particle states in Eq. 15.48 become

$$\epsilon_{s,\uparrow} \rightarrow \epsilon_s + \Delta, \tag{15.45}$$

$$\epsilon_{s,\downarrow} \rightarrow \epsilon_s - \Delta, \tag{15.46}$$

where

$$\Delta = \boldsymbol{m} \cdot \mathcal{B}_0, \tag{15.47}$$

with $\boldsymbol{m}$ the electron magnetic moment. In that case

$$\mathcal{Z}_{gr}^{FD} = \prod_s \left[1 + e^{-\beta(\epsilon_s + \Delta - \mu)}\right]\left[1 + e^{-\beta(\epsilon_s - \Delta - \mu)}\right] \tag{15.48}$$

and

$$\boldsymbol{\Omega}_{gr}^{FD} = -\frac{1}{\beta}\sum_s \left\{\ln\left[1 + e^{-\beta(\epsilon_s + \Delta - \mu)}\right] + \ln\left[1 + e^{-\beta(\epsilon_s - \Delta - \mu)}\right]\right\}, \tag{15.49}$$

where the sum $\sum_s$ is over all states $|\bar{\epsilon}_s\rangle$. Using the single-particle density of states $\mathscr{D}(\omega)$, the sums are replaced by integrals[13]

$$\boldsymbol{\Omega}_{gr}^{FD} = -\frac{1}{\beta}\left\{\int_0^\infty d\omega \mathscr{D}(\omega) \ln\left[1 + e^{-\beta(\omega + \Delta - \mu)}\right] + \int_0^\infty d\omega \mathscr{D}(\omega) \ln\left[1 + e^{-\beta(\omega - \Delta - \mu)}\right]\right\}. \tag{15.50}$$

[12] The single-particle density of states defined in this work does not include the factor 2 for spin.

[13] The reason for not having incorporated the spin factor 2 into the density of states should now be apparent.

For weak fields, expansions in Δ produce

$$\begin{aligned}
&\ln\left[1+e^{-\beta(\omega+\Delta-\mu)}\right]+\ln\left[1+e^{-\beta(\omega-\Delta-\mu)}\right] \\
&\quad\approx 2\ln\left[1+e^{-\beta(\omega-\mu)}\right]+\frac{1}{4}\beta^2\Delta^2\,\mathrm{sech}^2\left[\frac{\beta}{2}(\omega-\mu)\right]
\end{aligned} \tag{15.51}$$

so that keeping only the term of order Δ^2

$$\boldsymbol{\Omega}_{gr}^{FD}\approx-\frac{1}{\beta}\left\{\frac{\beta^2}{4}\Delta^2\int_0^\infty d\omega\mathscr{D}(\omega)\,\mathrm{sech}^2\left[\frac{\beta}{2}(\omega-\mu)\right]\right\}. \tag{15.52}$$

The integral is done as discussed in Appendix I, giving in the degenerate limit $(\beta\mu \gg 1)$

$$\boldsymbol{\Omega}_{gr}^{FD}\approx-\Delta^2\mathscr{D}(\epsilon_F) \tag{15.53}$$

where from Eq. 15.39 only the leading term is used. With

$$M=-\left(\frac{\partial}{\partial\mathcal{B}_0}\boldsymbol{\Omega}_{gr}^{FD}\right)_{T,\mu} \tag{15.54}$$

$$=2\boldsymbol{m}^2\mathcal{B}_0\mathscr{D}(\epsilon_F) \tag{15.55}$$

the Pauli susceptibility, $\chi_M = M/\mathcal{B}_0$, is[14]

$$\chi_M=2\boldsymbol{m}^2\mathscr{D}(\epsilon_F)\,. \tag{15.56}$$

15.10 Electron gas model

The electron gas model is used to describe the behavior of simple metals. It assumes a macroscopic collection of non-interacting spin-half particles, neutralized only by a uniform positive background, with a single-particle hamiltonian for the ith particle

$$\mathbf{h}_{op}(i)=-\frac{\hbar^2}{2m}\nabla_i^2. \tag{15.57}$$

Choosing periodic boundary conditions (see Appendix G), the single-particle eigenenergies are

$$\epsilon(\mathbf{k})=\frac{\hbar^2|\mathbf{k}|^2}{2m}, \tag{15.58}$$

where k_j $(j = x, y, z)$ are quantum numbers which assume the values

$$k_j=\frac{2\pi}{L}\nu_j \tag{15.59}$$

[14] If the density of states is defined to include the factor 2 for electron spin, as some authors prefer, $\chi_M = \boldsymbol{m}^2\mathscr{D}(\epsilon_F)$.

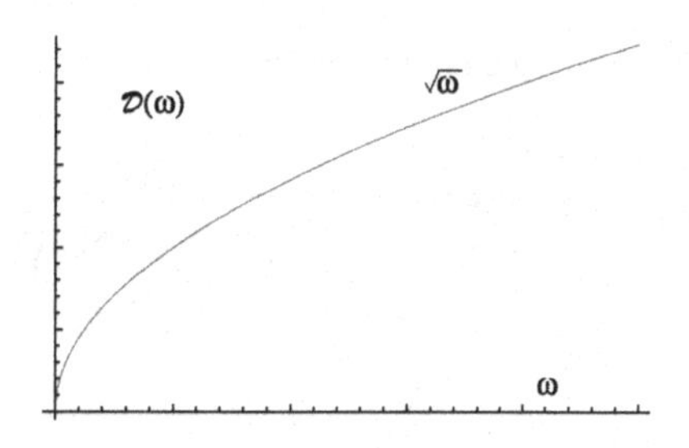

Fig. 15.2 Density of states for the three-dimensional electron gas model. (See Eq. 15.61).

where

$$\nu_j = 0, \pm 1, \pm 2, \pm 3, \ldots \tag{15.60}$$

Here L is the length of one side of a "notional" periodic cube with macroscopic volume $V = L^3$. From Appendix E the density of states, $\mathcal{D}(\omega)$, corresponding to this model (see Eq. 15.58) is

$$\begin{aligned}\mathcal{D}(\omega) &= \sum_{\mathbf{k}} \delta\left[\omega - \epsilon(\mathbf{k})\right] \\ &= \frac{V}{4\pi^2}\left(\frac{2m}{\hbar^2}\right)^{3/2}\sqrt{\omega},\end{aligned} \tag{15.61}$$

which is the distinctive three-dimensional electron gas result (see Figure 15.2).[15]

With Eq. 15.61 the grand potential (see Eq. 15.31) is

$$\boldsymbol{\Omega}_{gr}^{FD} = -\frac{V}{2\pi^2\beta}\left(\frac{2m}{\hbar^2}\right)^{3/2}\int_0^\infty d\omega\sqrt{\omega}\ln\left[1 + e^{-\beta(\omega-\mu)}\right], \tag{15.62}$$

which is integrated by parts to give

$$\boldsymbol{\Omega}_{gr}^{FD} = -\frac{V}{3\pi^2}\left(\frac{2m}{\hbar^2}\right)^{3/2}\int_0^\infty d\omega\frac{\omega^{3/2}}{e^{\beta(\omega-\mu)}+1}. \tag{15.63}$$

Since

$$p = -\left(\frac{\partial \boldsymbol{\Omega}_{gr}}{\partial V}\right)_{T,\mu} \tag{15.64}$$

[15] The mass can be replaced by an effective mass, $m \rightarrow m^*$, where m^* approximates the effect of a crystal lattice.

the fermion gas pressure is

$$p^{FD} = \frac{1}{3\pi^2}\left(\frac{2m}{\hbar^2}\right)^{3/2} \int_0^\infty d\omega \frac{\omega^{3/2}}{e^{\beta(\omega-\mu)}+1}. \tag{15.65}$$

In the degenerate case ($\beta\mu \gg 1$), Eq. 15.65 is expanded in powers of T by applying the steps in Appendix I, starting with an expansion of the integrand about $\mu = \epsilon_F$,

$$\int_0^\infty d\omega \frac{\omega^{3/2}}{e^{\beta(\omega-\mu)}+1} = \int_0^\infty d\omega \frac{\omega^{3/2}}{e^{\beta(\omega-\epsilon_F)}+1} + \frac{\beta(\mu-\epsilon_F)}{4}\int_0^\infty d\omega\, \omega^{3/2}\,\mathrm{sech}^2\left[\frac{\beta}{2}(\omega-\epsilon_F)\right]. \tag{15.66}$$

The first integral on the right-hand side is approximated using Sommerfeld's method

$$\int_0^\infty d\omega \frac{\omega^{3/2}}{e^{\beta(\omega-\epsilon_F)}+1} \approx \frac{2}{5}\epsilon_F^{5/2} + \frac{\pi^2}{4\beta^2}\epsilon_F^{1/2}, \tag{15.67}$$

while the second integral is

$$\int_0^\infty d\omega\, \omega^{3/2}\,\mathrm{sech}^2\left[\frac{\beta}{2}(\omega-\epsilon_F)\right] \approx \frac{4}{\beta}\epsilon_F^{3/2}. \tag{15.68}$$

Therefore Eq. 15.66, together with Eqs. 15.67, 15.68 and 15.39, gives the pressure of a degenerate ($\beta\mu \gg 1$) Fermi–Dirac gas model

$$p^{FD} = \frac{2}{15\pi^2}\left(\frac{2m}{\hbar^2}\right)^{3/2} \epsilon_F^{5/2}\left(1 + \frac{5\pi^2}{12\beta^2\epsilon_F^2}\right). \tag{15.69}$$

Unlike a classical ideal gas for which $p \approx T$, the pressure of a degenerate Fermi–Dirac gas is nearly independent of temperature.

Finally, in the Fermi gas model we have (see Eq. 15.37)

$$\frac{\langle \mathcal{N}_{op}\rangle}{V} = \frac{1}{2\pi^2}\left(\frac{2m}{\hbar^2}\right)^{3/2} \int_0^{\epsilon_F} d\omega \sqrt{\omega} \tag{15.70}$$

$$= \frac{1}{3\pi^2}\left(\frac{2m}{\hbar^2}\right)^{3/2} \epsilon_F^{3/2} \tag{15.71}$$

or

$$\epsilon_F = \frac{\hbar^2}{2m}\left(3\pi^2\frac{\langle\mathcal{N}_{op}\rangle}{V}\right)^{2/3} \tag{15.72}$$

permitting thermodynamic properties to be expressed in terms of the fermion density (e.g. Table 15.1).

15.11 White dwarf stars

The thermodynamics of stars is a study in self-gravitating systems – a titanic struggle between gravity, which compresses stellar matter, and outward gas pressure usually associated with high temperatures within stellar interiors.[16] As a result of nuclear and chemical evolution, equilibrium between these forces is in constant flux, resulting in continuous stellar transformations.

Stars that are not too massive, say of the order of our Sun's mass, eventually "burn" nearly all their core hydrogen resulting in a cooler stellar core richer in helium (and heavier nuclei). Consequently gas pressure is reduced which, seemingly, throws the dynamical advantage to gravitational forces, suggesting the star's continued collapse, perhaps without limit. Although ordinary gas pressure is no longer capable of resisting gravitational collapse, star stability can now be maintained by quantum behavior of degenerate (but not necessarily relativistic) electrons, whose source is the now fully ionized He atoms.

These stars can collapse to radii of the order of the Earth's while maintaining masses of the order of the Sun – a density of 10^6 times that of water. Such stars were discovered by British astronomer William Herschel (1783) who failed to realize their significance. These relatively near (on a galactic scale) but intrinsically dim stars are called *white dwarfs*. They are intermittently resurrected corpses in the evolutionary stellar graveyard for stars of about a solar mass and less.

15.11.1 White dwarf thermodynamics

A white dwarf star can be modeled as densely packed He nuclei at the star's hot central region. The He nuclei intermingle with a degenerate ($\beta\mu \gg 1$) electron gas whose origin is the fully ionized He atoms.

[16] High temperatures result from nuclear fusion at the star's center in which hydrogen "burns" to helium and, eventually, heavier nuclei including carbon and oxygen.

The degenerate electron gas behavior is governed by the PEP, so using Eqs. 15.69 and 15.72, the nearly temperature-independent outward pressure of the degenerate electron gas component is

$$p^e = \frac{(3\pi^2)^{2/3}}{5}\left(\frac{\hbar^2}{m_e}\right)\left(\frac{\langle\mathcal{N}^e_{op}\rangle}{V}\right)^{5/3}, \tag{15.73}$$

where m_e is the electron mass and $\frac{\langle\mathcal{N}^e_{op}\rangle}{V}$ is the electron number density. The stellar mass, M, is approximately

$$M = 2\langle\mathcal{N}^e_{op}\rangle m_{He}, \tag{15.74}$$

where m_{He} is the mass of a He nucleus.

The star's dense He core exerts inward gravitational pressure

$$p^{core} = -\left(\frac{\partial\mathbf{\Omega}^{core}_{gr}}{\partial V}\right)_{T,\mu} \tag{15.75}$$

where, for He nuclei (treated classically),

$$\mathbf{\Omega}^{core}_{gr} = -\frac{V}{3\pi^2}\left(\frac{2m_{He}}{\hbar^2}\right)^{3/2}\int_0^\infty d\omega\,\omega^{3/2}e^{-\beta(\omega-\mu+W)} \tag{15.76}$$

$$= \frac{V}{(2\pi^3)^{1/2}\beta^{5/2}}\left(\frac{m_{He}}{\hbar^2}\right)^{3/2}e^{\beta(\mu-W)}, \tag{15.77}$$

with W the mean field gravitational potential energy per He nucleus (see Chapter 12). Assuming a uniform, spherically distributed mass

$$\rho = \frac{M}{V} \tag{15.78}$$

$$= \frac{3M}{4\pi R^3}, \tag{15.79}$$

where R is the star radius, we find, as in Section 12.6.1,

$$W = -m_{He}\frac{3GM}{5R}, \tag{15.80}$$

where G is the universal gravitational constant.

Applying Eq. 15.75 the inward He nuclear core pressure is

$$p^{core} = \frac{1}{(2\pi^3)^{1/2}\beta^{5/2}}\left(\frac{m_{He}}{\hbar^2}\right)^{3/2}e^{\beta(\mu-W)} - \frac{1}{5\sqrt{\pi}}\frac{GM^2}{V^{1/3}}\left(\frac{m_{He}}{\beta\hbar^2}\right)^{3/2} \times\left(\frac{\sqrt{2}}{3\pi^2}\right)^{1/3}e^{\beta(\mu-W)}. \tag{15.81}$$

Next, applying Eq. 15.17 to Eq. 15.77, the average number of He nuclei is

$$\langle \mathcal{N}_{op}^{He} \rangle = \frac{V}{\sqrt{2}} \left(\frac{m_{He}}{\pi \beta \hbar^2} \right)^{3/2} e^{\beta(\mu - W)}. \tag{15.82}$$

Therefore, substituting for the chemical potential μ, the nuclear core pressure becomes

$$p^{core} = \frac{\langle \mathcal{N}_{op}^{He} \rangle}{\beta V} - \frac{1}{5} \left(\frac{4\pi}{3} \right)^{1/3} \frac{GM^2}{V^{4/3}} \tag{15.83}$$

where the first term is the classical ideal He gas pressure, which is negligible for this cool star. The total pressure – i.e. the equation of state – for the white dwarf star is therefore

$$p_{total} = \frac{(3\pi^2)^{2/3}}{5} \left(\frac{\hbar^2}{m_e} \right) \left(\frac{\langle \mathcal{N}_{op}^{e} \rangle}{V} \right)^{5/3} - \frac{1}{5} \left(\frac{4\pi}{3} \right)^{1/3} \frac{GM^2}{V^{4/3}}. \tag{15.84}$$

At mechanical equilibrium $p_{total} = 0$, giving

$$\frac{(3\pi^2)^{2/3}}{5} \left(\frac{\hbar^2}{m_e} \right) \left(\frac{\langle \mathcal{N}_{op}^{e} \rangle}{V} \right)^{5/3} = \frac{1}{5} \left(\frac{4\pi}{3} \right)^{1/3} \frac{GM^2}{V^{4/3}}, \tag{15.85}$$

which, with

$$\langle \mathcal{N}_{op}^{e} \rangle = 2 \langle \mathcal{N}_{op}^{He} \rangle \text{ (charge neutrality)} \tag{15.86}$$

$$= \frac{M}{2m_{He}}, \tag{15.87}$$

yields the remarkable (non-relativistic) white dwarf mass–radius condition

$$\left[\frac{3^{4/3} \pi^{2/3} \hbar^2}{8G (m_{He})^{5/3} m_e} \right] = RM^{1/3}. \tag{15.88}$$

Problems and exercises

15.1 Eventually the Sun will pass through a red giant phase and collapse into a white dwarf star ($\approx 6 \times 10^9$ years). Neglecting the shedding of mass in the red dwarf stage, what fraction of its present radius will the Sun be?

15.2 Consider a white dwarf star in an advanced stage of evolution with nearly all its fully degenerate electrons behaving ultra-relativistically

$$\hbar \omega \approx cp \tag{15.89}$$

where c is the speed of light and $p = \hbar \, |\mathbf{k}|$.

(a) Find the single-particle density of states $\mathscr{D}(\omega)$.
(b) Find an expression for the star's pressure.
(c) Find a critical mass (in solar mass units) beyond which the star is no longer stable and will gravitationally collapse into a neutron star or a black hole.

15.3 An uncharged degenerate, spin-half Fermi gas (say He^3) at pressure p and temperature T is confined to one half of a rigid insulated chamber, of total volume V. The other half of the chamber is empty (vacuum). The partition separating the two halves suddenly dissolves and the Fermi gas freely and adiabatically flows to occupy the entire chamber (free expansion). Find the change in temperature of the Fermi gas.

15.4 A cylinder is separated into two compartments by a sliding frictionless piston. Each compartment is occupied by an ideal Fermi gas, say A and B, which differ only in that A is composed of particles of spin $\frac{1}{2}$ while B is composed of particles of spin $\frac{3}{2}$. Find the relative equilibrium density of the two gases:

a. at $T = 0\,\text{K}$;
b. as $T \to \infty$.

15.5 A degenerate Fermi gas of non-interacting electrons moves in a uniform positive background confined to one dimension. (An example is the conducting polymer trans-polyacetylene.)

a. Find an expression for the single-particle density of states $\mathscr{D}(\omega)$ of independent particles in one dimension.
b. Find the leading non-zero term for the chemical potential of a one-dimensional conductor as a function of temperature in the limit of high degeneracy $\beta\mu \gg 1$.

15.6 In E. Fermi's model for ultra-high energy particle collisions[17] a microscopic "fireball" is created consisting of:

(a) ultra-relativistic nucleon-antinucleon pairs:
- protons (two up quarks and one down quark),
- antiprotons (two up antiquarks and one down antiquark),
- neutrons (one up quark and two down quarks),
- antineutrons (one up antiquark and two down antiquarks);

(b) Ultra-relativistic pions π^-, π^+ and π^0:
i. π^-, negatively charged (down quark and anti-up quark),
ii. π^+, positively charged (up quark and anti-down quark),
iii. π^0, uncharged and is its own antiparticle (up quark - anti-up quark – down quark - anti-down quark superposition).

[17] E. Fermi, "High energy nuclear events", *Prog. Theoret. Phys. (Japan)*, **5**, 570–583 (1950).

All nucleons are spin-half Fermi–Dirac particles. All pions are spin-zero Bose–Einstein particles.
At these high energies particles have negligible rest masses, i.e. they behave like photons (except for their statistics) with zero chemical potential and energies

$$E = \hbar c\,|\mathbf{k}| \tag{15.90}$$

with periodic boundary conditions

$$|\mathbf{k}|^2 = {k_x}^2 + {k_y}^2 + {k_z}^2, \tag{15.91}$$

where

$$k_j = \frac{2\pi}{L}\nu_j; \quad \nu_j = 0, \pm 1, \pm 2, \ldots \tag{15.92}$$

Dynamic assumptions are:

i. Interaction between pions and nucleons is so strong that statistical equilibrium is attained in times less than pion lifetimes.
ii. Energy and charge are conserved.
iii. The energy available in the center of mass system, $\mathscr{W}$, is released in a small volume $\mathscr{V} = \frac{4}{3}\pi R^3$, where R is the Compton wavelength $R \approx \frac{\hbar}{\mu c}$.
iv. Angular momentum conservation is ignored.
   a. Find an expression for the temperature T^* of the "fireball" in terms of $\mathscr{W}$ and $\mathscr{V}$.
   b. Find the average density of pions and of nucleons produced.
   c. Show that the internal energy per unit volume for the fermion "fireball" is exactly $\frac{7}{2}$ the internal energy per unit volume for the photon (EM radiation) "fireball". Consider two alternative scenarios:
      (1) The "fireball" is entirely composed of ultra-relativistic spin-half fermions.
      (2) The "fireball" is entirely composed of photons.

Note: The four distinct fermion varieties give rise to eight-fold degeneracy, i.e. each of the four created particles has two spin polarizations.
You may need the following integrals:

$$\int_0^\infty \mathrm{d}x x^3 \left(\frac{1}{e^x+1}\right) = \frac{7\pi^4}{120}; \quad \int_0^\infty \mathrm{d}x x^3 \left(\frac{1}{e^x-1}\right) = \frac{\pi^4}{15}. \tag{15.93}$$

16 Ideal Bose–Einstein system

From a certain temperature on, the molecules "condense" without attractive forces; that is, they accumulate at zero velocity. The theory is pretty, but is there also some truth to it?

Albert Einstein, Letter to Ehrenfest (Dec. 1924), Abraham Pais, *Subtle Is the Lord: The Science and the Life of Albert Einstein*, Oxford University Press, New York (1982)

16.1 Introduction

For over 50 years the low-temperature liquid state of uncharged, spinless He^4 was the only system in which a Bose–Einstein (BE) condensation was considered experimentally realized.[1,2,3] In that case, cold (<2.19 K) liquid He^4 passes into an extraordinary phase of matter called a *superfluid*, in which the liquid's viscosity and entropy become zero.

With advances in atomic cooling[4,5] (to $\approx 10^{-9}$ K) the number of Bose systems which demonstrably pass into a condensate has considerably increased. These include several isotopes of alkali gas atoms as well as fermionic atoms that pair into integer-spin (boson) composites.[6]

Although an ideal Bose gas does exhibit a low-temperature critical instability, the ideal BE gas theory is not, on its own, able to describe the BE condensate wave state.[7] In order for a theory of bosons to account for a condensate wave state, interactions between the bosons must be included. Nevertheless, considerable interesting physics is contained in the ideal Bose gas model.

[1] Peter Kapitza, "Viscosity of liquid helium below the λ-point", *Nature* **141**, 74 (1938).
[2] John F. Allen and Don Misener, "Flow of liquid helium II", *Nature* **141**, 75 (1938).
[3] Allan Griffin, "New light on the intriguing history of superfluidity in liquid ^{4}He", *J. Phys.: Condens. Matter* **21**, 164220 (2009).
[4] M.H. Anderson, J.R. Ensher, M.R. Matthews, C.E. Wieman and E.A. Cornell, "Observation of Bose–Einstein condensation in a dilute atomic vapor", *Science* **269**, 198 (1995).
[5] Eric A. Cornell and Carl E. Wieman, "The Bose–Einstein Condensate", *Scientific American* **278**, 40 (1998).
[6] C. A. Regal, M. Greiner and D. S. Jin, "Observation of resonance condensation of fermionic atom pairs", *Phys. Rev. Lett.* **92**, 040403 (2004).
[7] V.L. Ginzburg and L.D. Landau, *Zh. Eksp. Teor. Fiz.* **20**, 1064 (1950).

16.2 Ideal Bose gas

The Bose–Einstein (BE) system allows any occupation number, $n = 0, 1, 2, 3, \ldots, \infty$, so that the thermal Lagrangian is

$$\begin{aligned}\mathcal{L}^{BE} = &-k_B \sum_{n_1=0}^{\infty} \cdots \sum_{n_N=0}^{\infty} \hat{\mathbf{P}}^{BE}(n_1, n_2, \ldots) \ln \hat{\mathbf{P}}^{BE}(n_1, n_2, \ldots) \\ &- \frac{1}{T}\left\{ \sum_{n_1=0}^{\infty} \cdots \sum_{n_N=0}^{\infty} \hat{\mathbf{P}}^{BE}(n_1, n_2, \ldots) \sum_s (E_s - \mu)\, n_s \right\} \\ &- \lambda_0 \sum_{n_1=0}^{\infty} \cdots \sum_{n_N=0}^{\infty} \hat{\mathbf{P}}^{BE}(n_1, n_2, \ldots),\end{aligned} \tag{16.1}$$

where $\hat{\mathbf{P}}^{BE}(n_1, n_2, \ldots)$ is the probability[8] there are n_1 particles in the single-particle state $|E_1\rangle$, n_2 particles in the single-particle state $|E_2\rangle$, etc. Maximizing the thermal Lagrangian in the usual way gives the Bose–Einstein probabilities

$$\hat{\mathbf{P}}^{BE}(n_1, n_2, \ldots) = \frac{e^{-\beta \sum_s n_s (E_s - \mu)}}{\sum_{n_1=0}^{\infty} \cdots \sum_{n_N=0}^{\infty} e^{-\beta \sum_s n_s (E_s - \mu)}}. \tag{16.2}$$

The denominator is the BE grand partition function

$$\mathcal{Z}_{gr}^{BE} = \sum_{n_1=0}^{\infty} \cdots \sum_{n_N=0}^{\infty} e^{-\beta \sum_s n_s (E_s - \mu)} \tag{16.3}$$

$$= \prod_s \sum_{n_s=0}^{\infty} e^{-\beta n_s (E_s - \mu)}. \tag{16.4}$$

The infinite sum over n_s converges only for $e^{-\beta(E_s - \mu)} < 1$. Thus if the minimum value in the E_s energy spectrum is taken to be zero, convergence is guaranteed only if $\mu < 0$. As will soon be shown, this convergence criterion has astonishing physical implications!

Assuming convergence of the n_s sum, the BE grand partition function is

$$\mathcal{Z}_{gr}^{BE} = \prod_s \left(\frac{1}{1 - e^{-\beta(E_s - \mu)}} \right) \tag{16.5}$$

[8] These are "surrogate" probabilities.

and the corresponding grand potential is

$$\Omega^{BE} = \frac{1}{\beta}\sum_{s=0}^{\infty}\ln\left(1 - e^{\beta\mu}e^{-\beta E_s}\right). \tag{16.6}$$

16.3 Bose–Einstein thermodynamics

Applying the probabilities from Eq. 16.2:

1. Average number of particles in the pth single-particle state is

$$\langle n_p{}^{BE}\rangle = \frac{\sum_{n_p=0}^{\infty} n_p e^{-\beta n_p(E_p-\mu)}}{\sum_{n_p=0}^{\infty} e^{-\beta n_p(E_p-\mu)}} \tag{16.7}$$

or

$$\langle n_p{}^{BE}\rangle = \frac{1}{\beta}\left[\frac{\partial}{\partial\mu}\ln\sum_{n_p=0}^{\infty} e^{-\beta n_p(E_p-\mu)}\right]_T. \tag{16.8}$$

After carrying out the sum and differentiation in Eq. 16.8 the BE average occupation number for the single particle state $|E_p\rangle$ is

$$\langle n_p{}^{BE}\rangle = \frac{1}{e^{\beta(E_p-\mu)}-1}. \tag{16.9}$$

Similarly, the average total particle number is

$$\langle \mathcal{N}_{op}^{BE}\rangle = \sum_{p=0}\langle n_p{}^{BE}\rangle \tag{16.10}$$

$$= \frac{\partial}{\partial\mu}\Omega_{gr}^{BE} \tag{16.11}$$

$$= \sum_{p=0}\frac{1}{e^{\beta(E_p-\mu)}-1}. \tag{16.12}$$

But Eq. 16.9 presents a problem. Since the lowest value of E_p is taken to be *zero*, if $\mu > 0$ the average number of particles in state $|E_p\rangle$ is negative, which is nonsense. It clearly must be that

$$\mu^{BE} \leq 0. \tag{16.13}$$

This constraint (already mentioned above) has implications that will be discussed in the next sections.

2. Internal energy of the ideal BE system is the intuitive result

$$\mathcal{U}_{BE} = \sum_{p=0} E_p \langle n_p^{BE} \rangle. \tag{16.14}$$

16.3.1 The quasi-classical limit

The distinction between $\langle n_p^{BE} \rangle$ and $\langle n_p^{FD} \rangle$ – i.e. the sign in the denominator – has enormous physical consequences at low temperature. But in the limit $\mu \to -\infty$, i.e. the quasi-classical limit, where $n/n_Q \leq 1$, all distinctions disappear,

$$\langle n_p^{BE} \rangle \equiv \langle n_p^{FD} \rangle \to e^{\beta\mu} e^{-\beta E_p}. \tag{16.15}$$

16.4 The ideal BE gas and the BE condensation

Using the free-particle eigenstates and density of states previously applied in Fermi–Dirac thermodynamics the spin $= 0$ (no spin degeneracy) BE grand potential (see Eq. 16.6) is

$$\Omega_{gr}^{BE} = -\frac{1}{\beta}\left(\frac{V}{4\pi^2}\right)\left(\frac{2m}{\hbar^2}\right)^{3/2} \int_0^\infty \mathrm{d}\omega \sqrt{\omega} \ln\left(1 - e^{\beta\mu} e^{-\beta\omega}\right) \tag{16.16}$$

with $\mu \leq 0$. Integrating by parts

$$\Omega_{gr}^{BE} = -\frac{V}{6\pi^2}\left(\frac{2m}{\hbar^2}\right)^{3/2} \int_0^\infty \mathrm{d}\omega \frac{\omega^{3/2}}{e^{\beta(\omega-\mu)} - 1} \tag{16.17}$$

and then applying Eq. 16.11 gives

$$\langle \mathcal{N}_{op}^{BE} \rangle = \frac{V}{4\pi^2}\left(\frac{2m}{\hbar^2}\right)^{3/2} \int_0^\infty \mathrm{d}\omega \frac{\omega^{1/2}}{e^{\beta(\omega-\mu)} - 1}. \tag{16.18}$$

Now recalling the Fermi–Dirac expression

$$\langle \mathcal{N}_{op}^{FD} \rangle = \frac{V}{2\pi^2}\left(\frac{2m}{\hbar^2}\right)^{3/2} \int_0^\infty \mathrm{d}\omega \frac{\omega^{1/2}}{e^{\beta(\omega-\mu)} + 1}, \tag{16.19}$$

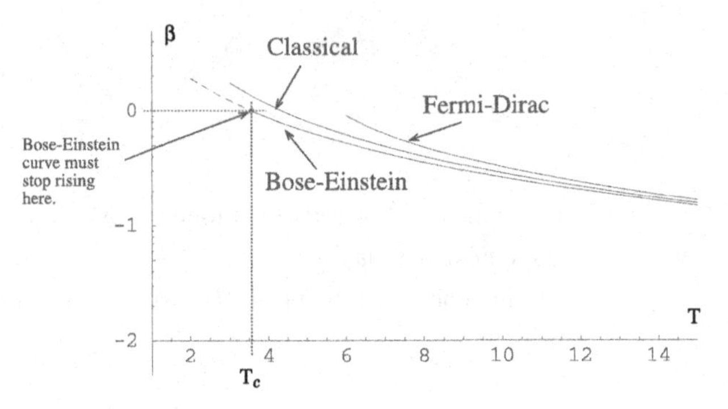

Fig. 16.1 Bose–Einstein, Fermi–Dirac and quasi-classical result for $\beta\mu$ vs. T.

and the quasi-classical expression

$$\langle \mathcal{N}_{op}^{QC} \rangle = \frac{V}{2\pi^2}\left(\frac{2m}{\hbar^2}\right)^{3/2} \int_0^\infty d\omega\, \omega^{1/2} e^{-\beta(\omega-\mu)}, \tag{16.20}$$

the variation of μ with T can, in each case, be plotted to show any distinctions among the three kinds of "statistics". At high temperature all three curves begin to merge as indicated in Eq. 16.15. As temperature falls the BE, FD and quasi-classical curves begin to diverge, as shown in Figure 16.1, with FD and quasi-classical curves rising above $\mu = 0$. However, the BE curve cannot rise above $\mu = 0$ and, since the curve is single-valued, there is some temperature T_c given by

$$\left(\frac{\langle \mathcal{N}_{op}^{BE} \rangle}{V}\right)_{T \to T_c^+} = \frac{1}{4\pi^2}\left(\frac{2m}{\hbar^2}\right)^{3/2} \int_0^\infty d\omega \frac{\omega^{1/2}}{e^{\beta_c \omega} - 1}, \tag{16.21}$$

where $\beta_c = \frac{1}{k_B T_c}$, at which it continues for $T < T_c$ as a horizontal line that coincides with $\mu = 0$ (see Figure 16.1) – although the BE curve is continuous, its slope is discontinuous. At temperatures $T < T_c$ the chemical potential assumes the fixed value $\mu = 0$ and the average BE particle density becomes

$$\left(\frac{\langle \mathcal{N}_{op}^{BE} \rangle}{V}\right)_{T < T_c} = \frac{1}{4\pi^2}\left(\frac{2m}{\hbar^2}\right)^{3/2} \int_0^\infty d\omega \frac{\omega^{1/2}}{e^{\beta \omega} - 1}. \tag{16.22}$$

Comparing Eq. 16.21 with Eq. 16.22 we have the result that

$$\left(\frac{\langle \mathcal{N}_{op}^{BE} \rangle}{V}\right)_{\substack{T < T_c \\ \mu = 0}} < \left(\frac{\langle \mathcal{N}_{op}^{BE} \rangle}{V}\right)_{\substack{T \geq T_c \\ \mu < 0}}. \tag{16.23}$$

Bosons are "*missing*"! Where are they? They have, in fact, thrown in their lot with an entirely new state of matter, having its own wavefunction, the *Bose–Einstein condensate*. For $T < T_c$, as $T \to T_c^-$, ordinary bosons begin to reappear. Thus for $T < T_c$ there are two coexisting phases – normal bosons and a BE condensate.

The transition temperature T_c at which the BE condensate begins to form is found by integrating Eq. 16.21. A procedure is to first expand the denominator

$$\int_0^\infty \mathrm{d}x \frac{x^s}{e^x - 1} = \int_0^\infty \mathrm{d}x x^s \sum_{n=0}^{\infty} e^{-(n+1)x} \tag{16.24}$$

and then integrate term by term

$$\int_0^\infty \mathrm{d}x \frac{x^s}{e^x - 1} = \Gamma(s+1) \sum_{\nu=1}^{\infty} \frac{1}{\nu^{s+1}}, \tag{16.25}$$

where the Γ function is defined by

$$\Gamma(r) = \int_0^\infty \mathrm{d}z\, z^{r-1} e^{-z}; \quad \mathcal{R}\mathrm{e}\, r > -1. \tag{16.26}$$

The remaining sum is the Riemann ζ function

$$\zeta(s) = \sum_{n=1}^{\infty} \frac{1}{n^s}, \tag{16.27}$$

which finally gives

$$\int_0^\infty \mathrm{d}x \frac{x^s}{e^x - 1} = \Gamma(s+1)\, \zeta(s+1) \tag{16.28}$$

and therefore

$$\left(\frac{\langle \mathcal{N}_{op}^{BE} \rangle}{V} \right)_{T \to T_c^+} = \frac{\Gamma(3/2)\, \zeta(3/2)}{4\pi^2} \left(\frac{2m}{\beta_c \hbar^2} \right)^{3/2}. \tag{16.29}$$

Solving for T_c,

$$T_c = \frac{\hbar^2}{2mk_B} \left[\frac{4\pi^2}{\Gamma(3/2)\, \zeta(3/2)} \right]^{2/3} \left(\frac{\langle \mathcal{N}_{op}^{BE} \rangle}{V} \right)^{2/3}_{T \to T_c^+} \tag{16.30}$$

where $\Gamma(3/2) = 0.886$ and $\zeta(3/2) = 2.612$. Therefore

$$T_c = 3.31 \left(\frac{\hbar^2}{mk_B} \right) \left(\frac{\langle \mathcal{N}_{op}^{BE} \rangle}{V} \right)^{2/3}. \tag{16.31}$$

For the case of He^4 (mass = 4.002 602 u,[9] $\langle \mathcal{N}_{op}^{BE} \rangle / V = 2.18 \times 10^{22}\,\mathrm{cm}^{-3}$) we find $T_c = 3.14\,\mathrm{K}$. This may be compared to the temperature $T_\lambda = 2.19\,\mathrm{K}$ (the λ point)

[9] $1\,\mathrm{u} = 1.660\,538\,782 \times 10^{-27}\,\mathrm{kg}$.

at which liquid He^4 is observed to undergo a transition to a remarkable superfluid phase.

16.4.1 The condensate

As noted above, at temperatures $0 < T < T_c$ the boson chemical potential remains $\mu = 0$, generating the missing boson "paradox". From what we know about the electromagnetic field's photons as well as the elastic field's phonons – whose chemical potential is inherently zero – these particles are created and destroyed without regard to conservation. Similarly, at temperatures $T < T_c$ Bose–Einstein "particles" assume the property that they can be created and destroyed without regard to number conservation. Here physical particles are freely disappearing into and emerging from a condensate that no longer keeps track of particle number – condensate particle number has large uncertainty, with the uncertainty relation[10] $\Delta n \Delta \cos\theta \approx \frac{1}{2}$ describing the trade-off. The condensate "wavefunction" is a superposition of states with different particle numbers, becoming, instead, a *coherent* phase of matter.

An emergent property of the condensate is classical phase. In analogy with laser light (and superconductors) – which surrender particle number certainty to acquire "good" phase – the condensate behaves like a classical macroscopic wave with definite (or nearly definite) phase. Also in analogy with the photon vacuum, the condensate acts like a Bose–Einstein particle vacuum, i.e. as temperature increases particles return from the condensate (vacuum) into the ideal BE gas phase and vice versa.

The density of "missing" particles (the number subsumed into the condensate) is

$$\left(\frac{\langle \mathcal{N}_{op}^{BE}\rangle}{V}\right)_{superfluid} = \left(\frac{N}{V}\right)_{T>T_c} - \left(\frac{N}{V}\right)_{T<T_c}. \tag{16.32}$$

Therefore integrating Eq. 16.22 to get

$$\left(\frac{\langle \mathcal{N}_{op}^{BE}\rangle}{V}\right)_{T<T_c} = \frac{\Gamma(3/2)\,\zeta(3/2)}{4\pi^2}\left(\frac{2m}{\beta\hbar^2}\right)^{3/2}, \tag{16.33}$$

then taking its ratio with Eq. 16.29 gives

$$\left(\frac{\langle \mathcal{N}_{op}^{BE}\rangle}{V}\right)_{T<T_c} \Bigg/ \left(\frac{\langle \mathcal{N}_{op}^{BE}\rangle}{V}\right)_{T>T_c} = (T/T_c)^{3/2}. \tag{16.34}$$

The mystery of the "missing" bosons has been "solved". Specifically,

$$\left(\frac{\langle \mathcal{N}_{op}^{BE}\rangle}{V}\right)_{superfluid} = \left(\frac{\langle \mathcal{N}_{op}^{BE}\rangle}{V}\right)_{T>T_c}\left[1 - \left(\frac{T}{T_c}\right)^{3/2}\right]. \tag{16.35}$$

[10] L. Susskind and J. Glogower, "Quantum mechanical phase and time operator", *Physica* 1, 49–61 (1964).

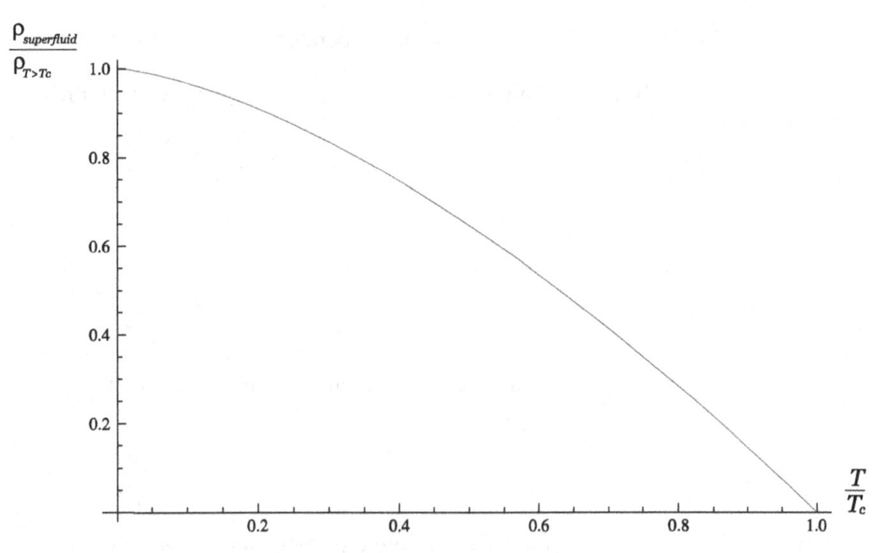

Fig. 16.2 Fractional condensate occupation (Eq. 16.35) vs. T/T_c.

At $T = 0$ all the bosons are subsumed by the condensate, whereas at $T = T_c$ the condensate is empty (there is no condensate.) (See Figure 16.2.)

A behavior such as Eq. 16.35 occurs in certain phase transitions. (See e.g. Chapter 12.) The exponent 3/2 in the temperature dependence is this phase transition's critical exponent.

Problems and exercises

16.1 When an ideal Bose–Einstein gas passes through the critical temperature T_c the heat capacity $\mathcal{C}_V$ remains constant but its temperature derivative is discontinuous, with

$$\left(\frac{\mathrm{d}\mathcal{C}_V}{\mathrm{d}T}\right)_{T_c^+} - \left(\frac{\mathrm{d}\mathcal{C}_V}{\mathrm{d}T}\right)_{T_c^-} = \gamma\beta_c. \tag{16.36}$$

Find the value of γ.

16.2 Consider an ultra-relativistic Bose–Einstein gas with single-particle eigen-energies

$$\epsilon = \hbar c\,|\mathbf{k}|\,. \tag{16.37}$$

a. Show that the single-particle density of states is

$$\mathscr{D}(\epsilon) = \frac{V}{2\pi^2\hbar^3c^3}\epsilon^2. \tag{16.38}$$

b. Find thc critical temperature T_c of the BE gas.

16.3 The volume of an n-dimensional sphere with "radius" k is

$$V_n = \frac{2\pi^{n/2}}{n\Gamma(n/2)} k^n, \tag{16.39}$$

where

$$k^2 = \sum_{i=1}^{n} k_i^2. \tag{16.40}$$

a. Assuming an n-dimensional Bose–Einstein gas with single-particle eigenvalues

$$\epsilon = \frac{\hbar^2 k^2}{2m}, \tag{16.41}$$

show that the single-particle density of states is

$$\mathscr{D}(\epsilon) = \frac{n\pi^{n/2}}{\Gamma(n/2)} \left(\frac{2m}{\hbar^2}\right) \epsilon^{(n/2)-1}. \tag{16.42}$$

b. Find the critical temperature T_c for an n-dimensional BE gas.

c. Show that the two-dimensional BE gas exhibits no Bose–Einstein condensation.

17 Thermodynamics and the cosmic microwave background

We are quantifying the cosmos in a different way to open up a new window for understanding the universe in its earliest times. Researchers have long been working to test cosmological models for how the universe developed, but previously have not been able to collect data that could provide a picture this clear.

Lyman Page, Princeton University, WMAP co-investigator (2006)

17.1 Introduction

The once super-hot but now cold afterglow from the creation of the universe has been studied by the satellites COBE[1] and WMAP[2]. WMAP's continuing task, and that of the European Space Agency's Planck satellite, is to measure small-scale temperature anisotropies in the cosmic microwave background (CMB) in order to better determine cosmological parameters which are a test of inflationary scenarios.[3]

The satellite moves through the microwave background with a local group velocity $|\mathbf{v}| = 627 \pm 22\,\mathrm{km\,s^{-1}}$ in the galactic coordinate[4] direction $\left(l{:}\,276^\circ \pm 3^\circ, b{:}\,33^\circ \pm 3^\circ\right)$ relative to the CMB rest frame, so that WMAP's local velocity parameter is $\left|\frac{\mathbf{v}}{c}\right| = 0.002$. Relativistic effects will therefore produce measurable temperature variations that must be subtracted from the total signal.[5] This effect was first

[1] COsmic Background Explorer.

[2] Wilkinson Microwave Anisotropy Probe.

[3] M. White and J. D. Cohn, "Resource letter: TACMB-1: The theory of anisotropies in the cosmic microwave background", *Am. J. Phys.* **70**, 106–118 (2002).

[4] A celestial coordinate system centered on the Sun and aligned with the apparent center of the Milky Way galaxy. The "equator" is aligned to the galactic plane. *l* represents galactic longitude; *b* is galactic latitude.

[5] M. Kamionkowski and L. L. Knox, "Aspects of the cosmic microwave background dipole", *Phys. Rev. D* **67**, 63001–63005 (2003).

calculated exactly by Peebles and Wilkinson (PW)[6] only a few years after the discovery of the CMB.[7]

The intent in this chapter is not to re-derive the well-known PW result, but is solely to discuss relativity in blackbody thermodynamics by applying the "thermal Lagrangian" formulation[8] and, by happenstance, deriving the PW result.

To find thermodynamic properties measured in a frame of reference moving with velocity $\mathbf{v}$ through a blackbody radiation rest frame (CMB), we start with a "thermodynamic" hamiltonian which contains, in addition to the formal electromagnetic quantum hamiltonian, the energy (work) required to transport the satellite through the radiation field,

$$\mathcal{H}_{op} = \sum_{\substack{\mathbf{k} \\ \lambda=1,2}} \hbar\omega_{\mathbf{k},\lambda}\left[a^{\dagger}_{\mathbf{k},\lambda}a_{\mathbf{k},\lambda} + \frac{1}{2}\right] - \sum_{\substack{\mathbf{k} \\ \lambda=1,2}} \mathbf{v}\cdot\mathbf{p}_{op}, \tag{17.1}$$

where

$$\mathbf{p}_{op} = \hbar\mathbf{k}a^{\dagger}_{\mathbf{k},\lambda}a_{\mathbf{k},\lambda}, \tag{17.2}$$

and

$$a^{\dagger}_{\mathbf{k},\lambda}a_{\mathbf{k},\lambda} = N_{op}(\mathbf{k},\lambda) \tag{17.3}$$

is the photon number operator whose eigenvectors and eigenvalues are given by

$$a^{\dagger}_{\mathbf{k},\lambda}a_{\mathbf{k},\lambda}\left|n_{k,\lambda}\right\rangle = n_{k,\lambda}\left|n_{k,\lambda}\right\rangle. \tag{17.4}$$

Here $\mathbf{k}$ is the photon wavevector, $\lambda = 1, 2$ covers the two transverse photon polarizations, and $n_{\mathbf{k},\lambda} = 0, 1, 2, \ldots$ is the occupation number of photons with wavevector $\mathbf{k}$ and polarization λ.

The eigen-energies of the thermodynamic hamiltonian (including the contribution from the satellite's uniform velocity $\mathbf{v}$ relative to the CMB rest frame) are, from Eq. 17.1,

$$E(n) = \sum_{\substack{\mathbf{k} \\ \lambda=1,2}} \hbar\omega_{\mathbf{k},\lambda}\left[n_{\mathbf{k},\lambda} + \frac{1}{2}\right] - \sum_{\substack{\mathbf{k} \\ \lambda=1,2}} \mathbf{v}\cdot\hbar\mathbf{k}n_{\mathbf{k},\lambda}. \tag{17.5}$$

[6] P. J. E. Peebles and D. T. Wilkinson, "Comment on the anisotropy of the primeval fireball", *Phys. Rev.* **174**, 2168 (1968).

[7] A. A. Penzias and R. W. Wilson, "A measurement of excess antenna temperature at 4080 Mc/s", *Astrophysical Journal* **142**, 419–421 (1965).

[8] See Appendix B.

17.2 Thermodynamic method

Using the eigen-energies of Eq. 17.5, the moving frame thermal Lagrangian is

$$\mathcal{L} = -k_B \sum_{n_{\mathbf{k},\lambda}} \mathbf{P}\left(n_{\mathbf{k},\lambda}\right) \ln \mathbf{P}\left(n_{\mathbf{k},\lambda}\right)$$
$$-\frac{1}{\tau}\sum_{n_{\mathbf{k},\lambda}} \mathbf{P}\left(n_{\mathbf{k},\lambda}\right)\left\{\sum_{\substack{\mathbf{k}\\ \lambda=1,2}} \hbar\omega_{\mathbf{k},\lambda}\left[n_{\mathbf{k},\lambda}+\frac{1}{2}\right] - \sum_{\substack{\mathbf{k}\\ \lambda=1,2}} \mathbf{v}\cdot\hbar\mathbf{k}n_{\mathbf{k},\lambda}\right\}$$
$$-\lambda_0 \sum_{n_{\mathbf{k},\lambda}} \mathbf{P}\left(n_{\mathbf{k},\lambda}\right). \tag{17.6}$$

The $\mathbf{P}\left(n_{\mathbf{k},\lambda}\right)$ are probabilities that $n_{\mathbf{k},\lambda}$ photons are in the mode $\mathbf{k}$ with polarization λ. Here τ is the temperature in the moving frame. Maximizing the Lagrangian of Eq. 17.6 with respect to $\mathbf{P}\left(n_{\mathbf{k},\lambda}\right)$ (following by now familiar steps) gives, in the moving frame, the probabilities

$$\mathbf{P}\left(n_{\mathbf{k},\lambda}\right) = \frac{\exp\left\{-\tilde{\beta}\sum\limits_{\substack{\mathbf{k}\\ \lambda=1,2}}\left[\hbar\omega_{\mathbf{k},\lambda}\left(n_{\mathbf{k},\lambda}+\frac{1}{2}\right) - \hbar n_{\mathbf{k},\lambda}\left|\mathbf{v}\right|\left|\mathbf{k}\right|\cos\theta\right]\right\}}{\sum\limits_{n_{\mathbf{k},\lambda}}\exp\left\{-\tilde{\beta}\sum\limits_{\substack{\mathbf{k}\\ \lambda=1,2}}\left[\hbar\omega_{\mathbf{k},\lambda}\left(n_{\mathbf{k},\lambda}+\frac{1}{2}\right) - \hbar n_{\mathbf{k},\lambda}\left|\mathbf{v}\right|\left|\mathbf{k}\right|\cos\theta\right]\right\}}, \tag{17.7}$$

where $\tilde{\beta} = (k_B\tau)^{-1}$ and θ is the angle (in the moving frame) between the local velocity vector $\mathbf{v}$ and the direction of CMB observation. The frequency $\omega_{\mathbf{k},\lambda}$ is also in the moving frame. The denominator in Eq. 17.7 is identified as the moving frame blackbody partition function.

17.3 Moving frame thermodynamic potential $\tilde{\Omega}$

The thermodynamic entropy $\mathcal{S}$ is

$$\mathcal{S} = -k_B \sum_{n_{\mathbf{k},\lambda}} \mathbf{P}\left(n_{\mathbf{k},\lambda}\right) \ln \mathbf{P}\left(n_{\mathbf{k},\lambda}\right), \tag{17.8}$$

which after substituting the probabilities of Eq. 17.7 is evaluated to give

$$\tau\mathcal{S} = \tilde{\mathcal{U}} - \mathbf{v}\cdot\langle\tilde{\mathbf{p}}\rangle_{op} + \frac{1}{\tilde{\beta}}\ln\tilde{\mathcal{Z}}, \tag{17.9}$$

where $\tilde{\mathcal{U}}$ is the moving frame blackbody internal energy, $\langle \tilde{\mathbf{p}}_{op} \rangle$ is the moving frame average radiation momentum and $\tilde{\mathcal{Z}}$ is the moving frame blackbody partition function. We can now define a moving frame thermodynamic potential function

$$\tilde{\Omega} = -\frac{1}{\tilde{\beta}} \ln \tilde{\mathcal{Z}} \tag{17.10}$$

where, from the denominator of Eq. 17.7,

$$\tilde{\mathcal{Z}} = \prod_{\mathbf{k}} \left\{ \sum_{n_{\mathbf{k}}} \exp \left[-\hbar \tilde{\beta} n_{\mathbf{k}} \left(\omega_{\mathbf{k}} - |\mathbf{v}| \, |\mathbf{k}| \cos \theta \right) \right] \right\}^2 \tag{17.11}$$

with "squaring" accounting for the two transverse photon polarizations. Combining Eqs. 17.9 and 17.10 we can write

$$\tilde{\Omega} = \tilde{U} - \tau \mathcal{S} - \mathbf{v} \cdot \langle \tilde{\mathbf{p}}_{op} \rangle. \tag{17.12}$$

Moreover, by direct differentiation of Eq. 17.8, we find a corresponding thermodynamic identity, i.e. a moving frame "fundamental equation"

$$\tau \, d\mathcal{S} = d\tilde{\mathcal{U}} + \tilde{p} \, d\tilde{V} - \mathbf{v} \cdot d\langle \tilde{\mathbf{p}}_{op} \rangle, \tag{17.13}$$

where $\tilde{p}$ is the moving frame radiation pressure. Combining this with the total differential of Eq. 17.12,

$$d\tilde{\Omega} = -\tilde{p} \, d\tilde{V} - \mathcal{S} \, d\tau - \langle \tilde{\mathbf{p}}_{op} \rangle \cdot d\mathbf{v}, \tag{17.14}$$

from which follow the thermodynamic results

$$\begin{aligned} \mathcal{S} &= -\left(\frac{\partial \tilde{\Omega}}{\partial \tau} \right)_{\tilde{V}, \mathbf{v}}, \\ \tilde{p} &= -\left(\frac{\partial \tilde{\Omega}}{\partial \tilde{V}} \right)_{\tau, \mathbf{v}}, \\ \langle \tilde{\mathbf{p}}_{op} \rangle &= -\left(\frac{\partial \tilde{\Omega}}{\partial \mathbf{v}} \right)_{\tau, \tilde{V}}. \end{aligned} \tag{17.15}$$

17.3.1 Moving frame blackbody thermodynamics

The partition function of Eq. 17.11 is evaluated by first summing over $n_{\mathbf{k}}$ to get

$$\tilde{\mathcal{Z}} = \prod_{\mathbf{k}} \left\{ \frac{1}{1 - \exp\left\{-\tilde{\beta}\left[\hbar c\,|\mathbf{k}|\,(1 - |\mathbf{v}/c|\cos\theta)\right]\right\}} \right\}^2, \tag{17.16}$$

where $\omega = c\,|\mathbf{k}|$ has been used. In the CMB rest frame $\mathbf{v} = 0$ and $\tilde{\beta} \to \beta$, where $\beta = (k_B T_0)^{-1}$ with T_0 the CMB rest frame temperature.

The moving frame thermodynamic potential is

$$\tilde{\Omega} = -\frac{1}{\tilde{\beta}} \ln \tilde{\mathcal{Z}} \tag{17.17}$$

$$= \frac{2}{\tilde{\beta}} \sum_{\mathbf{k}} \ln\left\{1 - \exp\left\{-\tilde{\beta}\left[\hbar c\,|\mathbf{k}|\,(1 - |\mathbf{v}/c|\cos\theta)\right]\right\}\right\}. \tag{17.18}$$

In the large volume limit this is expressed as the integral

$$\tilde{\Omega} = \frac{\tilde{V}}{2\pi^2\tilde{\beta}} \int_0^{2\pi} d\varphi \int_0^{\pi} d\theta \sin\theta \int_0^{\infty} d\,|\mathbf{k}|\,|\mathbf{k}|^2 \times \ln\left\{1 - \exp\left[-\tilde{\beta}\hbar c\,|\mathbf{k}|\,(1 - |\mathbf{v}/c|\cos\theta)\right]\right\}, \tag{17.19}$$

which, with the change in variable

$$z = \tilde{\beta}\hbar c\,(1 - |\mathbf{v}/c|\cos\theta), \tag{17.20}$$

$$dz = \tilde{\beta}\hbar c\,|\mathbf{v}/c|\sin\theta d\theta, \tag{17.21}$$

followed by an integration by parts, is

$$\tilde{\Omega} = -\frac{\pi^2 k_B^4}{45\hbar^3 c^3} \times \frac{\tilde{V}\tau^4}{\left(1 - |\mathbf{v}/c|^2\right)^2}. \tag{17.22}$$

Using this result with Eq. 17.15, the moving frame radiation pressure is

$$\tilde{p} = \frac{\pi^2 k_B^4}{45\hbar^3 c^3} \times \frac{\tau^4}{\left(1 - |\mathbf{v}/c|^2\right)^2}. \tag{17.23}$$

Again applying Eq. 17.15, we find an expression for the "moving frame" entropy

$$\mathcal{S} = \frac{4}{3} \times \frac{\pi^2 k_B^4}{45\hbar^3 c^3} \times \frac{\tilde{V}\tau^3}{\left(1 - |\mathbf{v}/c|^2\right)^2}. \tag{17.24}$$

However, in the CMB rest frame

$$\mathcal{S}_0 = \frac{4}{3} \times \frac{\pi^2 k_B^4}{45\hbar^3 c^3} \times V_0 T_0^3, \tag{17.25}$$

where V_0 is a rest frame volume and T_0 is the rest frame temperature. But entropy is not altered by changing reference frames,[9] so that $\mathcal{S} = \mathcal{S}_0$. In the moving frame the rest frame volume is Lorentz-contracted,

$$\tilde{V} = V_0\sqrt{1 - |\mathbf{v}/c|^2}, \tag{17.26}$$

so that equating Eqs. 17.24 and 17.25 gives

$$\frac{1}{\left[1 - |\mathbf{v}/c|^2\right]^{3/2}}\tau^3 = T_0^3 \tag{17.27}$$

so that

$$\tau = T_0\left[1 - |\mathbf{v}/c|^2\right]^{1/2}, \tag{17.28}$$

which is the Lorentz transformation for temperature.[10]

Completing this part of the discussion, an application of Eq. 17.15 gives

$$\langle\tilde{\mathbf{p}}_{op}\rangle = -\frac{4k_B^4\pi^2\mathbf{v}\tau^4\tilde{V}}{45\hbar^3c^5\left(1 - |\mathbf{v}/c|^2\right)^3}. \tag{17.29}$$

Finally, with Eqs. 17.29, 17.24 and 17.22, we get from Eq. 17.12 the moving frame radiation field internal energy $\tilde{\mathcal{U}}$

$$\tilde{\mathcal{U}} = \frac{\pi^2k_B^4\tilde{V}}{15\hbar^3c^3}\frac{\left(1 + \frac{1}{3}|\mathbf{v}/c|^2\right)}{\left(1 - |\mathbf{v}/c|^2\right)^3}\tau^4 \tag{17.30}$$

or

$$\tilde{\mathcal{U}} = \mathcal{U}_0\frac{\left(1 + \frac{1}{3}|\mathbf{v}/c|^2\right)}{\left(1 - |\mathbf{v}/c|^2\right)^{1/2}}, \tag{17.31}$$

where

$$\mathcal{U}_0 = \frac{\pi^2k_B^4T_0^4V_0}{15\hbar^3c^3} \tag{17.32}$$

is the rest frame radiation field energy (see Eq. 14.37).[11] Finally, since entropy is Lorentz invariant, for quasi-static heat transfer

$$d\mathcal{S} = \frac{\widetilde{đQ}}{\tau} = \frac{đQ}{T_0} \tag{17.33}$$

[9] N. G. Van Kampen, "Relativistic thermodynamics of moving systems", *Phys. Rev.* **173**, 295 (1968).

[10] D. Mi, H. Y. Zhong and D. M. Tong, "Different proposals for relativistic temperature transforms: Whys and wherefores", *Mod. Phys. Lett.* **24**, 73–80 (2009).

[11] C. Møller, *The Theory of Relativity*, Clarendon Press, Oxford (1952).

so that

$$\widetilde{dQ} = \text{đ}Q\left(1 - \frac{v^2}{c^2}\right)^{1/2}. \tag{17.34}$$

17.4 Radiation energy flux

That part of the CMB anisotropy (frequency $\omega_{\mathbf{k},\lambda}$ and velocity $\mathbf{v}$ at a particular angle θ with respect to the local group velocity) which arises solely from satellite motion with respect to the microwave CMB rest frame provides a "base" anisotropy to be subtracted[12] from the microwave signal in order to obtain the anisotropy of the 13.7 billion year old CMB fireball itself.

For this we write the Poynting vector $\mathscr{S}(\theta, \mathbf{v})$ (energy per unit time per unit detector area) observed from the angle θ with respect to the moving frame velocity $\mathbf{v}$. For a single mode $\mathbf{k}$ and one polarization λ,

$$\frac{\mathrm{d}\left|\mathscr{S}_{\mathbf{k},\lambda}(\theta, \mathbf{v})\right|}{\mathrm{d}A} = \left(1/\tilde{V}\right) c\hbar\omega_{\mathbf{k},\lambda} \langle \tilde{n}(\mathbf{k}, \lambda; \theta, \mathbf{v})\rangle \frac{\mathbf{k}}{|\mathbf{k}|} \cdot \frac{\mathbf{v}}{|\mathbf{v}|}, \tag{17.35}$$

where $\mathscr{S}_{\mathbf{k},\lambda}(\theta, \mathbf{v})$ is the radiation flux for the mode $\mathbf{k}$ with polarization λ, and $\langle \tilde{n}(\mathbf{k}, \lambda; \theta, \mathbf{v})\rangle$ is the moving frame average photon occupation number. Using Eq. 17.7 we find directly

$$\langle \tilde{n}(\mathbf{k}, \lambda; \theta, \mathbf{v})\rangle = \frac{1}{e^{\beta\hbar\left(\omega_{\mathbf{k},\lambda} - |\mathbf{v}|\,|\mathbf{k}|\cos\theta\right)} - 1}, \tag{17.36}$$

which is a Bose–Einstein average occupation number. With $\omega_{\mathbf{k},\lambda} = c\,|\mathbf{k}|$ and in the large volume limit the sum in Eq. 17.35 – including both polarizations – becomes, per unit solid angle $\mathrm{d}\Omega$, the integral

$$\frac{\mathrm{d}^2\left|\mathscr{S}_{\mathbf{k},\lambda}(\theta, \mathbf{v})\right|}{\mathrm{d}A\ \mathrm{d}\Omega} = \frac{2\tilde{V}}{(2\pi)^3}\left(1/\tilde{V}\right) c^2\hbar \int_0^\infty \mathrm{d}\,|\mathbf{k}| \frac{|\mathbf{k}|^3}{e^{\tilde{\beta}\hbar c\,|\mathbf{k}|\left[1 - (|\mathbf{v}|/c)\cos\theta\right]} - 1}. \tag{17.37}$$

Integrating over all modes $\mathbf{k}$ for a particular radiation reception angle θ gives

$$\frac{\mathrm{d}^2\left|\mathscr{S}_{\mathbf{k},\lambda}(\theta, \mathbf{v})\right|}{\mathrm{d}A\,\mathrm{d}\Omega} = \frac{\pi}{60c^2\hbar^3\left(1 - |\mathbf{v}/c|\cos\theta\right)^4 \tilde{\beta}^4} \tag{17.38}$$

$$= \frac{\pi k_B^4 T_0^4 \left(1 - |\mathbf{v}/c|\right)^2}{60c^2\hbar^3\left(1 - |\mathbf{v}/c|\cos\theta\right)^4}. \tag{17.39}$$

[12] Along with other known galactic radiation corrections.

In other words, the moving observer's Planck radiation function is

$$u\left(\omega, T_0; \theta, \mathbf{v}\right) = \left\{\exp\left[\frac{\hbar\omega\left(1 - |\mathbf{v}/c|\cos\theta\right)}{k_B T_0\sqrt{1 - |\mathbf{v}/c|^2}}\right] - 1\right\}^{-1} \tag{17.40}$$

and the moving frame temperature for the admittance angle θ is

$$\tau\left(\theta, \mathbf{v}\right) = \frac{T_0\sqrt{1 - |\mathbf{v}/c|^2}}{\left(1 - |\mathbf{v}/c|\cos\theta\right)} \tag{17.41}$$

which is the PW result.

The sensitivity of modern microwave satellite detectors assures that relativistic temperature anisotropy effects will be observable up to several multipole orders. These have to be carefully subtracted in order to extract the significant underlying cosmological information. Eq. 17.41 is essential to such analyses.

Problems and exercises

17.1 Find an expressions in terms of rest values H_0, F_0, G_0 for:

(a) moving frame blackbody enthalpy $\tilde{H}$;
(b) moving frame blackbody Helmholtz potential $\tilde{F}$;
(c) moving frame blackbody Gibbs potential $\tilde{G}$.

Appendix A How pure is pure? An inequality

A.1 $0 < \mathcal{T}r\left(\rho_{op}^{\tau}\right)^2 \leq 1$

For the general mixed state

$$\mathcal{T}r\left(\rho_{op}^{\tau}\right)^2 = \sum_{i,j,n} \langle n \mid \psi_i\rangle\, w_i \left\langle \psi_i \mid \psi_j\right\rangle w_j \left\langle \psi_j \mid n\right\rangle, \tag{A.1}$$

which is rearranged to read

$$\mathcal{T}r\left(\rho_{op}^{\tau}\right)^2 = \sum_{i,j,n} w_i \left\langle \psi_i \mid \psi_j\right\rangle w_j \left\langle \psi_j \mid n\right\rangle \langle n \mid \psi_i\rangle\,. \tag{A.2}$$

Explicitly carrying out the trace operation (sum over n)

$$\mathcal{T}r\left(\rho_{op}^{\tau}\right)^2 = \sum_{i,j} w_i \left\langle \psi_i \mid \psi_j\right\rangle w_j \left\langle \psi_j \mid \psi_i\right\rangle. \tag{A.3}$$

"Sharing" the classical (real, positive) probabilities w_k among the terms

$$\mathcal{T}r\left(\rho_{op}^{\tau}\right)^2 = \sum_{i,j} \left\langle \psi_i\sqrt{w_i} \mid \psi_j\sqrt{w_j}\right\rangle \left\langle \psi_j\sqrt{w_j} \mid \psi_i\sqrt{w_i}\right\rangle. \tag{A.4}$$

By the Schwartz inequality

$$\begin{aligned}&\sum_{i,j} \left\langle \psi_i\sqrt{w_i} \mid \psi_j\sqrt{w_j}\right\rangle \left\langle \psi_j\sqrt{w_j} \mid \psi_i\sqrt{w_i}\right\rangle \\ &\quad \leq \left\{\sum_{i} \left\langle \psi_i\sqrt{w_i} \mid \psi_i\sqrt{w_i}\right\rangle\right\} \left\{\sum_{j} \left\langle \psi_j\sqrt{w_j} \mid \psi_j\sqrt{w_j}\right\rangle\right\}\end{aligned} \tag{A.5}$$

or

$$0 < \mathcal{T}r\left(\rho_{op}^{\tau}\right)^2 \leq \left(\mathcal{T}r\rho_{op}^{\tau}\right)^2 = 1. \tag{A.6}$$

Appendix B Bias and the thermal Lagrangian

B.1 Properties of $\mathcal{F}$

We seek a functional[1] of probabilities $\mathcal{F}(P_1, P_2, P_3, \ldots)$ that serves as a mathematical measure of "bias" (uncertainty). The possibilities for $\mathcal{F}$ can be considerably narrowed by imposing reasonable properties:[2]

1. Probabilities P_s are normalized to unity.
2. $\mathcal{F}(P_1, P_2, P_3, \ldots)$ is a continuous function of the P_s.
3. Any incremental changes in any of the P_s produce only incremental changes in $\mathcal{F}(P_1, P_2, P_3, \ldots)$.
4. $\mathcal{F}(P_1, P_2, P_3, \ldots)$ is a symmetric function of its arguments. No change in ordering of the arguments can change the value of the function.
5. $\mathcal{F}(P_1, P_2, P_3, \ldots)$ is a minimum when all probabilities but one are zero (only one possible outcome – completely biased).
6. $\mathcal{F}(P_1, P_2, P_3, \ldots)$ is a maximum when all probabilities are equal (completely unbiased).
7. If an event of zero probability is added to the set, the value of the function remains unchanged, i.e.

$$\mathcal{F}(P_1, P_2, P_3, \ldots, 0) = \mathcal{F}(P_1, P_2, P_3, \ldots)\,. \tag{B.1}$$

8. For the case in which all probabilities P_s are equal, i.e.

$$P_1 = p, P_2 = p, P_3 = p, \ldots, P_n = p, \tag{B.2}$$

a function $f(n)$ is defined

$$\mathcal{F}(p, p, p, \ldots, P_n = p) \equiv f(n) \tag{B.3}$$

such that $f(n)$ is a monotonically increasing function of n. (More distinct outcomes mean decreasing bias; fewer mean increased bias.)

[1] The abbreviation $P(\varepsilon_s) \equiv P_s$ is used.

[2] J. N. Kapur, *Maximum Entropy Models in Science and Engineering*, John Wiley and Sons, New York (1989).

9. There is an addition rule for combining "bias" functions of independent events. For example, consider two independent probability distributions corresponding to events $\boldsymbol{P}$ and $\boldsymbol{Q}$,

$$\boldsymbol{P} = \{P_1, P_2, P_3, \ldots, P_n\}, \tag{B.4}$$

$$\boldsymbol{Q} = \{Q_1, Q_2, Q_3, \ldots, Q_m\}, \tag{B.5}$$

with "bias" functions

$$\mathcal{F}_P(P_1, P_2, P_3, \ldots, P_n), \tag{B.6}$$

$$\mathcal{F}_Q(Q_1, Q_2, Q_3, \ldots, Q_m). \tag{B.7}$$

Let there be a set of joint events

$$\boldsymbol{P} \cup \boldsymbol{Q} = \{P_1Q_1, P_1Q_2, \ldots, P_2Q_1, P_2Q_2, P_2Q_3, \ldots, P_mP_n\} \tag{B.8}$$

with "bias" function

$$\mathcal{F}_{P\cup Q}(P_1Q_1, P_1Q_2, \ldots, P_2Q_1, P_2Q_2, P_2Q_3, \ldots, P_mP_n). \tag{B.9}$$

The addition rule is

$$\begin{aligned}&\mathcal{F}_{P\cup Q}(P_1Q_1, P_1Q_2, \ldots, P_2Q_1, P_2Q_2, P_2Q_3, \ldots, P_mP_n)\\&\quad = \mathcal{F}_P(P_1, P_2, P_3, \ldots, P_n) + \mathcal{F}_Q(Q_1, Q_2, Q_3, \ldots, Q_m).\end{aligned} \tag{B.10}$$

The addition rule excludes, for example, forms like purity $\mathcal{I}$.

10. In the special case that all probabilities are equal, applying Eq. B.3 together with the addition rule Eq. B.10 gives

$$f(n \times m) = f(n) + f(m) \tag{B.11}$$

and

$$f(m^y) = yf(m); \quad y > 0. \tag{B.12}$$

B.2 The "bias" function

From Eq. B.11 we can guess a solution

$$f(m) = \kappa \ln m \tag{B.13}$$

and since $f(m)$ is a monotonically increasing function of m, κ is a positive constant. In particular, since there are m events with equal probability p, that probability must be $p = 1/m$, so

$$\mathcal{F}(p, p, p, \ldots) = -\kappa m p \ln p, \tag{B.14}$$

which, in the present restricted case $p_1 = p_2 = p_3 = \ldots = p_m$, may be written

$$\mathcal{F}(p, p, p, \ldots) = -\kappa \sum_{r=1}^{m} p_r \ln p_r. \tag{B.15}$$

Relaxing this restriction to include p_r that differ in value, say P_r, the result becomes a general "bias" function

$$\mathcal{F}(P_1, P_2, P_3, \ldots, P_m) = -\kappa \sum_{r=1}^{m} P_r \ln P_r \tag{B.16}$$

which has the form of *Gibbs–Shannon entropy*.

B.3 A thermal Lagrangian

The method chosen for determining the "best" $\mathbf{P}(\varepsilon_s)$ is to:

1. Minimize the average total energy

$$\langle \mathcal{H} \rangle = \sum_{r=1}^{n} \tilde{\mathbf{P}}(\varepsilon_r)\, \varepsilon_r \tag{B.17}$$

   with respect to variations in the $\tilde{\mathbf{P}}$. Here $\boldsymbol{\varepsilon}_r$ are macroscopic eigen-energies of the hamiltonian,

$$\mathcal{H} = \mathbf{h}_0 - \sum_{m=1} \boldsymbol{X}_m \boldsymbol{x}_m \tag{B.18}$$

   and where $\boldsymbol{X}_m$ and $\boldsymbol{x}_m$ are conjugate intensive and extensive variables, respectively (see Eq. 6.12).
2. Simultaneously minimize bias among the $\tilde{\mathbf{P}}$ by maximizing the "bias" functional (see Eq. B.16)

$$\mathcal{F}\left[\tilde{\mathbf{P}}\right] = -\kappa \sum_{r=1}^{n} \tilde{\mathbf{P}}(\varepsilon_r) \ln \tilde{\mathbf{P}}(\varepsilon_r). \tag{B.19}$$

3. Maintain normalization

$$\sum_{r=1}^{n} \tilde{\mathbf{P}}(\varepsilon_r) = 1. \tag{B.20}$$

Forming $\Lambda\left[\tilde{\mathbf{P}}\right]$, a functional combining "bias" and average energy with a Lagrange multiplier λ_0 to enforce normalization of the $\tilde{\mathbf{P}}(\varepsilon_r)$,

$$\Lambda\left[\tilde{\mathbf{P}}\right] = \left\{-\kappa\sum_{r=1}^{n}\tilde{\mathbf{P}}(\varepsilon_r)\ln\tilde{\mathbf{P}}(\varepsilon_r) - \sum_{r=1}^{n}\tilde{\mathbf{P}}(\varepsilon_r)\,\varepsilon_r\right\} - \lambda_0\sum_{r=1}^{n}\tilde{\mathbf{P}}(\varepsilon_r)\,. \tag{B.21}$$

$\Lambda\left[\tilde{\mathbf{P}}\right]$ is then maximized with respect to the $\tilde{\mathbf{P}}(\varepsilon_r)$ by setting the functional derivatives $\delta\Lambda/\delta\tilde{\mathbf{P}}$ equal to zero:

$$\frac{\delta\Lambda}{\delta\tilde{\mathbf{P}}} = 0 = -\,\kappa\left(\ln\tilde{\mathbf{P}}(\varepsilon_r) + 1\right) - \varepsilon_r - \lambda_0;\quad r = 1, 2, \ldots, n, \tag{B.22}$$

along with

$$\sum_{r=1}^{n}\tilde{\mathbf{P}}(\varepsilon_r) = 1. \tag{B.23}$$

The "best" $\mathbf{P}(\varepsilon_r)$ are then

$$\mathbf{P}(\varepsilon_r) = \frac{\exp\left(-\kappa^{-1}\varepsilon_r\right)}{\mathcal{Z}}, \tag{B.24}$$

where

$$\mathcal{Z} = \sum_{s=1}^{n}\exp\left(-\kappa^{-1}\varepsilon_s\right), \tag{B.25}$$

with κ still to be determined.

Taking the constant $\kappa \rightarrow k_B T$ the functional $\Lambda\left[\tilde{\mathbf{P}}\right]$ becomes

$$\mathcal{L}\left[\mathbf{P}\right] = -k_B\sum_{r=1}^{n}\mathbf{P}(\varepsilon_r)\ln\mathbf{P}(\varepsilon_r) - \frac{1}{T}\sum_{r=1}^{n}\mathbf{P}(\varepsilon_r)\,\boldsymbol{\varepsilon}_r - \lambda_0\sum_{r=1}^{n}\mathbf{P}(\varepsilon_r)\,. \tag{B.26}$$

Now, identifying the entropy

$$\mathcal{S} = -k_B\sum_{r=1}^{n}\mathbf{P}(\varepsilon_r)\ln\mathbf{P}(\varepsilon_r) \tag{B.27}$$

and $\beta = (k_B T)^{-1}$,

$$\mathbf{P}(\varepsilon_r) = \frac{\exp(-\beta\varepsilon_r)}{\mathcal{Z}}, \tag{B.28}$$

with

$$\mathcal{Z} = \sum_{s=1}^{n}\exp(-\beta\varepsilon_s)\,, \tag{B.29}$$

this becomes identical to determining the equilibrium thermodynamic configuration by maximizing a negative Helmholtz potential functional $-F\left[\mathbf{P}\right]$ (see Section 5.1.3).

C

Appendix C Euler's homogeneous function theorem

C.1 The theorem

If $f(x_1, x_2, x_3, \ldots, x_N)$ is a function with the property

$$f(\lambda x_1, \lambda x_2, \lambda x_3, \ldots, \lambda x_N) = \lambda^n f(x_1, x_2, x_3, \ldots, x_N), \tag{C.1}$$

it is said to be *homogeneous of order n*. Then, according to Euler's homogeneous function theorem,

$$\sum_{i=1}^{N} x_i \left(\frac{\partial f}{\partial x_i}\right) = nf(x_1, x_2, \ldots, x_N). \tag{C.2}$$

C.2 The proof

Differentiating Eq. C.1 with respect to λ

$$\frac{d}{d\lambda} f(\lambda x_1, \lambda x_2, \lambda x_3, \ldots, \lambda x_N) = \frac{d}{d\lambda} \lambda^n f(x_1, x_2, x_3, \ldots, x_N). \tag{C.3}$$

But

$$\frac{d}{d\lambda} f(\lambda x_1, \lambda x_2, \ldots, \lambda x_N) = \sum_{i=1}^{N} \frac{\partial f}{\partial(\lambda x_i)} \frac{d(\lambda x_i)}{d\lambda} \tag{C.4}$$

$$= \sum_{i=1}^{N} \frac{\partial f}{\partial(\lambda x_i)} x_i \tag{C.5}$$

and

$$\frac{d}{d\lambda} \lambda^n f(x_1, x_2, x_3, \ldots, x_N) = n\lambda^{n-1} f(x_1, x_2, x_3, \ldots, x_N). \tag{C.6}$$

Now, setting $\lambda = 1$,

$$\sum_{i=1}^{N} x_i \left(\frac{\partial f}{\partial x_i}\right) = nf(x_1, x_2, \ldots, x_N). \tag{C.7}$$

D Appendix D Occupation numbers and the partition function

Explicit determination of the "eigenstate" degeneracy $g(E)$[1] has been used in arriving at Eq. 8.6. In some circumstances this is too limited or inconvenient for calculating a partition function.

A different method is demonstrated by reconsidering the two-level Schottky problem of Section 8.2. The macroscopic eigen-energies may be written with a different emphasis as

$$E\left\{\begin{bmatrix} n_1^1 \\ n_2^1 \end{bmatrix};\begin{bmatrix} n_1^2 \\ n_2^2 \end{bmatrix};\begin{bmatrix} n_1^3 \\ n_2^3 \end{bmatrix};\ldots\begin{bmatrix} n_1^N \\ n_2^N \end{bmatrix}\right\} = \sum_{j=1}^{N}\left(\epsilon_1 n_1^j + \epsilon_2 n_2^j\right), \tag{D.1}$$

where

$$\begin{bmatrix} n_1^j \\ n_2^j \end{bmatrix} \tag{D.2}$$

represents jth atom occupation number *pairs*, which for the two-level example of Section 8.2 – where only ϵ_1 or ϵ_2 can be occupied on each atom – are

$$\begin{pmatrix} n_1^j = 1 \\ n_2^j = 0 \end{pmatrix} \quad \textit{and} \quad \begin{pmatrix} n_1^j = 0 \\ n_2^j = 1 \end{pmatrix}. \tag{D.3}$$

The sum over j in Eq. D.1 is over all atom occupation *pairs*. Using Eq. D.1 the partition function is

$$\mathcal{Z} = \sum_{\begin{bmatrix} n_1^1 \\ n_2^1 \end{bmatrix},\begin{bmatrix} n_1^2 \\ n_2^2 \end{bmatrix},\ldots,\begin{bmatrix} n_1^N \\ n_2^N \end{bmatrix}} \exp\left\{-\beta\sum_{j=1}^{N}\left(\epsilon_1 n_1^j + \epsilon_2 n_2^j\right)\right\} \tag{D.4}$$

$$= \left\{\sum_{\begin{bmatrix} n_1 \\ n_2 \end{bmatrix}} \exp\left\{-\beta\left(\epsilon_1 n_1 + \epsilon_2 n_2\right)\right\}\right\}^N, \tag{D.5}$$

[1] Similar to Boltzmann's "microstate" count, i.e. his classical idea of an "internal energy degeneracy".

where the remaining sum is over the occupation pair

$$\begin{bmatrix} n_1 \\ n_2 \end{bmatrix} \equiv \begin{pmatrix} 1 \\ 0 \end{pmatrix} \quad \textit{and} \quad \begin{pmatrix} 0 \\ 1 \end{pmatrix}, \tag{D.6}$$

i.e. there are only two terms, which gives

$$\mathcal{Z} = \left(e^{-\beta\varepsilon_1} + e^{-\beta\varepsilon_2}\right)^N. \tag{D.7}$$

E

Appendix E Density of states

E.1 Definition

The *density of states* $\mathscr{D}(\omega)$ of a quantum system is simply the number of states at the energy level ω. This number is determined by sifting through all the quantum eigen-energies ϵ_s of the system and counting how many belong in the energy bin $\epsilon = \omega$. If there is degeneracy among the eigen-energies, e.g. $\epsilon_1 = \epsilon_2 = \ldots = \epsilon_g = \omega$, all g of these degenerate energies must find their way into the energy bin ω.

Finding a density of states $\mathscr{D}(\omega)$ is a mathematical counting process defined by

$$\mathscr{D}(\omega) = \sum_s \delta(\omega - \epsilon_s), \tag{E.1}$$

where $\delta(\omega - \varepsilon_s)$ is the Dirac delta function which carries out the task of adding ϵ_s to the number of states accumulating in bin ω.

The Dirac delta function has the property of a well-chosen unity in the sense that

$$1 = \int_{-\infty}^{\infty} \mathrm{d}\omega\, \delta(\omega - \epsilon_s). \tag{E.2}$$

In calculating a density of states the eigen-energy subscript s is usually replaced by quantum numbers that identify the eigen-energies, so that counting in Eq. E.1 can take place over quantum numbers.

E.2 Examples

Example 1

Consider a one-dimensional system of independent particles with only kinetic energy

$$\epsilon(k) = \frac{\hbar^2 k^2}{2m}, \tag{E.3}$$

where

$$-\infty < k < \infty. \tag{E.4}$$

Here ϵ_s in Eq. E.1 is replaced by $\epsilon(k)$, where k is the free particle quantum number and the density of single-particle states is written

$$\mathscr{D}(\omega) = \sum_k \delta[\omega - \epsilon(k)] \tag{E.5}$$

$$= \sum_k \delta\left[\omega - \frac{\hbar^2 k^2}{2m}\right] \tag{E.6}$$

with the sum over ϵ_s replaced by a sum over k.

The property of the Dirac delta function that is used in density of states calculations is

$$\delta[f(k)] = \sum_{k_0} \frac{\delta(k - k_0)}{\left|\frac{\mathrm{d}}{\mathrm{d}k} f(k)\right|_{k=k_0}} \tag{E.7}$$

where k_0 denotes the zeros of $f(k)$

$$f(k_0) = 0 \tag{E.8}$$

and the sum in Eq. E.7 is over k_0, all the zeros of $f(k)$.

In this one-dimensional case where

$$\mathscr{D}(\omega) = \sum_k \delta[f(k)] \tag{E.9}$$

with

$$f(k) = \omega - \frac{\hbar^2 k^2}{2m}, \tag{E.10}$$

locating its zeros we find the pair

$$k_{01} = +\sqrt{\frac{2m\omega}{\hbar^2}} \tag{E.11}$$

and

$$k_{02} = -\sqrt{\frac{2m\omega}{\hbar^2}}. \tag{E.12}$$

Next, the denominator in Eq. E.7 is evaluated,

$$\left|\frac{\mathrm{d}}{\mathrm{d}k} f(k)\right|_{k=k_{01}} = \frac{\hbar^2 |k_{01}|}{m}, \tag{E.13}$$

$$\left|\frac{\mathrm{d}}{\mathrm{d}k} f(k)\right|_{k=k_{02}} = \frac{\hbar^2 |k_{02}|}{m}, \tag{E.14}$$

so that, finally, the density of states is written

$$\mathscr{D}(\omega) = \sum_k \left\{ \frac{\delta(k - k_{01})}{\frac{\hbar^2}{m}\sqrt{\frac{2m\omega}{\hbar^2}}} + \frac{\delta(k - k_{02})}{\frac{\hbar^2}{m}\sqrt{\frac{2m\omega}{\hbar^2}}} \right\}. \tag{E.15}$$

In the case of a macroscopically long line of particles the k-values become closely spaced so that in one dimension the "sum" becomes the integral

$$\sum_k \rightarrow \frac{L}{2\pi} \int_{-\infty}^{\infty} \mathrm{d}k, \tag{E.16}$$

where L is the line length.[1,2]

The integral is easily evaluated to give

$$\mathscr{D}(\omega) = \frac{L}{\pi}\sqrt{\frac{m}{2\hbar^2\omega}}. \tag{E.19}$$

The density of single-particle states is customarily given per unit length,

$$\frac{\mathscr{D}(\omega)}{L} = \frac{1}{\pi}\sqrt{\frac{m}{2\hbar^2\omega}}. \tag{E.20}$$

If there is spin degeneracy a factor g may be included,

$$\frac{\mathscr{D}(\omega)}{L} = g\frac{1}{\pi}\sqrt{\frac{m}{2\hbar^2\omega}}, \tag{E.21}$$

where $g = 2$ in the case $spin = \frac{1}{2}$.

Example 2

In the case of the three-dimensional Debye model[3]

$$\omega = \langle c_s \rangle |\mathbf{k}|. \tag{E.22}$$

[1] In two dimensions

$$\sum_{\mathbf{k}} \rightarrow \frac{A}{(2\pi)^2} \int_{-\infty}^{\infty}\int_{-\infty}^{\infty} \mathrm{d}k_x\, \mathrm{d}k_y \tag{E.17}$$

with A the two-dimensional area.

[2] In three dimensions

$$\sum_{\mathbf{k}} \rightarrow \frac{V}{(2\pi)^3} \int_{-\infty}^{\infty}\int_{-\infty}^{\infty}\int_{-\infty}^{\infty} \mathrm{d}k_x\, \mathrm{d}k_y\, \mathrm{d}k_z \tag{E.18}$$

with V the three-dimensional volume.

[3] In this example the distinction between longitudinal (1 mode) and transverse (2 modes) phonons is ignored, using instead an average speed of sound $\langle c_s \rangle$ and a degeneracy $g = 3$.

Therefore, since the zero of Eq. E.22 is

$$|k|_0 = \omega/\langle c_s\rangle, \tag{E.23}$$

we can write

$$\mathscr{D}(\omega) = \frac{V}{(2\pi)^3}\left[4\pi\int_0^\infty \frac{1}{\langle c_s\rangle}\delta\left(|k| - \omega/\langle c_s\rangle\right)|k|^2 \,\mathrm{d}\,|k|\right] \tag{E.24}$$

to finally give

$$\mathscr{D}(\omega) = \frac{3V}{2\pi^2}\frac{\omega^2}{\langle c_s\rangle^3}, \tag{E.25}$$

where the approximated mode degeneracy (1 longitudinal + 2 transverse) is accounted for by the inserted factor $g = 3$.

F

Appendix F A lab experiment in elasticity

F.1 Objectives

Here we will explore the possibility of finding the *microscopic* interchain interaction energy ϵ for a particular elastomer sample using macroscopic thermodynamic measurements. The experiment also serves to emphasize the physical meaning and importance of the partial derivatives that are crucial in applying thermodynamics.

The experiment is based on the theoretical discussion in Section 10.5 on a non-ideal elastomer model. We start by writing elastic tension τ as a function of length $L = \langle L_z \rangle$ and temperature T, i.e. $\tau = \tau(L, T)$. The corresponding tension differential is

$$d\tau = \left(\frac{\partial \tau}{\partial L}\right)_T dL + \left(\frac{\partial \tau}{\partial T}\right)_L dT. \tag{F.1}$$

If we perform an experiment in which a rubber band's length L is held fixed, the tension differential becomes

$$d\tau = \left(\frac{\partial \tau}{\partial T}\right)_L dT. \tag{F.2}$$

Furthermore, applying a Maxwell relation

$$\left(\frac{\partial \mathcal{S}}{\partial L}\right)_T = -\left(\frac{\partial \tau}{\partial T}\right)_L, \tag{F.3}$$

we can replace Eq. F.2 by

$$d\tau = -\left(\frac{\partial \mathcal{S}}{\partial L}\right)_T dT, \tag{F.4}$$

which demonstrates that $-(\partial \mathcal{S}/\partial L)_T$, a quantity in which we have some interest (see the discussion below), is the slope on a τ vs. T curve acquired at fixed elastomer length.

Now consider the Helmholtz potential $\mathcal{F}$

$$\mathcal{F} = \mathcal{U} - T\mathcal{S} \tag{F.5}$$

and its derivative with respect to L at constant T

$$\left(\frac{\partial \mathcal{F}}{\partial L}\right)_T = \left(\frac{\partial \mathcal{U}}{\partial L}\right)_T - T\left(\frac{\partial \mathcal{S}}{\partial L}\right)_T. \tag{F.6}$$

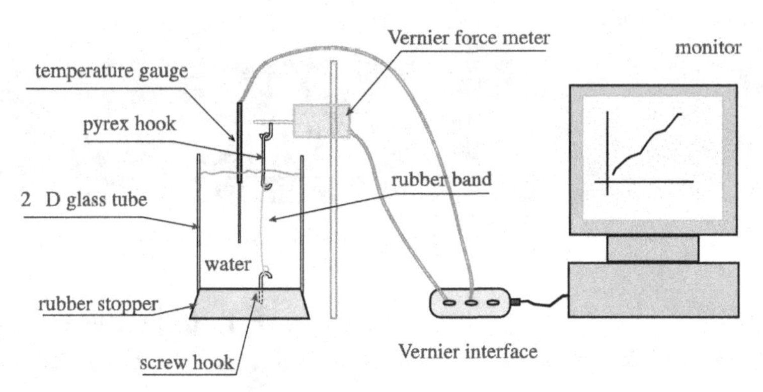

Fig. F.1 Experimental arrangement for measurement of τ vs. T. Add boiling water to the glass tube or fill the tube with acetone and dry ice $\approx -78\ ^\circ$C. Then allow the system to approach room temperature.

Taking the total differential of Eq. F.5 and combining it with the First Law we have

$$d\mathcal{F} = -\mathcal{S}dT + \tau dL \tag{F.7}$$

from which we note

$$\left(\frac{\partial \mathcal{F}}{\partial L}\right)_T = \tau. \tag{F.8}$$

Then we can rewrite Eq. F.6 as

$$\tau = -T\left(\frac{\partial \mathcal{S}}{\partial L}\right)_T + \left(\frac{\partial U}{\partial L}\right)_T, \tag{F.9}$$

which is equivalent to having integrated Eq. F.4.

According to Eq. F.9, if we fix the length of a rubber band and measure the tension τ while varying temperature T, the slope of the curve at any temperature is $-(\partial\mathcal{S}/\partial L)_T$ while the τ-intercept (at $T = 0$) is $(\partial U/\partial L)_T$, which is another quantity of interest. In fact comparing

$$\left|T\left(\frac{\partial \mathcal{S}}{\partial L}\right)_T\right|, \tag{F.10}$$

obtained from the slope, with

$$\left|\left(\frac{\partial U}{\partial L}\right)_T\right|, \tag{F.11}$$

obtained from the τ-intercept, it is possible to assess the relative importance of entropy vs. internal energy to an elastomer's elasticity. This is one motivation for the proposed experiment, although not the prime objective. A possible apparatus is illustrated in Figure F.1.

F.2 Results and analysis

1. Using the Vernier instrumentation and Logger-Pro computer software, obtain best τ vs. T linear fits to the data. Identify the values of the slope and the τ-intercept.
2. Determine if the elasticity of this rubber band is entropy or internal energy driven.
3. Now we must see what the experiment can say about the parameter ϵ. According to the model's thermodynamics as expressed in Eqs. 10.78 and 10.80 of Chapter 10,

$$\left(\frac{\partial \mathcal{S}}{\partial L}\right)_T = -k_B\beta\left\{\tau + \frac{2\epsilon\ \sinh(\beta a\tau)}{a\left[e^{\beta\epsilon} + 2\cosh(\beta a\tau)\right]}\right\}. \tag{F.12}$$

Verify this result.

Also, from Eqs. 10.78 and 10.81,

$$\left(\frac{\partial U}{\partial L}\right)_T = -\frac{2\epsilon\ \sinh(\beta a\tau)}{a\left[e^{\beta\epsilon} + 2\cosh(\beta a\tau)\right]}. \tag{F.13}$$

Verify this result.

Now expand Eqs. F.12 and F.13 for small τ. The resulting series have only odd powers of τ. Do the same thing with the equation of state and use that result to eliminate τ from each of the expansions obtained from F.12 and F.13. This gives

$$\left(\frac{\partial \mathcal{S}}{\partial L}\right)_T \cong -k_B\left[1 + 2e^{-\beta\epsilon}(1+\beta\epsilon)\right]\left(L/Na^2\right). \tag{F.14}$$

Verify this result.

It also gives

$$\left(\frac{\partial U}{\partial L}\right)_T \cong -2\epsilon\, e^{-\beta\epsilon}\left(L/Na^2\right). \tag{F.15}$$

Verify this result.

4. Evaluate Eqs. F.14 and F.15 in the limit $\epsilon \to 0$. Compare these results with ideal gas counterparts, i.e. $(\partial\mathcal{S}/\partial V)_T$ and $(\partial U/\partial V)_T$.
5. To this order the ratio of Eq. F.14 and Eq. F.15 does not contain the length parameter L/Na^2 and can therefore be used to estimate the interaction energy ϵ.
6. Use the experimental results to evaluate ϵ and $\left(L/Na^2\right)$. (You may have to use Maple or Mathematica to solve the equations.)

G

Appendix G Magnetic and electric fields in matter

G.1 Introduction

Magnetic and electric fields and their response functions (magnetization and polarization) are generally defined, consistent with Maxwell's equations, as local averages over microscopic regions. Except for special sample geometries and field orientations, these fields within matter:

- differ from external fields;
- are non-uniform inside matter and in the surrounding free space.

This can make thermodynamic calculations and experimental interpretation a non-trivial problem.[1]

G.2 Thermodynamic potentials and magnetic fields

With magnetic work, as in Eq. 11.10, the fundamental equation becomes

$$T\delta\mathcal{S} = \delta\mathcal{U} + p\mathrm{d}V - \frac{1}{4\pi}\int_V \mathcal{H}\cdot\delta\mathcal{B}\mathrm{d}V. \tag{G.1}$$

The Helmholtz potential differential δF is

$$\delta F = -\mathcal{S}\mathrm{d}T - p\mathrm{d}V + \frac{1}{4\pi}\int_V \mathcal{H}\cdot\delta\mathcal{B}\mathrm{d}V. \tag{G.2}$$

[1] Experimentalists strive to fabricate samples which conform to special geometries for which internal demagnetizing fields are usually – but not always – negligible.

The enthalpy differential δH is

$$\delta H = T\delta \mathcal{S} + V\mathrm{d}p - \frac{1}{4\pi}\int_V \mathcal{B}\cdot\delta\mathscr{H}\,\mathrm{d}V. \tag{G.3}$$

The Gibbs potential differential δG is

$$\delta G = -\mathcal{S}\mathrm{d}T + V\mathrm{d}p - \frac{1}{4\pi}\int_V \mathcal{B}\cdot\delta\mathscr{H}\,\mathrm{d}V. \tag{G.4}$$

These results are based on average local fields (Maxwell's equations) and are valid for any mechanism of magnetization. They include, in principle, internal interactions.

G.3 Work and uniform fields

Microscopic-based theory and experiment can be properly compared only if magnetic models use practical expressions for thermodynamic work in the case of matter being inserted into previously uniform external fields ($\mathcal{B}_0$, $\mathcal{E}_0$).[2,3,4]

To the experimentalist this means:

- constraining free currents that create the external field $\mathscr{H}_0$ to remain unchanged when matter is inserted into the field, so that $\nabla\times\mathscr{H} = \nabla\times\mathscr{H}_0$;
- constraining free charges that create the external field $\mathcal{D}_0$ to remain unchanged when matter is inserted into the field, so that $\nabla\cdot\mathcal{D} = \nabla\cdot\mathcal{D}_0$.

Because magnetization $\mathcal{M}$ and polarization $\mathcal{P}$ are both zero in the absence of matter, the defining constitutive relations for $\mathscr{H}$ and $\mathcal{D}$,

$$\mathscr{H} = \mathcal{B} - 4\pi\mathcal{M}, \tag{G.5}$$

$$\mathcal{D} = \mathcal{E} + 4\pi\mathcal{P}, \tag{G.6}$$

specify that before the sample is introduced $\mathscr{H}_0 = \mathcal{B}_0$ and $\mathcal{D}_0 = \mathcal{E}_0$. After the sample is introduced $\mathcal{B}$, $\mathscr{H}$, $\mathcal{E}$ and $\mathcal{D}$ become the local average fields.

[2] L.D. Landau, L.P. Pitaevskii and E.M. Lifshitz, *Electrodynamics of Continuous Media*, 2nd edition, Elsevier, Amsterdam (1984).

[3] V. Heine, "The thermodynamics of bodies in static electromagnetic fields", *Proc. Camb. Phil. Soc.* **52**, 546–552 (1956).

[4] E. A. Guggenheim, *Thermodynamics: An Advanced Treatment for Chemists and Physicists*, North-Holland, Amsterdam (1967).

From this point we examine only magnetic contributions, the electric field case being formulated by parallel arguments.

For the uniform pre-existent field (prior to insertion of matter) $\mathcal{B}_0 = \mathscr{H}_0$ and we can construct the identity

$$\frac{1}{4\pi}\int_V \mathcal{B}_0 \cdot \delta\mathcal{B} \mathrm{d}V - \frac{1}{4\pi}\int_V \mathscr{H}_0 \cdot \delta\mathcal{B} \mathrm{d}V = 0, \tag{G.7}$$

and use it to construct yet another identity for the difference between magnetic work with average Maxwellian fields (see Eq. 11.9) and magnetic energy of the uniform field before matter is introduced, namely

$$\begin{aligned}&\frac{1}{4\pi}\int_V \mathscr{H} \cdot \delta\mathcal{B} \mathrm{d}V - \frac{1}{4\pi}\int_V \mathcal{B}_0 \cdot \delta\mathcal{B}_0 \mathrm{d}V \\ &\quad = \frac{1}{4\pi}\int_V (\mathscr{H} - \mathscr{H}_0) \cdot \delta\mathcal{B} \mathrm{d}V + \frac{1}{4\pi}\int_V \mathcal{B}_0 \cdot (\delta\mathcal{B} - \delta\mathcal{B}_0)\, \mathrm{d}V.\end{aligned} \tag{G.8}$$

Using the constitutive equation relating $\mathcal{B}$ and $\mathscr{H}$

$$(\mathscr{H} - \mathscr{H}_0) = (\mathcal{B} - \mathcal{B}_0) - 4\pi \mathcal{M} \tag{G.9}$$

and

$$(\delta\mathscr{H} - \delta\mathscr{H}_0) = (\delta\mathcal{B} - \delta\mathcal{B}_0) - 4\pi \delta\mathcal{M}, \tag{G.10}$$

where $\mathcal{B}_0$ and $\mathscr{H}_0$ are the prior existing fields, allows the right-hand side of Eq. G.8 to be re-expressed in terms of magnetization as

$$\begin{aligned}&\frac{1}{4\pi}\int_V \mathscr{H} \cdot \delta\mathcal{B} \mathrm{d}V - \frac{1}{4\pi}\int_V \mathcal{B}_0 \cdot \delta\mathcal{B}_0 \mathrm{d}V \\ &= \frac{1}{4\pi}\int_V (\mathscr{H} - \mathscr{H}_0) \cdot \delta\mathcal{B} \mathrm{d}V + \frac{1}{4\pi}\int_V \mathcal{B}_0 \cdot [(\delta\mathscr{H} - \delta\mathscr{H}_0) + 4\pi \delta\mathcal{M}]\, \mathrm{d}V \\ &= \frac{1}{4\pi}\int_V (\mathscr{H} - \mathscr{H}_0) \cdot \delta\mathcal{B}\, \mathrm{d}V + \frac{1}{4\pi}\int_V \mathcal{B}_0 \cdot (\delta\mathscr{H} - \delta\mathscr{H}_0)\, \mathrm{d}V + \int_V \mathcal{B}_0 \cdot \delta\mathcal{M}\, \mathrm{d}V.\end{aligned} \tag{G.11}$$

Applying

$$\mathcal{B} = \nabla \times \mathcal{A}, \tag{G.12}$$

$$\delta\mathcal{B} = \nabla \times \delta\mathcal{A}, \tag{G.13}$$

$$\mathcal{B}_0 = \nabla \times \mathcal{A}_0, \tag{G.14}$$

$$\delta\mathcal{B}_0 = \nabla \times \delta\mathcal{A}_0, \tag{G.15}$$

where $\mathcal{A}$ and $\mathcal{A}_0$ are static magnetic vector potentials, the first term on the final line of Eq. G.11 is

$$\frac{1}{4\pi}\int_V (\mathcal{H} - \mathcal{H}_0) \cdot \nabla \times \delta\mathcal{A}\, \mathrm{d}V \tag{G.16}$$

and the second term is

$$\frac{1}{4\pi}\int_V (\nabla \times \mathcal{A}_0) \cdot (\delta\mathcal{H} - \delta\mathcal{H}_0)\, \mathrm{d}V. \tag{G.17}$$

Then, with the vector identity

$$U \cdot \nabla \times V = V \cdot \nabla \times U + \nabla \cdot (V \times U) \tag{G.18}$$

the integral G.16 is transformed

$$\begin{aligned}&\frac{1}{4\pi}\int_V (\mathcal{H} - \mathcal{H}_0) \cdot \nabla \times \delta\mathcal{A}\, \mathrm{d}V\\ &\quad\rightarrow \frac{1}{4\pi}\int_V \delta\mathcal{A} \cdot \nabla \times (\mathcal{H} - \mathcal{H}_0)\, \mathrm{d}V - \frac{1}{4\pi}\int_V \nabla \cdot [\delta\mathcal{A} \times (\mathcal{H} - \mathcal{H}_0)]\, \mathrm{d}V.\end{aligned} \tag{G.19}$$

Similarly, the integral G.17 is transformed

$$\begin{aligned}&\frac{1}{4\pi}\int_V (\nabla \times \mathcal{A}_0) \cdot (\delta\mathcal{H} - \delta\mathcal{H}_0)\, \mathrm{d}V\\ &\quad\rightarrow \frac{1}{4\pi}\int_V \mathcal{A}_0 \cdot \nabla \times (\delta\mathcal{H} - \delta\mathcal{H}_0)\, dV - \frac{1}{4\pi}\int_V \nabla \cdot [\mathcal{A} \cdot \nabla \times (\delta\mathcal{H} - \delta\mathcal{H}_0)]\, \mathrm{d}V.\end{aligned} \tag{G.20}$$

Both resulting integrals, Eqs. G.19 and G.20, are zero since:

- The currents that create the external field are constrained to remain unchanged when introducing the sample into the field. Therefore in the first terms of each of these integrals, $\nabla \times (\mathcal{H} - \mathcal{H}_0) = 0 = \nabla \times (\delta\mathcal{H} - \delta\mathcal{H}_0)$.
- The second term of each integral can be turned into a surface integral (Gauss' theorem) on a very distant surface where static (non-radiative) fields fall off faster than $\frac{1}{r^2}$.

With these simplifications Eq. G.11 is rearranged to become

$$\frac{1}{4\pi}\int_V \mathcal{H} \cdot \delta\mathcal{B}\mathrm{d}V = \frac{1}{4\pi}\int_V \mathcal{B}_0\, \delta\mathcal{B}_0\, \mathrm{d}V + \int_{V'} \mathcal{B}_0 \cdot \delta\mathcal{M}\mathrm{d}V. \tag{G.21}$$

The left-hand side is the magnetic work with average local fields while the right-hand side involves:

- the total magnetic field energy before insertion of matter;
- the energy associated with the magnetic material in the prior existing external field, $\mathcal{B}_0$.

Since $\mathcal{B}_0$ is constant and $\mathcal{M} = 0$ outside matter, the volume in the second term, V', covers magnetized material only. Therefore, whether or not $\mathcal{M}$ is uniform,

$$\mathbf{M} = \int_{V'} \mathcal{M} \mathrm{d}V, \tag{G.22}$$

where $\mathbf{M}$ is the total magnetization. The final result is therefore

$$\frac{1}{4\pi} \int_V \mathscr{H} \cdot \delta\mathcal{B} \mathrm{d}V = \frac{V}{8\pi} \delta \left(\mathcal{B}_0\right)^2 + \mathcal{B}_0 \cdot \mathrm{d}\mathbf{M}. \tag{G.23}$$

Applying Eq. G.1 the resulting thermodynamic relations are

$$T\mathrm{d}\mathcal{S} = \mathrm{d}\mathcal{U}^* + p\mathrm{d}V - \mathcal{B}_0 \cdot \mathrm{d}\mathbf{M}, \tag{G.24}$$

$$T\mathrm{d}\mathcal{S} = \mathrm{d}H^* - V\mathrm{d}p + \mathbf{M} \cdot \mathrm{d}\mathcal{B}_0, \tag{G.25}$$

$$\mathrm{d}F^* = -\mathcal{S}\mathrm{d}T - p\mathrm{d}V + \mathcal{B}_0 \cdot \mathrm{d}\mathbf{M}, \tag{G.26}$$

$$\mathrm{d}G^* = -\mathcal{S}\mathrm{d}T + V\mathrm{d}p - \mathbf{M} \cdot \mathrm{d}\mathcal{B}_0, \tag{G.27}$$

each having the advantage of $\mathcal{B}_0$ being a simple independent variable rather than a function of position. (The field energies $V\delta\left(\mathcal{B}_0\right)^2/8\pi$ have been included in the internal energy and hence into the thermodynamic potentials, which are now distinguished by superscripted stars.)

G.4 Thermodynamics with internal fields

But of course there is no free lunch. Although the potentials F^* and G^*, as defined in Eqs. G.26 and G.27, are convenient thermodynamic state functions (with mathematical properties discussed in Chapter 4), they take no account of effective contributions to $\mathcal{B}$ from the remaining magnetized matter. However, they can be used in the usual thermodynamic calculations where, say, from Eq. G.27 the magnetization $\mathbf{M}$ is given by

$$\mathbf{M} = -\left(\frac{\partial G^*}{\partial \mathcal{B}_0}\right)_{p,T}, \tag{G.28}$$

but it is a magnetization that takes no account of any internal fields. They are not true free energies.

Nor do they carry information about currents and fields inside matter. For example, the kinetic energy in the hamiltonian cannot simply be $p^2/2m$ because that would ignore consequences of the external field within matter. The kinetic energy should be

$$\mathcal{H} = \frac{1}{2m}\left\{p - \frac{e}{c}(\mathcal{A}_0 + \mathcal{A})\right\}^2, \tag{G.29}$$

where $\mathcal{A}_0$ is an external vector potential and $\mathcal{A}$ is the internal vector potential whose source is the currents induced by the external field and which interacts with internal currents via the interaction hamiltonian $\mathcal{A} \cdot \mathbf{p}$.

Specifically, the final term in the mean field decomposition in Chapter 11

$$\begin{aligned}\mathbf{m}_{op}(i')\mathbf{m}_{op}(i) &= [\mathbf{m}_{op}(i') - \langle\mathbf{m}_{op}\rangle][\mathbf{m}_{op}(i) - \langle\mathbf{m}_{op}\rangle] \\ &\quad + \mathbf{m}_{op}(i')\langle\mathbf{m}_{op}\rangle + \langle\mathbf{m}_{op}\rangle\mathbf{m}_{op}(i) - \langle\mathbf{m}_{op}\rangle\langle\mathbf{m}_{op}\rangle\end{aligned} \tag{G.30}$$

constitutes an internal magnetic contribution to the Gibbs potential (see Eqs. 11.97 and 11.98)

$$\mathbf{W}_{\text{MFA}} = -\frac{1}{2}\mathcal{B}_{int} \cdot \mathbf{M} \tag{G.31}$$

so that the Gibbs potential G^* becomes

$$\tilde{G}^* = G^* - \frac{1}{2}\mathcal{B}_{int} \cdot \mathbf{M} \tag{G.32}$$

$$= G^* - \frac{z\mathcal{K}}{2N}\mathbf{M} \cdot \mathbf{M}, \tag{G.33}$$

where $\tilde{G}^*$ denotes this additional dependence on the effective field, Eq. 11.97. Furthermore

$$d\tilde{G}^* = dG^* - \frac{z\mathcal{K}}{N}\mathbf{M}\, d\mathbf{M} \tag{G.34}$$

$$= -SdT - \mathbf{M}d\mathcal{B}_0 - \frac{z\mathcal{K}}{N}\mathbf{M}\,d\mathbf{M} \tag{G.35}$$

$$= -SdT - \mathbf{M}d\left(\mathcal{B}_0 + \frac{z\mathcal{K}}{N}\mathbf{M}\right) \tag{G.36}$$

$$= -SdT - \mathbf{M}d\mathcal{B}^*, \tag{G.37}$$

which means that as a result of the internal field the magnetization is now

$$\mathbf{M} = -\left(\frac{\partial\tilde{G}^*}{\partial\mathcal{B}^*}\right)_T. \tag{G.38}$$

Therefore, re-examining Eq. 11.99 we see a proper role for $\mathcal{B}^*$ in the following way:

$$\mathbf{M} = \frac{1}{\beta}\left(\frac{\partial}{\partial\mathcal{B}^*}\ln\mathcal{Z}_{M^*}\right)_T \tag{G.39}$$

$$= N\mu_{\frac{1}{2}}\tanh(\beta\mu_{\frac{1}{2}}\mathcal{B}^*). \tag{G.40}$$

Another internal field of practical importance is the so-called "demagnetization" field. Demagnetization is defined through geometrically based factors η_j (demagnetization factors), which describe a uniform field $\mathcal{H}_{int,j}$ inside an ellipsoid arising from magnetization

$$\mathcal{H}_{int,j} = \mathcal{H}_{0,j} - 4\pi\eta_{j,k}\mathbf{M}_k, \tag{G.41}$$

with $j = x, y, z$. Formally, its source is the fictitious, classical magnetic surface "poles" arising from the applied field $\mathcal{H}_0$. Demagnetization factors (actually tensors) $\eta_{j,k}$ depend on sample geometry and are obviously responsible for magnetic anisotropy. Calculating demagnetization "factors" for different geometries remains an on-going activity in the field of practical magnetics.[5]

Applying the constitutive equation

$$\mathcal{H}_{int,j} = \mathcal{B}_{int,j} - 4\pi\mathbf{M}_j \tag{G.42}$$

to Eq. G.41, we have (ignoring, for simplicity, the tensor character of η) the effective field

$$\mathcal{B}_{int,j} = \mathcal{B}_{0,j} + 4\pi(1-\eta)\mathbf{M}_j. \tag{G.43}$$

[5] M. Beleggia, M. DeGraff and Y. Millev, "Demagnetization factors of the general ellipsoid", *Phil. Mag* **86**, 2451 (2006).

H Appendix H Maxwell's equations and electromagnetic fields

H.1 Maxwell's equations

Electric fields $\mathcal{E}(\mathbf{x}, \mathbf{t})$ and magnetic fields $\mathcal{B}(\mathbf{x}, \mathbf{t})$ in free space satisfy the differential Maxwell relations[1]

$$\nabla \cdot \mathcal{E}(\mathbf{x}, \mathbf{t}) = 0, \tag{H.1}$$

$$\nabla \cdot \mathcal{B}(\mathbf{x}, \mathbf{t}) = 0, \tag{H.2}$$

$$\nabla \times \mathcal{E}(\mathbf{x}, \mathbf{t}) = -\frac{1}{c}\frac{\partial \mathcal{B}(\mathbf{x}, \mathbf{t})}{\partial t}, \tag{H.3}$$

$$\nabla \times \mathcal{B}(\mathbf{x}, \mathbf{t}) = \frac{1}{c}\frac{\partial \mathcal{E}(\mathbf{x}, \mathbf{t})}{\partial t}. \tag{H.4}$$

Unlike Schrödinger's wave mechanics, which retains some suggestion of classical particle dynamics, a quantum theory of radiation is guided by Maxwell's field relationships and Planck's photon postulate, which Dirac used in his theory of quantum[2] fields.

H.2 Electromagnetic waves

In classical theory time-varying electric and magnetic fields propagate in free space as waves with velocity c. This can be seen by combining the last of Maxwell's equations, Eq. H.4, with Maxwell's first equation, Eq. H.1, to give the wave equation for an electric field

$$\nabla^2 \mathcal{E}(\mathbf{x}, t) = \frac{1}{c^2}\frac{\partial^2 \mathcal{E}(x, t)}{\partial t^2}. \tag{H.5}$$

[1] In rationalized cgs units.

[2] P. A. M. Dirac, "Quantum theory of emission and absorption of radiation", *Proc. Roy. Soc. (London)*, **A114**, 243 (1927).

Similarly, Eqs. H.2 and H.3 are combined for a corresponding wave equation for the magnetic field

$$\nabla^2 \mathcal{B}(\mathbf{x}, t) = \frac{1}{c^2} \frac{\partial^2 \mathcal{B}(x, t)}{\partial t^2}. \tag{H.6}$$

H.2.1 Solution in free space – periodic boundary conditions

Imagine a cube with side L and volume $V = L^3$ within which a solution to Eq. H.5 is assumed consisting of the sum of propagating waves,

$$\mathcal{E}_j(\mathbf{x}, t) = \frac{1}{\sqrt{V}} \int_{-\infty}^{\infty} \frac{d\omega}{2\pi} \sum_{\mathbf{k}} \mathbf{e}_j(\mathbf{k}, \omega)\, e^{i\mathbf{k}\cdot\mathbf{x}} e^{-i\omega t}, \tag{H.7}$$

where $\mathcal{E}_j$ is the jth component of the Maxwell electric field vector ($j = x, y, z$) and $\mathbf{e}_j$ is, correspondingly, the Fourier vector component of the $\mathbf{k}$th propagating wave amplitude. The vector $\mathbf{k}$ in Eq. H.7 defines the spatial direction in which that particular Fourier component propagates (its direction of propagation). Each propagation vector $\mathbf{k}$ is identified as a mode index.

The solutions within the notional cubic volume are supplemented by sensible (but synthetic) boundary conditions according to which translation in any direction x, y or z by the finite side length L results in exactly the same electric fields, i.e.

$$\mathcal{E}_j(\mathbf{x} + \mathbf{L}, t) = \frac{1}{\sqrt{V}} \int_{-\infty}^{\infty} \frac{d\omega}{2\pi} \sum_{\mathbf{k}} \mathbf{e}_j(\mathbf{k}, \omega)\, e^{i\mathbf{k}\cdot(\mathbf{x}+\mathbf{L})} e^{-i\omega t} = \mathcal{E}_j(\mathbf{x}, t). \tag{H.8}$$

These are called *periodic boundary conditions*. They are used in nearly all field theories, classical and quantum. When the cube is finally made large enough to really look like infinite free space it hardly matters what goes on at the boundaries since the cube's surface-to-volume ratio is rapidly diminishing to zero.

But in order for Eq. H.8 to hold, the following restrictions on $\mathbf{k}$ must apply:

$$k_x L = 2\pi n_x; \quad n_x = 0, \pm 1, \pm 2, \ldots, \tag{H.9}$$

$$k_y L = 2\pi n_y; \quad n_y = 0, \pm 1, \pm 2, \ldots, \tag{H.10}$$

$$k_z L = 2\pi n_z; \quad n_z = 0, \pm 1, \pm 2, \ldots, \tag{H.11}$$

i.e. under periodic boundary conditions the propagation vectors can have only discrete positive and negative values

$$k_j = \frac{2\pi n_j}{L}. \tag{H.12}$$

However, as can be seen from Eq. H.12, in the limit of a large volume, i.e. $L \to \infty$, the spacing between successive values of k_j becomes infinitesimal. This important result allows the sums in Eq. H.19 (and other similar sums) to be treated as integrals.

Properties of wave solutions

Substituting Eq. H.7 into the wave equation Eq. H.5,

$$\nabla^2 \left\{ \frac{1}{\sqrt{V}} \int_{-\infty}^{\infty} \frac{d\omega}{2\pi} \sum_{\mathbf{k}} \mathbf{e}_j (\mathbf{k}, \omega) e^{i\mathbf{k}\cdot\mathbf{x}} e^{-i\omega t} \right\} = \frac{1}{\sqrt{V}} \int_{-\infty}^{\infty} \frac{d\omega}{2\pi} \sum_{\mathbf{k}} [-\mathbf{k}\cdot\mathbf{k}]\, \mathbf{e}_j \times (\mathbf{k}, \omega) e^{i\mathbf{k}\cdot\mathbf{x}} e^{-i\omega t} \tag{H.13}$$

and

$$\frac{1}{c^2} \frac{\partial^2}{\partial t^2} \left\{ \frac{1}{\sqrt{V}} \int_{-\infty}^{\infty} \frac{d\omega}{2\pi} \sum_{\mathbf{k}} \mathbf{e}_j (\mathbf{k}, \omega) e^{i\mathbf{k}\cdot\mathbf{x}} e^{-i\omega t} \right\} = \frac{1}{\sqrt{V}} \int_{-\infty}^{\infty} \frac{d\omega}{2\pi} \sum_{\mathbf{k}} \left[-\frac{\omega^2}{c^2} \right] \mathbf{e}_j \times (\mathbf{k}, \omega) e^{i\mathbf{k}\cdot\mathbf{x}} e^{-i\omega t}, \tag{H.14}$$

showing that Eq. H.7 is a wave equation solution only if

$$k^2 = \frac{\omega^2}{c^2}, \tag{H.15}$$

where

$$k^2 = {k_x}^2 + {k_y}^2 + {k_z}^2, \tag{H.16}$$

i.e. the magnitude of the propagation vector $\mathbf{k}$ is proportional to the frequency ω.

Of course, the solution must also satisfy the first of Maxwell's equations, i.e. Eq. H.1, which also has important implications. Taking the divergence of Eq. H.7 and setting it equal to *zero*,

$$\nabla \cdot E = \frac{1}{\sqrt{V}} \int_{-\infty}^{\infty} \frac{d\omega}{2\pi} \sum_{\mathbf{k}} \left[k_x \mathbf{e}_x (\mathbf{k}, \omega) + k_y \mathbf{e}_y (\mathbf{k}, \omega) + k_z \mathbf{e}_z (\mathbf{k}, \omega) \right] e^{i\mathbf{k}\cdot\mathbf{x}} e^{-i\omega t} = 0, \tag{H.17}$$

from which is concluded

$$\mathbf{k} \cdot \mathbf{e} (\mathbf{k}, \omega) = 0, \tag{H.18}$$

i.e. the Fourier vector amplitude $\mathbf{e}(\mathbf{k}, \omega)$ must be perpendicular to the propagation vector $\mathbf{k}$. The Fourier vector amplitude $\mathbf{e}(\mathbf{k}, \omega)$ can then be expressed in terms of two mutually perpendicular unit vectors, $\hat{\varepsilon}(\mathbf{k}, 1)$ and $\hat{\varepsilon}(\mathbf{k}, 2)$ – a pair for each mode $\mathbf{k}$ – both of which are perpendicular to the propagation vector $\mathbf{k}$, i.e. $\hat{\varepsilon}(\mathbf{k}, \lambda)\cdot\mathbf{k} = 0$ where

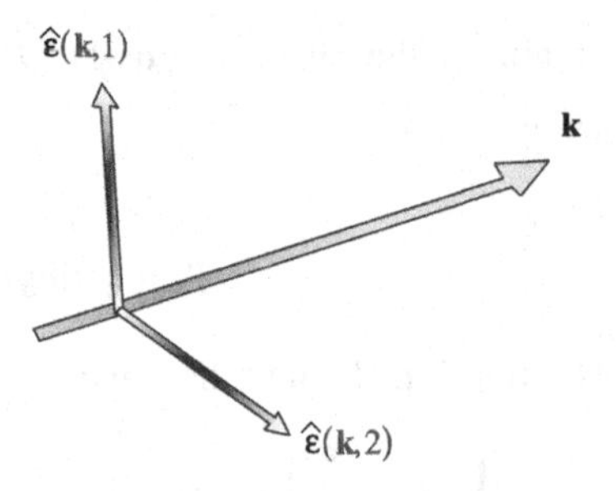

Fig. H.1 Propagation vector k and two transverse polarization unit vectors $\hat{\varepsilon}\,(\mathbf{k},1)$ and $\hat{\varepsilon}\,(\mathbf{k},2)$. Each cartesian component of the Fourier vector amplitude $\mathbf{e}_j\,(\mathbf{k},\omega)$ can be expressed as a linear combination of $\hat{\varepsilon}\,(\mathbf{k},1)$ and $\hat{\varepsilon}\,(\mathbf{k},2)$.

$\lambda = 1, 2$. These two transverse unit vectors are called *electric field polarization vectors.*[3]

In terms of unit polarization vectors the electric field is written

$$\mathcal{E}_j\,(\mathbf{x},t) = \frac{1}{\sqrt{V}}\int_{-\infty}^{\infty}\frac{d\omega}{2\pi}\sum_{\substack{\mathbf{k}\\ \lambda=1,2}}\hat{\varepsilon}\,(\mathbf{k},\lambda)\left|\mathbf{e}_{j,\lambda}\right|(\mathbf{k},\omega)\,e^{i\mathbf{k}\cdot\mathbf{x}}e^{-i\omega t}, \tag{H.19}$$

where $\left|\mathbf{e}_{j,\lambda}\right|$ is the projection of the jth cartesian component ($j = x,\ y,\ z$) of the field vector's **k**th Fourier component along the λth polarization vector direction $\hat{\varepsilon}\,(\mathbf{k},\lambda)$. (See Figure H.1.)

Each photon therefore requires two identifiers:

- **k**, the mode;
- $\hat{\varepsilon}\,(\mathbf{k},\lambda)$, the polarization ($\lambda = 1$ or 2).

H.3 Electromagnetic vector potential

The energy eigenvalues of an electromagnetic field in free space are expressed in terms of photons – particle-like excitations first hypothesized by Planck. Quantum electrodynamics, the 20th-century refinement of Planck's theory, describes photons as quantum excitations of the vector potential field $\mathcal{A}\,(\mathbf{x},t)$, where

$$\mathcal{B}\,(\mathbf{x},t) = \nabla\times\mathcal{A}\,(\mathbf{x},t) \tag{H.20}$$

[3] Vibrational modes in solids (phonons) have no governing relation similar to Maxwell's first law, and so maintain three independent polarization directions.

and

$$\mathcal{E}(\mathbf{x},t) = -\nabla\varphi - \frac{1}{c}\frac{\partial}{\partial t}\mathcal{A}(\mathbf{x},t), \tag{H.21}$$

where ϕ is a scalar potential. Because the vector potential is not unique, gauge fixing is required. One common choice is

$$\nabla \cdot A(\mathbf{x},t) = 0, \tag{H.22}$$

which is called the *Coulomb gauge*[4] or, alternatively, the *transverse gauge* (since $\mathcal{A} \rightarrow \mathcal{A}_{\perp}$).[5,6] In the Coulomb gauge and in the absence of sources we can take $\phi \equiv 0$.

In this gauge the classical wave equation becomes

$$\nabla^2 \mathcal{A}_{\perp} - \frac{\partial^2 \mathcal{A}_{\perp}}{\partial t^2} = 0. \tag{H.23}$$

The (transverse) vector potential can be expanded in a Fourier series of plane waves

$$\mathcal{A}_{\perp}(\mathbf{x},t) = \sum_{\substack{k \\ \lambda=1,2}} \left(\frac{1}{2V\omega_{\mathbf{k}}}\right)^{\frac{1}{2}} \left[\hat{\varepsilon}(\mathbf{k},\lambda)\, a_{\mathbf{k},\lambda}(t)\, e^{i\mathbf{k}\cdot\mathbf{x}} + \hat{\varepsilon}^*(\mathbf{k},\lambda)\, a^*_{\mathbf{k},\lambda}(t)\, e^{-i\mathbf{k}\cdot\mathbf{x}}\right], \tag{H.24}$$

where $a_{\mathbf{k},\lambda}$ and $a^*_{\mathbf{k},\lambda}$ are complex coefficients, $\omega_{\mathbf{k}} = c\,|\mathbf{k}|$ is the vacuum electromagnetic dispersion relation and the coefficient $(2V\omega_{\mathbf{k}})^{-\frac{1}{2}}$ is chosen with an eye towards simplifying the final result.

It follows from the Coulomb gauge condition that

$$\mathbf{k} \cdot \hat{\varepsilon}(\mathbf{k},\lambda) = 0, \quad \lambda = 1, 2, \tag{H.25}$$

where the unit polarization vectors themselves are chosen to be mutually perpendicular and therefore satisfy the orthogonality condition

$$\hat{\varepsilon}(\mathbf{k},\mu) \cdot \hat{\varepsilon}^*(\mathbf{k},\gamma) = \delta_{\mu,\gamma}. \tag{H.26}$$

The classical (transverse) electric field is from Eqs. H.21 and H.24 (ignoring circular or elliptical polarization)

$$\mathcal{E}_{\perp}(\mathbf{x},t) = \frac{1}{c} \sum_{\substack{k \\ \lambda=1,2}} \left(\frac{1}{2V\omega_{\mathbf{k}}}\right)^{1/2} \hat{\varepsilon}(\mathbf{k},\lambda) \left[\frac{\partial a_{\mathbf{k},\lambda}(t)}{\partial t} e^{i\mathbf{k}\cdot\mathbf{x}} + \frac{\partial a^*_{\mathbf{k},\lambda}(t)}{\partial t} e^{-i\mathbf{k}\cdot\mathbf{x}}\right]. \tag{H.27}$$

[4] Because Coulomb's law is satisfied in this gauge.

[5] This gauge restricts the vector potential to have only transverse components.

[6] This gauge has the same consequence as Maxwell's first law, i.e. Coulomb's law.

However, by inserting the expansion Eq. H.24 into the wave equation Eq. H.23 the following simplifying results are obtained:

$$\frac{\partial a_{\mathbf{k},\lambda}}{\partial t} = -i\omega_{\mathbf{k}} a_{\mathbf{k},\lambda} \tag{H.28}$$

and

$$\frac{\partial a^*_{\mathbf{k},\lambda}}{\partial t} = i\omega_{\mathbf{k}} a^*_{\mathbf{k},\lambda} \tag{H.29}$$

so that

$$\mathcal{E}_\perp(\mathbf{x}, t) = \frac{i}{c} \sum_{\substack{\mathbf{k} \\ \lambda=1,2}} \left(\frac{\omega_{\mathbf{k}}}{2V}\right)^{1/2} \hat{\varepsilon}(\mathbf{k}, \lambda) \left[a_{\mathbf{k},\lambda}(t) e^{i\mathbf{k}\cdot\mathbf{x}} - a^*_{\mathbf{k},\lambda}(t) e^{-i\mathbf{k}\cdot\mathbf{x}}\right]. \tag{H.30}$$

The transverse magnetic field is, similarly,

$$\begin{aligned} \mathcal{B}_\perp(\mathbf{x}, t) &= \nabla \times \mathcal{A}_\perp \\ &= i \sum_{\substack{\mathbf{k} \\ \lambda=1,2}} \left(\frac{1}{2V\omega_{\mathbf{k}}}\right)^{1/2} \left[\mathbf{k} \times \hat{\varepsilon}(\mathbf{k}, \lambda)\right] \left[a_{\mathbf{k},\lambda}(t) e^{i\mathbf{k}\cdot\mathbf{x}} - a^*_{\mathbf{k},\lambda}(t) e^{-i\mathbf{k}\cdot\mathbf{x}}\right]. \end{aligned} \tag{H.31}$$

H.4 Quantized electromagnetic hamiltonian

The field theoretic canonical quantization procedure postulates the complex coefficients in the expansions for Eqs. H.30 and H.31 "morph" into Fock space operators

$$a_{\mathbf{k},\lambda} \Longmapsto a_{op}(\mathbf{k}, \lambda), \tag{H.32}$$

$$a^*_{\mathbf{k},\lambda} \Longmapsto a^\dagger_{op}(\mathbf{k}, \lambda), \tag{H.33}$$

where the operators fulfill equal time *commutator* relations

$$\left[a_{op}(\mathbf{k}, \mu), a^\dagger_{op}(\mathbf{k}', \mu')\right] = \delta_{\mathbf{k},\mathbf{k}'}\delta_{\mu,\mu'}, \tag{H.34}$$

$$\left[a_{op}{}^\dagger(\mathbf{k}, \mu), a_{op}{}^\dagger(\mathbf{k}', \mu')\right] = 0, \tag{H.35}$$

$$\left[a_{op}(\mathbf{k}, \mu), a_{op}(\mathbf{k}', \mu')\right] = 0, \tag{H.36}$$

where

$$a^\dagger_{op}(\mathbf{k}, \mu)\, a_{op}(\mathbf{k}, \mu) = n_{op}(\mathbf{k}, \mu) \tag{H.37}$$

with $n_{op}(\mathbf{k},\mu)$ the photon number operator. Now we have the vector potential operator $\mathcal{A}_{op}(\mathbf{x},t)$

$$\mathcal{A}_{op}(\mathbf{x},t)=\sum_{\substack{k\\ \lambda=1,2}}\left(\frac{1}{2V\omega_{\mathbf{k}}}\right)^{\frac{1}{2}}\hat{\varepsilon}(\mathbf{k},\lambda)\left[a_{op}(\mathbf{k},\lambda)e^{i\mathbf{k}\cdot\mathbf{x}}+a_{op}^{\dagger}(\mathbf{k},\lambda)e^{-i\mathbf{k}\cdot\mathbf{x}}\right], \quad \text{(H.38)}$$

the electric field operator $\mathcal{E}_{op}(\mathbf{x},\mathbf{t})$

$$\mathcal{E}_{op}(\mathbf{x},t)=\frac{i}{c}\sum_{\substack{\mathbf{k}\\ \lambda=1,2}}\left(\frac{\omega_{\mathbf{k}}}{2V}\right)^{\frac{1}{2}}\hat{\varepsilon}(\mathbf{k},\lambda)\left[a_{op}(\mathbf{k},\lambda)e^{i\mathbf{k}\cdot\mathbf{x}}-a_{op}^{\dagger}(\mathbf{k},\lambda)e^{-i\mathbf{k}\cdot\mathbf{x}}\right] \quad \text{(H.39)}$$

and a magnetic field operator $\mathcal{B}_{op}(\mathbf{x},t)$

$$\mathcal{B}_{op}(\mathbf{x},t)=i\sum_{\substack{\mathbf{k}\\ \lambda=1,2}}\left(\frac{1}{2V\omega_{\mathbf{k}}}\right)^{\frac{1}{2}}\left[\mathbf{k}\times\hat{\varepsilon}(\mathbf{k},\lambda)\right]\left[a_{op}(\mathbf{k},\lambda)e^{i\mathbf{k}\cdot\mathbf{x}}-a_{op}^{\dagger}(\mathbf{k},\lambda)e^{-i\mathbf{k}\cdot\mathbf{x}}\right]. \quad \text{(H.40)}$$

The electromagnetic hamiltonian is

$$\mathcal{H}_{op}^{EM}=\frac{1}{2}\int_V d\mathbf{x}\left[\mathcal{E}_{op}^2(\mathbf{x},t)+\mathcal{B}_{op}^2(\mathbf{x},t)\right]. \quad \text{(H.41)}$$

Now substituting Eqs. H.39 and H.40, carrying out the required integrations, using the commutator relations and applying the results

$$\frac{1}{V}\int_V d\mathbf{x}\, e^{i\mathbf{k}\cdot\mathbf{x}}e^{i\mathbf{k}\cdot\mathbf{x}}=\delta_{\mathbf{k}+\mathbf{p},0}, \quad \text{(H.42)}$$

$$\hat{\varepsilon}(\mathbf{k},\lambda)\cdot\hat{\varepsilon}(-\mathbf{k},\mu)=\hat{\varepsilon}(\mathbf{k},\lambda)\cdot\hat{\varepsilon}(\mathbf{k},\mu) \quad \text{(H.43)}$$

$$=\delta_{\lambda,\mu} \quad \text{(H.44)}$$

and

$$\left[\mathbf{k}\times\hat{\varepsilon}(\mathbf{k},\lambda)\right]\cdot\left[-\mathbf{k}\times\hat{\varepsilon}(-\mathbf{k},\mu)\right]=-\left[\mathbf{k}\times\hat{\varepsilon}(\mathbf{k},\lambda)\right]\cdot\left[\mathbf{k}\times\hat{\varepsilon}(\mathbf{k},\mu)\right] \quad \text{(H.45)}$$

$$=-|k|^2\delta_{\lambda,\mu}, \quad \text{(H.46)}$$

wc find

$$\mathcal{H}_{op}^{EM} = \frac{1}{2} \sum_{\substack{\mathbf{k} \\ \lambda=1,2}} \omega_{\mathbf{k}} \left[a_{op}(\mathbf{k},\lambda)\, a_{op}^{\dagger}(\mathbf{k},\lambda) + a_{op}^{\dagger}(\mathbf{k},\lambda)\, a_{op}(\mathbf{k},\lambda) \right] \tag{H.47}$$

$$= \sum_{\substack{\mathbf{k} \\ \lambda=1,2}} \omega_{\mathbf{k}} \left[a_{op}^{\dagger}(\mathbf{k},\lambda)\, a_{op}(\mathbf{k},\lambda) + \frac{1}{2} \right] \tag{H.48}$$

$$= \sum_{\substack{\mathbf{k} \\ \lambda=1,2}} \omega_{\mathbf{k}} \left[n_{op}(\mathbf{k},\lambda) + \frac{1}{2} \right]. \tag{H.49}$$

I Appendix I Fermi–Dirac integrals

I.1 Introduction

The typical Fermi–Dirac integral in the theory of metals and semiconductors has the form

$$\mathcal{I}^{FD}(\mu) = \int_0^\infty d\omega s(\omega) \mathcal{D}(\omega) \frac{1}{e^{\beta(\omega-\mu)}+1}, \tag{I.1}$$

where $\mathcal{D}(\omega)$ is the density of states and $s(\omega)$ is some function of ω, often a fractional or integer power, such that $\mathcal{D}(\omega) s(\omega)$ is analytic (differential) near $\omega \approx \mu$. (When this is not the case, special techniques can be applied.[1])

Analytic evaluation of these integrals is not possible so several approximations and numerical methods have appeared over the years.[2,3,4,5]

I.2 Expansion method: $\beta\mu \gg 1$

The Sommerfeld asymptotic expansion, which is the earliest published result,[6] exploits the approximate property that

$$\frac{d}{d\omega}\left[\frac{1}{e^{\beta(\omega-\mu)}+1}\right] \simeq -\frac{1}{\beta}\delta(\omega-\mu), \quad \beta\mu \gg 1. \tag{I.2}$$

[1] A. Wasserman, T. Buckholtz and H. E. Dewitt, "Evaluation of some Fermi–Dirac integrals", *J. Math. Phys.* **11**, 477 (1970).

[2] W. B. Joyce and R. W. Dixon, "Analytic approximations for the Fermi energy of an ideal Fermi gas", *Appl. Phys. Lett.* **31**, 354 (1977).

[3] A. Wasserman, "Fermi–Dirac integrals", *Phys. Letters* **27A**, 360 (1968).

[4] T. M. Garoni, N. E. Frankel and M. L. Glasser, "Complete asymptotic expansions of the Fermi–Dirac integrals", *J. Math. Phys.* **42**, 1860 (2001).

[5] Raseong Kim and Mark Lundstrom, *Notes on Fermi–Dirac Integrals* (3rd edition) (2008) (http://nanohub.org/resources/5475).

[6] A. Sommerfeld, "Zur Elekronentheorie der Metalle auf Grund der Fermischen Statisik", *Z. Phys.* **47**, 1 (1928).

A logical first step in Sommerfeld's method is to integrate Eq. I.1 by parts to get

$$\mathscr{I}^{FD}(\mu) = \frac{\beta}{4}\int_0^\infty \mathrm{d}\omega \left\{\int_0^\omega \mathrm{d}\omega' s(\omega')\,\mathscr{D}(\omega')\right\} \mathrm{sech}^2\left[\frac{\beta}{2}(\omega-\mu)\right]; \tag{I.3}$$

where for $\beta\mu \gg 1$ the function $\mathrm{sech}^2\left[\frac{\beta}{2}(\omega-\mu)\right]$ peaks sharply at $\omega \approx \mu$. Now, Taylor-expanding the bracketed integral

$$\int_0^\omega \mathrm{d}\omega' s(\omega')\,\mathscr{D}(\omega') \tag{I.4}$$

about $\omega = \mu$,

$$\begin{aligned}\int_0^\omega \mathrm{d}\omega' s(\omega')\,\mathscr{D}(\omega') &= \int_0^\mu \mathrm{d}\omega'\, s(\omega')\mathscr{D}(\omega') + (\omega-\mu)\frac{\mathrm{d}}{\mathrm{d}\omega}\left[\int_0^\omega \mathrm{d}\omega'\, s(\omega')\mathscr{D}(\omega')\right]_{\omega=\mu} \\ &\quad + \frac{1}{2}(\omega-\mu)^2\frac{\mathrm{d}^2}{\mathrm{d}\omega^2}\left[\int_0^\omega \mathrm{d}\omega'\, s(\omega')\mathscr{D}(\omega')\right]_{\omega=\mu} + \ldots \\ &= \int_0^\mu \mathrm{d}\omega'\, s(\omega')\mathscr{D}(\omega') + (\omega-\mu)\, s(\mu)\,\mathscr{D}(\mu) \\ &\quad + \frac{1}{2}(\omega-\mu)^2\left\{\frac{\mathrm{d}}{\mathrm{d}\omega'}\left[s(\omega')\,\mathscr{D}(\omega')\right]\right\}_{\omega'=\mu} + \ldots\end{aligned} \tag{I.5}$$

After substituting $z = \beta(\omega-\mu)$ we have the approximate result

$$\begin{aligned}4\mathscr{I}^{FD}(\mu) &= \int_0^\mu \mathrm{d}\omega'\, s(\omega')\mathscr{D}(\omega')\int_{-\infty}^\infty \mathrm{d}z\,\mathrm{sech}^2\left(\frac{z}{2}\right) + s(\mu)\,\mathscr{D}(\mu)\int_{-\infty}^\infty \mathrm{d}z\, z\,\mathrm{sech}^2\left(\frac{z}{2}\right) \\ &\quad + \frac{1}{2\beta^2}\left\{\frac{\mathrm{d}}{\mathrm{d}\omega'}\left[s(\omega')\,\mathscr{D}(\omega')\right]\right\}_{\omega'=\mu}\int_{-\infty}^\infty \mathrm{d}z z^2\,\mathrm{sech}^2\left(\frac{z}{2}\right) + \ldots\end{aligned} \tag{I.6}$$

where in the "degenerate case" ($\beta\mu \gg 1$) lower limits on the ω integrals have been extended, $-\beta\mu \to -\infty$. Since the function $\mathrm{sech}^2\left(\frac{z}{2}\right)$ peaks sharply at $z = 0$, this introduces negligible error.[7] Evaluating the integrals:

$$\int_{-\infty}^\infty \mathrm{d}z\,\mathrm{sech}^2\left(\frac{z}{2}\right) = 4, \tag{I.7}$$

$$\int_{-\infty}^\infty \mathrm{d}z\, z\,\mathrm{sech}^2\left(\frac{z}{2}\right) = 0, \tag{I.8}$$

[7] This approximation makes strict convergence questionable and the approximation may become only "asymptotic".

$$\int_{-\infty}^{\infty} \mathrm{d}z\, z^2 \operatorname{sech}^2\left(\frac{z}{2}\right) = \frac{4\pi^2}{3}, \tag{I.9}$$

$$\int_{-\infty}^{\infty} \mathrm{d}z\, z^3 \operatorname{sech}^2\left(\frac{z}{2}\right) = 0, \tag{I.10}$$

we have the approximate result

$$\mathscr{I}^{FD}(\mu) = \int_0^{\mu} \mathrm{d}\omega'\, s(\omega')\mathscr{D}(\omega') + \frac{\pi^2}{6\beta^2}\left\{\frac{\mathrm{d}}{\mathrm{d}\omega'}\left[s(\omega')\,\mathscr{D}(\omega')\right]\right\}_{\omega'=\mu} + \ldots \tag{I.11}$$

Appendix J Bose–Einstein integrals

J.1 BE integrals: $\mu = 0$

In the Bose–Einstein (BE) problem we encounter two kinds of integrals. The first appears when $T < T_c$ so that $\mu = 0$. These have the form

$$\mathscr{I}_{BE} = \int_0^\infty \mathrm{d}x \frac{x^s}{e^x - 1}. \tag{J.1}$$

They can be done by first expanding the denominator

$$\int_0^\infty \mathrm{d}x \frac{x^s}{e^x - 1} = \int_0^\infty \mathrm{d}x\, x^s \sum_{n=0}^\infty e^{-(n+1)x} \tag{J.2}$$

and then integrating term by term

$$\int_0^\infty \mathrm{d}x \frac{x^s}{e^x - 1} = \Gamma(s+1) \sum_{\nu=1}^\infty \frac{1}{\nu^{s+1}} \tag{J.3}$$

where the Γ-function is defined by

$$\Gamma(r) = \int_0^\infty \mathrm{d}z\, z^{r-1} e^{-z}, \quad \mathcal{R}\mathrm{e}(r) > -1. \tag{J.4}$$

The remaining sum is the Riemann ζ function

$$\zeta(s) = \sum_{n=1}^\infty \frac{1}{n^s} \tag{J.5}$$

which finally gives

$$\int_0^\infty \mathrm{d}x \frac{x^s}{e^x - 1} = \Gamma(s+1)\,\zeta(s+1). \tag{J.6}$$

J.2 BE integrals: $\mu < 0$

The second class of integrals is best demonstrated by the specific example

$$\left\langle \mathcal{N}_{op}^{BE} \right\rangle = \frac{V}{4\pi^2} \left(\frac{2m}{\hbar^2} \right)^{3/2} \int_0^\infty d\omega \omega^{1/2} \left(\frac{1}{e^{\beta(\omega-\mu)} - 1} \right) \tag{J.7}$$

$$= \frac{V}{4\pi^2} \left(\frac{2m}{\hbar^2} \right)^{3/2} \int_0^\infty d\omega \omega^{1/2} \left(\frac{\lambda e^{-\beta\omega}}{1 - \lambda e^{-\beta\omega}} \right), \tag{J.8}$$

where $\lambda = e^{\beta\mu}$ and therefore $\lambda e^{-\beta\omega} < 1$. Expanding the denominator in a binomial series

$$\left\langle \mathcal{N}_{op}^{BE} \right\rangle = \frac{V}{4\pi^2} \left(\frac{2m}{\hbar^2} \right)^{3/2} \int_0^\infty d\omega^{1/2} \lambda e^{-\beta\omega} \sum_{n=0}^\infty \lambda^n e^{-n\beta\omega} \tag{J.9}$$

and integrating term by term

$$\left\langle \mathcal{N}_{op}^{BE} \right\rangle = \frac{V}{4\pi^2} \left(\frac{2m}{\hbar^2} \right)^{3/2} \left[\frac{\Gamma(3/2)}{\beta^{3/2}} \right] \left(\lambda + \frac{\lambda^2}{2^{3/2}} + \frac{\lambda^3}{3^{3/2}} + \cdots \right) \tag{J.10}$$

$$= \frac{V}{4\pi^2} \left(\frac{2m}{\hbar^2} \right)^{3/2} \left[\frac{\Gamma(3/2)}{\beta^{3/2}} \right] \varsigma(3/2, \lambda), \tag{J.11}$$

where

$$\varsigma(s, \lambda) = \lambda + \frac{\lambda^2}{2^s} + \frac{\lambda^3}{3^s} + \cdots \tag{J.12}$$

is a generalized ς function. The function as studied by mathematicians (and in the computational arsenal of, say, Mathematica) is Lerch's Φ function (also called Hurwitz–Lerch ς function) defined as

$$\Phi(z, s, a) = \sum_{n=0}^\infty \frac{z^n}{(n+a)^s}, \tag{J.13}$$

where any term for which the denominator is "zero" is excluded. But the generalized ς function appearing in Bose–Einstein integrals is

$$\varsigma(s, z) = \sum_{n=1}^\infty \frac{z^n}{n^s} \tag{J.14}$$

which is related to Lerch's Φ function by

$$\varsigma(s, z) = z\Phi(z, s, 1). \tag{J.15}$$

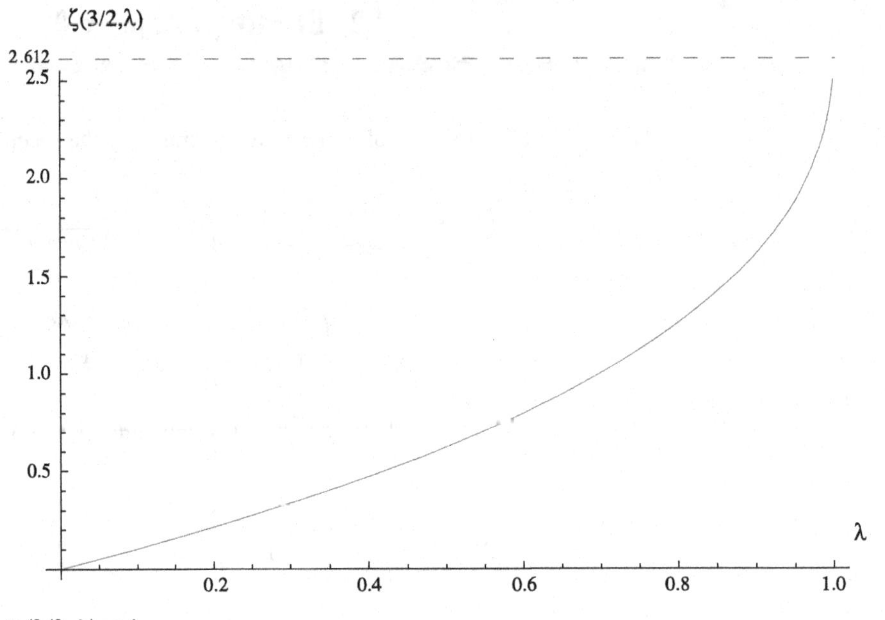

Fig. J.1 $\zeta\,(3/2,\,\lambda)$ vs. λ.

Therefore

$$\left\langle \mathcal{N}_{op}^{BE} \right\rangle = \frac{\lambda V}{4\pi^2}\left(\frac{2m}{\hbar^2}\right)^{3/2}\left[\frac{\Gamma\,(3/2)}{\beta^{3/2}}\right]\Phi\,(\lambda,\,3/2,\,1)\,. \tag{J.16}$$

Using Mathematica it is a simple matter to plot $\zeta\,(3/2,\,\lambda) = \lambda\Phi\,(\lambda,\,3/2,\,1)$ vs. λ (see Figure J.1).

It is useful to note that $\zeta\,(3/2,\,\lambda) \approx \lambda$ when $\lambda << 1$ so that

$$\left\langle \mathcal{N}_{op}^{BE} \right\rangle \cong \frac{\lambda}{4\pi^2}\left(\frac{2m}{\beta\hbar^2}\right)^{3/2}\Gamma\,(3/2)\,, \tag{J.17}$$

which is the quasi-classical result.

Index

Printed in the United States
By Bookmasters